Management Information Systems 2004

GIS and Remote Sensing

FOURTH INTERNATIONAL CONFERENCE ON MANAGEMENT INFORMATION SYSTEMS INCORPORATING GIS & REMOTE SENSING

MANAGEMENT INFORMATION SYSTEMS 2004

CONFERENCE CHAIRMAN

C.A. Brebbia
Wessex Institute of Technology, UK

INTERNATIONAL SCIENTIFIC ADVISORY COMMITTEE

Organised by

Wessex Institute of Technology, UK

Management Information Systems 2004

GIS and Remote Sensing

Editor:

C.A. Brebbia

Wessex Institute of Technology, UK

C.A. Brebbia
Wessex Institute of Technology, UK

Published by

WIT Press
Ashurst Lodge, Ashurst, Southampton, SO40 7AA, UK
Tel: 44 (0) 238 029 3223; Fax: 44 (0) 238 029 2853
E-Mail: witpress@witpress.com
http://www.witpress.com

For USA, Canada and Mexico

WIT Press
25 Bridge Street, Billerica, MA 01821, USA
Tel: 978 667 5841; Fax: 978 667 7582
E-Mail: infousa@witpress.com
http://www.witpress.com

British Library Cataloguing-in-Publication Data

A Catalogue record for this book is available
from the British Library

ISBN: 1-85312-728-0
ISSN: 1470-6326

The texts of the papers in this volume were set individually by the authors or under their supervision. Only minor corrections to the text may have been carried out by the publisher.

Printed in Great Britain by Athenaeum Press Ltd., Gateshead.

Preface

This volume of the Transactions of Wessex Institute contains the papers presented at the 4th International Conference on Management Information Systems (MIS) held in Malaga, Spain.

The book covers different aspects of MIS including GIS, GPS and Remote Sensing. MIS are now used in a wide range of fields and their applications cover topics as varied as physical sciences, business and economy amongst many others. The technology also plays an essential role in bringing different disciplines together and in this way, helps to set up new development strategies.

The book examines areas in which MIS are increasingly used as well as new applications. The papers in this volume are classified into the following sections:

- Applications of GIS and GPS
- Applications of MIS
- Remote sensing
- Information system strategies
- Hydro and Geo informatics
- Transportation

The Editor is indebted to all authors for their contributions and most specially to the members of the International Scientific Advisory Committee, who helped to review the papers contained in this Volume.

Carlos A Brebbia
Malaga, 2004

Contents

Section 3: Remote sensing

Section 4: Information system strategies and methodologies

Section 5: Hydro and Geo informatics

Section 6: Transportation

Section 1
Applications of GIS and GPS

An integrated GPS-GIS surface movement ground control system

M. N. Postorino & V. Barrile
"Mediterranea" University of Reggio Calabria, Engineering Faculty, DIMET

Abstract

The need to guarantee a high safety level in airport areas and the high congestion at the airports, due both to the increase of air traffic demand and the lack of significant intervention on the infrastructures, ask for more intensive studies concerning the problem of the ground circulation in an airport area. In particular, high traffic airport areas must be continuously monitored in order to know at each time the position of all the moving objects (land vehicles and aircrafts) on the ground (aprons, taxiways, runways). Recent solutions used to resolve the problem of object positioning move towards the use of GPS apparatus. In this work, an integrated GPS-GIS system is proposed for monitoring the airport areas, following the new requirements of Intelligent Transport System (ITS) applications.
Keywords: airport ground movements; land vehicles and aircrafts location; GIS and GPS technologies.

1 Introduction

The development of the air transport system in the last decades, due to a more and more increasing transport demand level, has a constraint in the bounded capacity of the system, that depends on the air control system and the actual specific set of rules.

The bottlenecks of the system can be identified in the finite capacity of both airways and airports, these latter even more constraining.

The ground movement control system plays an important role for the estimation of the airport capacity; in fact, it establishes the time and/or space

Management Information Systems, C. A. Brebbia (Editor)

minimum separations that has to be guaranteed between two aircrafts during the landing/take-off operations.

With reference to the take-off operations (similar considerations can be made for the landing operations), the management of ground movements requires the exact position of the aircraft to be known by the tower operators during the movement from the taxiway to the runway as well as its waiting point before the take-off. In order to guarantee safety conditions during the ground vehicle movements (not only aircrafts, but also land vehicles as bus for passenger transfers, luggage and refuelling vehicles, and so on) suitable separations has to be ensured among them as during the flight.

With reference to this aspect, the FAA (Federal Aviation Administration) defines the runway incursion at an airport as "any occurrence at an airport involving an aircraft, vehicle, person or object on the ground that creates a collision hazard or results in a loss of separation reduction with an aircraft taking-off, intending to take-off, landing or intending to land" [1]. A similar definition has been developed by the European Organization for the Safety of the Air Navigation (within Eurocontol, the European organization that deals with the air traffic management and control), that defines the runway incursion as "any unauthorized presence on a runway of an aircraft, vehicle person that creates a collision hazard or results in a potential loss of separation" [1].

Even of the air transport system can be considered a safe system [7], however, the risk linked to interferences among aircrafts and/or vehicles during ground movements is still high with respect to the suitable standards for the system. The safety problem of the ground operations has been studied also by means of a simulation approach [9], in order to verify the incidence of the airport configuration on the ground movement operations.

To assure safety conditions to the operations occurring in the airport area (as landing, take-off, ground circulation) different systems have been proposed and partly realized; they are known as Surface Movement Ground Control Systems, (SMGCS). They can be used not only to control the ground circulation in order to guarantee suitable safety standards, but also to optimal manage the ground movements. In fact, the exact knowledge of the actual position of the aircrafts (as well as the land vehicles) in real time could allow the system be managed in order to increase the airport capacity, particularly by both reducing possible excessive spaces among aircrafts and addressing the moving vehicles along the best path in order to optimise the whole system in terms of capacity.

At this time, the systems used to control the ground aircraft movement are based mainly on radar systems and/or underground detectors. Limits of radar systems are linked to the wave propagation system and to the reflections generated by adverse meteorological conditions that reduce their efficacy; as regards the detector systems, they can detect the movements of the vehicles (both aircrafts ad land vehicles) only at specific points, because they are located in prefixed points on the ground.

Even if technological progresses have given a good support to the management of the airport ground movements, the problem of the aircraft and land vehicle surveillance cannot be resolved by means of only one kind of

Management Information Systems, C. A. Brebbia (Editor)

technology given the continuous occurring of accidents due to insufficiency of the technological systems and wrong communications between pilots and tower operators.

In fact, the actual system cannot neither locate and identify the moving vehicles with the needed accuracy nor convert the signals acquired by detectors in analogical data to be used in an automatic computing system.

For this reason different systems have been studied in order to integrate the radar control with other apparatus, also by using completely different technologies to guarantee both the ground movement safety and the airport capacity increase.

In this paper, an integrated approach GPS-GIS will be describer; its goal is to provide a surveillance and guidance system for the airport ground movement; the GPS component is used to establish the aircrafts/vehicles position in real time while the GIS component is used to manage and depict the geographical information of the aircrafts/vehicles position.

The paper is organized as follows: in section 2 the ground movement control problem is discussed as well as the benefits deriving from an optimal management of the ground movements; in section 3 the GPS technology is briefly described; in section 4 the GPS-GIS system for the airport ground movement control is discussed, with particular reference to the needed apparatus and the data transmission system. Finally, section 5 reports some short conclusions.

2 Ground air traffic control system

The Advanced Surface Ground Movement Control System (A-SMGCS), as defined by ICAO [6], is a modular system able to support the ground movements of land vehicles and aircrafts at the airport area in a safe, orderly and quick way, whichever the meteorological conditions, traffic density and airport complexity are.

The need to have an A-SMGCS arises from the observation of the increasing number of accidents occurring in an airport area, the augmented complexity of the airport structure, the increased number of operations and the desire to ensure the operative capacity in all weather conditions.

Functions required to an A-SMGCS can be synthesised as follows:

- surveillance function
- routing function
- guidance function
- control function.

The surveillance function must be directed to the identification and location of aircrafts and land vehicles within the airport area. In most cases, the surveillance procedures are based on the criterion "see and be seen", in the sense that the separations among all the moving vehicles are visual and are directly managed by the pilots or by the land vehicle drivers. This procedure should be unsafe when the external conditions do not permit a good visibility (e.g., in presence of fog) or the traffic density is quite high. In these situations there is a

Management Information Systems, C. A. Brebbia (Editor)

real need for systems able to support or substitute the procedure "see and be seen", whichever the whether conditions are, in order to locate moving/stationary vehicles and verify the absence of interferences among moving vehicles.

The routing function must be directed to plan and assign a path to each moving aircraft and land vehicles in the airport area in order to assure a safe, quick and efficient movement from its current position (origin) to its final position (destination). At this time, the routing is provided via radio communications to pilots (or to drivers of land vehicles) in terms of sequence of taxiways that must be run to move from the origin position to the destination position.

The usual procedure is based on the use of a standard terminology (in order to avoid wrong interpretations) and confirmation in terms of message repetition by pilots. Paths are established by the tower operators (Air Traffic Control, ATC), on the basis of the visual observations from the tower and the knowledge of the airport landside. An advanced function of routing must consider the opportunity to change the path at each time, plan all the paths for aircrafts and land vehicles for each traffic condition, interact with the tower to minimize all possible conflicts at the intersections, answer quickly to all users path requests.

The guidance function must be directed to provide pilots (and drivers) with suitable, unequivocal and continuous information about the path to be followed and the speed to be maintained to continue safely. At this time, the control function is realized by means of visual aids, in other words the pilot moves along taxiways and aprons following the painted markings (particularly the centerlines) and in-pavement lights. Furthermore, pilots have got airport maps that can be used to have additional information. Sometimes, ATC operators can provide pilots with information or instructions about the taxiways. An A-SMGCS must provide guidance for all possible assigned paths, give clear and unambiguous information to pilots and drivers such that they can follow the assigned path, make them aware of their position along the assigned path, accept path modifications at any time.

Finally, the control function must guarantee from each possible collision and runway incursion, and assure safe, quick and efficient movements in the airport surface area. At this time, the control function is under the responsibility of both pilots and controllers; in fact, the pilot moves following the rule "see and avoid", while the controller reduces the number of movements in order to avoid possible conflicts. An advanced control function must be able to support the ground movements in any traffic condition and to support the required movements for up to one hour; it must be able to find conflicts and to provide fast solutions to avoid them; it must be able to verify that the required safety distances are kept and advise if they reduce by a minimum value. Similarly, it must advise in case of runway or other restricted area incursions by means of suitable alarm systems; it must coordinate driver and pilot actions; spacing aircrafts to ensure minimum delay and maximum utilization of the airport capacity; finally, it must separate movements from obstacles, secure areas and restricted areas.

Till today, the most used systems to control the surface air traffic refer to two projects, identified by the acronyms ASDE-3 (Airport Surface Detection

Management Information Systems, C. A. Brebbia (Editor)

Equipment) and AMASS (Airport Movement Area Safety System), based on the use of radars to locate aircrafts (Ground Movement Radar systems). The first one is formed by a GMR system whose aim is to help the tower operators to identify the surface aircraft position. The second one is a development of the first one and it is addressed to send alarm signals when possible surface conflicts can develop.

As known, the radar working principle is based on the following: radio-waves, generally impulse modulated, are sent towards the object to be looked for and then the waves reflected by the objects itself (radar echo) are received by the apparatus. The object distance can be deduced as a measure of the time length between the emission and reception of the wave; in other words, the time spent by a radio-frequency impulse, generated by the radar transmitter, to reach the object, be reflected by it and then return to the receiver can be used to measure the distance between the radar position and the object. The orientation of the electromagnetic beam generated by the transmitter gives the direction of the object.

Generally, only one transmitter is used both as transmitter and receiver; the reception is temporarily interrupted during the impulse generation and then it is suddenly activated to capture echoes. The apparatus used to this aim is a kind of switch activated by the transmitted impulses. The radar antenna generally spins with a constant speed in order to explore the whole horizon; this kind of exploration is called azimuthal because it allows the angle between the reflecting object and the local meridian (azimuth angle) to be identified.

An optimization of the system is due to the visualization of the object on a screen. In fact, the reflected signal, captured by the received and suitably amplified, is transformed in a video impulse on a screen where the signal is reproduced as a high brilliance trace and its position with respect to the screen centre reproduces, in the given scale, the distance between the radar (supposed located at the centre of the screen) and the object that generated the radar echo. At each time the angular direction with which the electron beam realizes the scanning corresponds to the direction of the radar electromagnetic beam.

Other systems used to follow the surface movements of aircrafts are under-pavement detectors whose aim is to control the high traffic density airport areas (Airport Surface Monitoring Equipment, ASME). Among these, ASME-4 assures the following four main functions: surveillance, control, guidance and assistance. The network of under-pavement detectors can distinguish different kinds of traffic by identifying the moving vehicles. A further detector network on the outside of the paved areas allows a possible aircraft out-of-way to be quickly located and assisted. The system can also be connected to the way lights, so, following the acquired information and those introduced by the operators using the keyboard, a suitable system activates the lights preceding the aircraft and switches off those behind it, in order to generate a light path that guides the pilot from the taxi/run-way to the assigned apron and vice versa.

Detectors that can be used refer mainly to three classes:

- magnetic turns under-pavement, that use the physical principle of the magnetic induction on metallic wire;

Management Information Systems, C. A. Brebbia (Editor)
 ISBN 1-85312-728-0

- piezoelectric detectors, that use the physical principle of the pressure forces generated over the ways;
- pyroelectric detectors, that use the physical principle of the infra-red radiation emitted by hot bodies.

Among these, the pyroelectric detectors are not intrusive because they do not require the pavement perforation and the following rebuilding of the pavement as for the first two systems.

3 Global Positioning System (GPS)

The NavSTAR GPS (Navigation Satellite Timing And Ranging Global Positioning System) system was originally borne in USA for military purposes; it allows the three-dimensional positioning of objects (also moving) to be identified by means of information coming from a geostationary satellite system by using distance-measuring spatial intersections (ground receiver - orbit satellite).

Even if the new European satellite system GALILEO is coming, for the next ten years the GPS will be still the dominant system in the application field including the dynamic control of the territory, the emergency and above all the navigation, the goods positioning in terms of container vehicle fleet, goods wagon and anti-collision systems.

Mainly two kinds of GPS measures can be used [2], [3], [5], [8]: the pseudo-range and phase measures. The first ones are used above all for navigation assistance, while the second ones are used in all applications for which a greater accuracy is required, as the land deformation control.

The differential (RT-DGPS) and relative (RTK) positioning are particularly useful to continuously control in real time the moving means of transport (cinematic GPS); in this case, the measure (Cartesian coordinates of the baseline vector between the known master station and the rover station referred to the vehicle to be positioned) must be initialized and the satellite visibility must be guaranteed (by means of the monitoring of the DOP value) during all the time of the positioning measure. The acquisition system must be equipped with a firmware for the on-the-fly (OTF) initialisation, in case of temporary loss of the signal, and suitable links among the acquisition stations (master and rover) to obtain in real time the correct transmission of the data following given protocols.

4 A GPS-GIS approach for the air traffic ground control problem

In this section an integrated GPS-GIS system is proposed for monitoring the airport areas, following the new requirements of Intelligent Transport System (ITS) applications. ITS applications use telematics technologies and data processing methods to obtain an “intelligent” management of the transport system in terms of efficiency, safety and environmental aspects.

Geographical Information Systems (GIS) are an essential support to plan, manage and depict all useful information related to the examined system, and

Management Information Systems, C. A. Brebbia (Editor)

particularly many applications have been made (and some other can be thought) also in the transport field [4].

The system proposed here involves both the GPS and GIS systems in order to obtain an efficient, safe and quick management of the vehicle movements in an airport area (both aircrafts and land vehicles), in order to cover the surveillance and control functions of an A-SMGCS for taking-off aircrafts moving from the apron along the taxiways and for vehicles operating in the airport areas.

In facts, thanks to GPS systems, the actual trace of the aircraft (or land vehicles) can be followed (by means of transmission and updating of the navigation data in real time), while the GIS system allows the geographic information and the position to be managed and visualised (by means of the implementation of suitable software functions to resolve the problems of perception and management of the information and to update in real time the acquired data).

Particularly, the proposed system considers:

- location and positioning of aircrafts and the other moving vehicles in the aprons by GPS and transfer of data referred to its position (other data measured by different sensors can also be considered);
- management and data processing operating in a GIS system;
- information re-transmission to the aircraft by local transmitters, internet and WAP systems.

This latter point is structured as follows: a master station, located in a point of known coordinates, sends positioning information to the remote station (aircrafts or vehicles, all equipped with firmware for RTK-OTF and/or DGPS) that, fixed the ambiguity flight (OTF initialization) if it is the case, processes immediately the self-acquired data and those sent by the master station giving as output its positioning in real time (it has to be noticed that the possible presence of radio repeaters and/or GSM technologies could increase the range of transmission); this positioning information is sent (via GSM transmission – Internet – radio modem) to the control centre (server) formed, for example, by the control station where a GIS tool allows the view of the monitored vehicle/aircraft (as suitable trace) in real time and continuously during the time. The GIS system has to be suitably modelled in order to transform the positioning information into geometrical coordinates in turn transformed in graphical information on a screen.

The main components of the system are:

- a master station of known coordinates (that can or cannot coincide with the control centre;
- a device (firmware RTK-OTF and/or DGPS), located on the rover (moving vehicles: aircrafts and land vehicles), that has to detect in real time their position and must be able to communicate with the master station in real time; it is equipped with a display, connected to a network of data transmission, where on a cartographic support is showed the path to be followed;
- radio transmitting and receiving the RCTM protocol or, alternatively, GSM modem;

- a control centre equipped with a data reception system, a GIS system for visualising and processing the data and with a data transmission system that returns the information to the aircraft (figures 1 and 2).

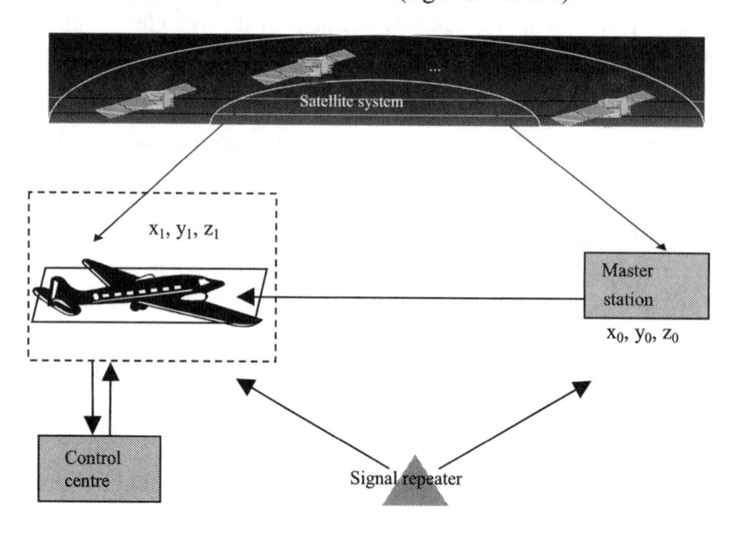

Figure 1: The GPS system to detect the moving vehicles on the ground.

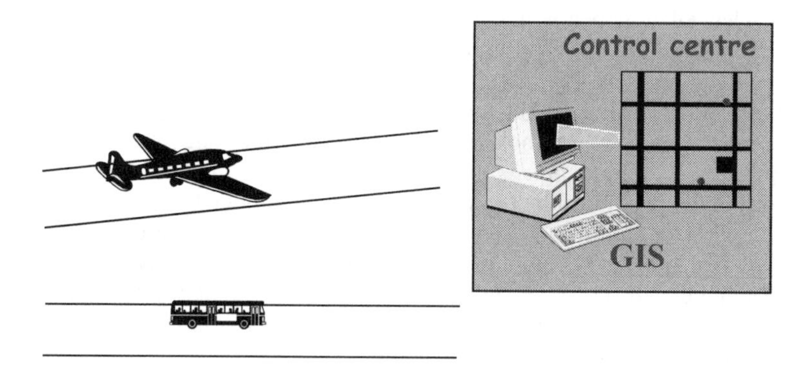

Figure 2: Identification and control of moving vehicles on the ground by means of a GIS system.

The device, autonomous and fed by suitable batteries, is characterized by a system able to receive and apply a real-time correction in terms of code range and/or phase (DGPS), initialize the measure (OTF firmware to initialize quickly or during the movement), compute the object position in real time, memorize and store the coordinates and transmit the memorized data to the control centre according to international protocols (NMEA).

Management Information Systems, C. A. Brebbia (Editor)

The control centre, whose aim is the real-time view of the movements of the vehicles/aircrafts on ground, is characterized by a GIS modulus realized to manage the monitored information; it works on a cartographic map that reproduces the airport ground area and is equipped in order to acquire, re-process, return and re-transmit positioning information on a digital base.

The system is then based on a client-server type, with moving clients (aircrafts and vehicles) and a control centre as server able to measure in real time the client position, showing the position to the server, re-transmit the view to the client. The system architecture is characterized by the following technologies:

- local technology: GPS system;
- commercial technology: GSM, Internet, radio modem;
- memorizing technology;
- view and analysis technology: GIS

Specific attention should be addressed to the transmission modalities of the positioning information. The radio-modem transmission, given the actual legal rules, does not allow the radio signal transmission with high power and then the range is limited to few kilometres; it is also dependent on the obstacles in the zone. Generally, the used radio modem apparatus transmit at 400-500 Mhz, that correspond to a 2-2.5 Km range.

To reduce these inconvenients, a repeater could be added to the master and rover stations (figure 1), according to two possible systems:

- radio repeater in a fixed location on the ground;
- mobile radio repeater on the ground.

The presence of the repeater increases the range of the RTK system and if there is a mobile repeater also the system flexibility increases. The repeater allows more precise positioning measures to be obtained than in the case where it is not considered; furthermore, more rover stations linked to the same master station could work at the same time.

Alternatively, the correction signal can be sent by using the GSM technology (phone system). The GSM transmission system is reliable in terms of signal transmission, provided that the signal is covered and the passage between two cells does not involve the telephonic link interruption. For both systems it is convenient operate in a range of 15-20 Km.

The last solution could be considered the most advantageous; in fact, the control unit of the RTK system, equipped with a firmware able to dial the phone number of the master station, while dialling the rover opens the telephonic link with the master station itself and receive the data of the differential correction. The monetary cost of the phone call is charged on the rover station suitable equipped with a special SIM card. The master station is equipped with a data transmission SIM card. The data transmission speed is 9600 bit/sec.

5 Conclusions

The system described before has been tested using a small private aeroplane, during the taking-off operations. Even if it is still under test, because more moving vehicles, higher taking-off speeds and different solutions to

Management Information Systems, C. A. Brebbia (Editor)

communicate data between the different apparatus are now testing, however the results are very encouraging.

Some problems could verify during the landing operations, given the possible GPS signal latency due to the high involved speeds and then the impossibility to visualize the aircraft in real time.

In any case, more tests and more work need in this field due to the importance of the topic involving the human safety.

References

[1] AA.VV., Methods of preventing runway collisions evolve in Europe and in the United States, *Flight Safety Foundation, Airport Operations*, **26(4)**, 2000.

[2] Crespi M., Manzoni G., Strumenti e metodi innovativi nel rilevamento, *Proc. of the IV Conferenza Nazionale ASITA*, Genova, 2000.

[3] Crocetto N., Gatti M., Neuroni F., Uso di ripetitori radio nel rilevamento GPS in tempo reale, *Proc. of the IV Conferenza Nazionale ASITA* , Genova 2000.

[4] Golledge R., Loomis M., A GIS for a GPS based personal guidance system, *Geographic information Science* **12**, pp. 727-749, 1998.

[5] Hoffman-Wellenhof et al, *GPS Theory and Practice,* Springer-Verlag New York, 1997.

[6] International Civil Aviation Organization, *European manual on Advanced Surface Movement Guidance and Control System*", draft version 07 – 9 January 2001.

[7] Janic M., An assessment of risk and safety in civil aviation, *Journal of Air Transport Management*, **6**, pp. 43-50, 2000.

[8] Laurini R., Thompson D., *Fundamental of spatial information system*, San Diego, California, 1992.

[9] Pitfiled D.E., Brooke A.S., Jerrard E.A., A Montecarlo simulation of potentially conflicting ground movement at a new international airport, *Journal of Air Transport Management*, **4**, pp. 3-9, 1998.

Management Information Systems, C. A. Brebbia (Editor)
© 2004 WIT Press, www.witpress.com, ISBN 1-85312-728-0

Web-GIS based urban management information system: the case of satellite cities in Istanbul

M. Ozturan, B. Egeli & F. Bacioglu
Department of Management Information Systems,
Bogazici University, Turkey

Abstract

A Geographic Information System (GIS) is defined as a computer based tool useful for capturing, storing, retrieving and manipulating, displaying and querying both spatial and non-spatial data about given areas to generate various planning scenarios for decision making. On the other hand, an Urban Management Information System (UMIS) is an economic resource that can benefit decision makers in the planning, development, and management of urban projects and resources using statistical data such as social, economic, demographic, housing and health data. Since Internet technology reduces the cost of data management and information distribution to mass usages as in UMIS, web based systems are mostly used for a UMIS applications throughout the world. Nowadays, the number of satellite cities is rapidly increasing in Turkey, especially in Istanbul. A satellite city for Istanbul can be defined as a mini city established outside Istanbul that has its own shopping centers, schools, social clubs, cinemas, restaurants, etc. Though it is not a municipality by itself, it needs to conform to the requirements of both the municipality it belongs to and its management. In this study, a prototype web-GIS based UMIS is developed for satellite cities. The system integrates the digital base map and statistical data of a satellite city that can be input through a user-interface and offers a variety of querying and reporting options. The developed system is tested for one satellite city in Istanbul and the results showed that it is promising as a starting system for other similar applications.
Keywords: geographic information system, urban management information system, satellite city.

Management Information Systems, C. A. Brebbia (Editor)

1 Introduction

Geographic Information System (GIS) is defined as a tool useful for capturing, storing, retrieving and manipulating, displaying and querying of both spatial and non-spatial data about given areas to generate various planning scenarios for decision making. It is a computer based tool for mapping and analyzing things that exist and events that happen on Earth. GIS technology integrates common database operations such as query and statistical analysis with the unique visualization and geographic analysis benefits offered by maps. These abilities distinguish GIS from other information systems and make it valuable to a wide range of public and private enterprises for explaining events, predicting outcomes, and planning strategies [1].

Urban Management Information System (UMIS) is an economic resource that can benefit decision makers in the planning, development, and management of urban projects and resources using statistical data such as social, economic, demographic, housing and health data. UMIS combines map images with other kinds of information (like tabular data) for the purpose of analyzing spatial relationships among data related to locations in the city [2]. It includes physical, technical, social, economic and administrative data, base maps, land use maps, master and development plans, the land information and cadastral maps. The system can make queries from legal documents, reports about the spaces, the citizens and result thematic maps like building, population and infrastructure maps [3]. The building maps show building occupancy, construction type or date, number of floors, license information, users information and so on. The system may contain transportation maps, traffic-density maps, shortest path for the firefighters or ambulance drivers. According to the population, the decisions can be made about the locations of the social facilities like schools, health centers or green areas, utility service areas, telecommunication, gas, electricity [4]. There are many UMIS applications around the world and most of them are web based since Internet technology reduces the cost of data management and information distribution to mass usages and also has become a tool to convert local government into an "open system" that makes the public services accessible [5]. But development of such systems needs a very careful study due to the importance of the integration of GIS, relational database and web programming issues.

Satellite cities, which are the target of this study, are also in need of UMIS since they are interpreted as mini cities established on county sites and have their own shopping centers, schools, social clubs, cinemas, restaurants, etc. Management of such cities needs to conform both the requirements of the municipality it belongs to and its internal management.

2 Prototype UMIS for satellite cities

The aim of this preliminary study is to gain experience and know how necessary for the development of web-GIS based UMIS. The prototype UMIS developed in this study is a part of a larger system (City Information System for Satellite

Management Information Systems, C. A. Brebbia (Editor)

Cities) that the authors are working on as an applied research project. The subsystems of the main system are given in fig. 1.

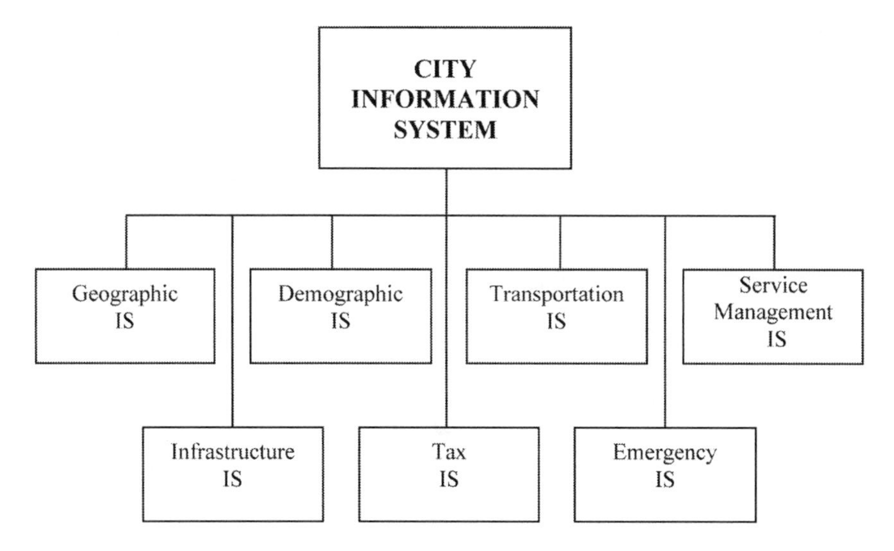

Figure 1: City Information system for satellite cities.

Utilities of these subsystems which are still under construction are given below:

Geographic Information Subsystem: Creation of maps by digitizing and/or coordinate entry, forming layers of infrastructure, natural resources and buildings; and related input forms, queries and reports.

Demographic Information Subsystem: Creation of city resident records, and buildings information; and related input forms, queries and reports.

Transportation Information Subsystem: Main roads, streets, parking places and traffic flow directions information; and related input forms, queries and reports.

Service Management Information Subsystem: Services given in the satellite city, financial issues and scheduling of these services; and related input forms, queries and reports.

Infrastructure Information Subsystem: Water, natural gas, waste water, telephone and electricity networks and their subscriptions; related input forms, queries and reports.

Tax Information Subsystem: Real estate and environment taxes; and related input forms, queries and reports.

Emergency Information Subsystem: Security, fire extinguishing, health, weather conditions, earthquake and similar emergency information; related input forms, queries and reports.

The prototype developed in this study integrates only some partial utilities of the geographic, demographic, transportation, service management and infrastructure subsystems of the main system.

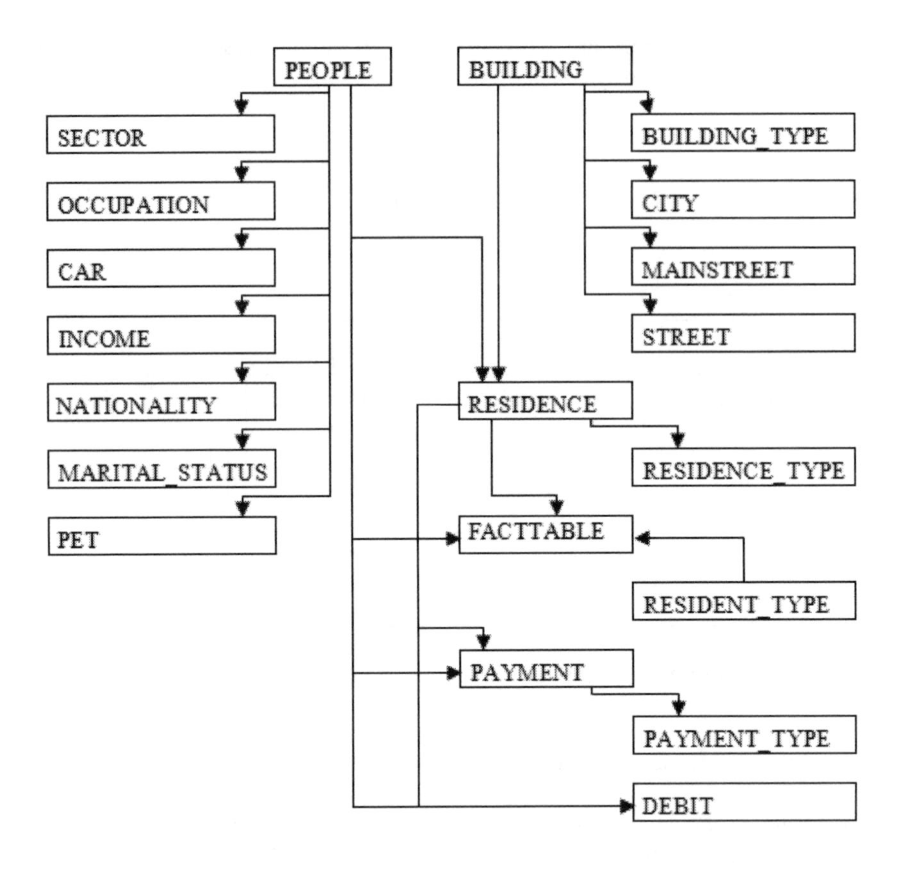

Figure 2: Relationship diagram of the database.

2.1 Prototype subsystems

The utilities taken into consideration for the subsystems included in this prototype are given below:

2.1.1 Geographic information subsystem

The base map is the basis of the GIS activities and involves the conversion of an existing available map into a digital format. In this study this map contains the city boundary layer and two more layers for buildings and for streets. Geographic references of each building and street are kept in the geodatabase of this subsystem.

2.1.2 Demographic information subsystem

This subsystem includes building information (building type, address, construction date, etc.), residence information (occupancy, pet, car, etc.) and people information (demographic data, professional data, etc.). It is possible to get several query reports from the system such as the buildings on a specific

Management Information Systems, C. A. Brebbia (Editor)

street where disabled people live and the building to where a special car belongs. The results of these queries are also shown on the base map.

2.1.3 Transportation information subsystem

Transportation information subsystem includes only the basic road and street information. There are a few query reports related to them such as buildings on a specific street. The query results can again be easily shown on the map.

2.1.4 Service management information subsystem

Includes payment requirements for the related services given by the satellite city administration and the related payments done by the residents. There are also some query reports in this subsystem such as residents that have debit accounts, payments to be done by the residents for special services and monthly payment requirements for a specific time period. There is also the possibility of showing the results of these queries on the base map.

2.1.5 Infrastructure information subsystem

Electric-gas-water meters information of the residents are kept in this subsystem. There is only one query report and it is for displaying and showing on map the residence that a specific meter belongs to.

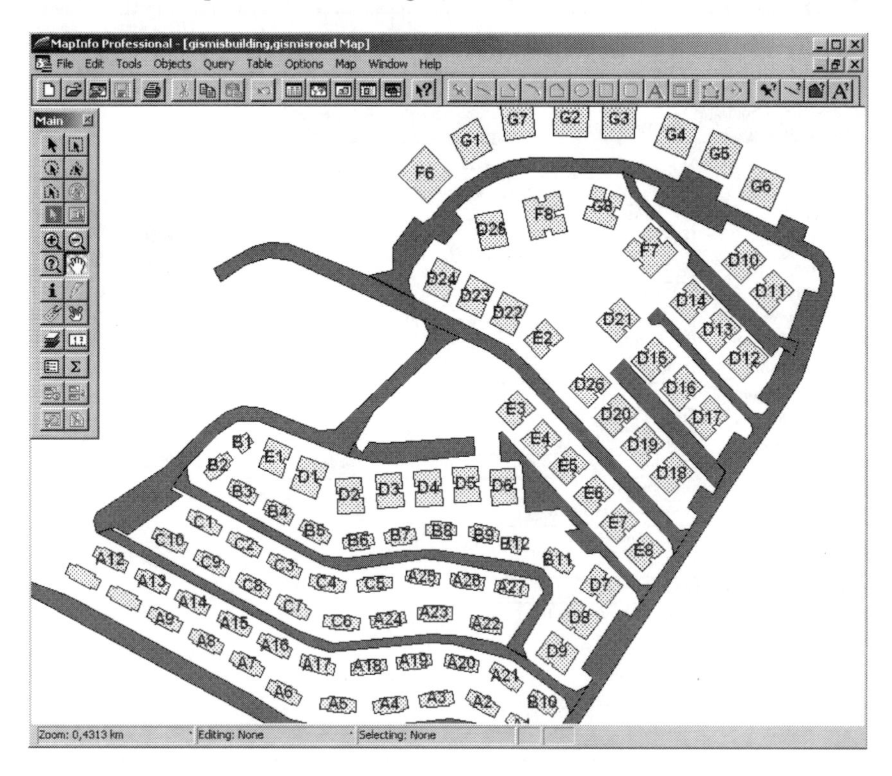

Figure 3: Base map of Hisarevleri satellite city.

Management Information Systems, C. A. Brebbia (Editor)
© 2004 WIT Press, www.witpress.com, ISBN 1-85312-728-0

2.2 System development environment

The system is developed using the following tools:

a. MapInfo Professional: Desktop map management software
b. MSAccess: Database management system
c. MapInfo MapXtreme: Internet based map server software
d. HyperText Markup Language (HTML): Web page creation tool
e. Active Server Pages (ASP): Internet programming tool

The geodatabase, which keeps spatial data of the map objects, is designed using MapInfo Professional and the relational database that keeps non spatial data described in the subsystems is designed using MSAccess for which the relationship diagram is given in fig. 2. Two databases are related to each other through ids of the map objects. The user interface, data input forms, query based reports and map displays are developed using HTML, ASP and MapXtreme environments.

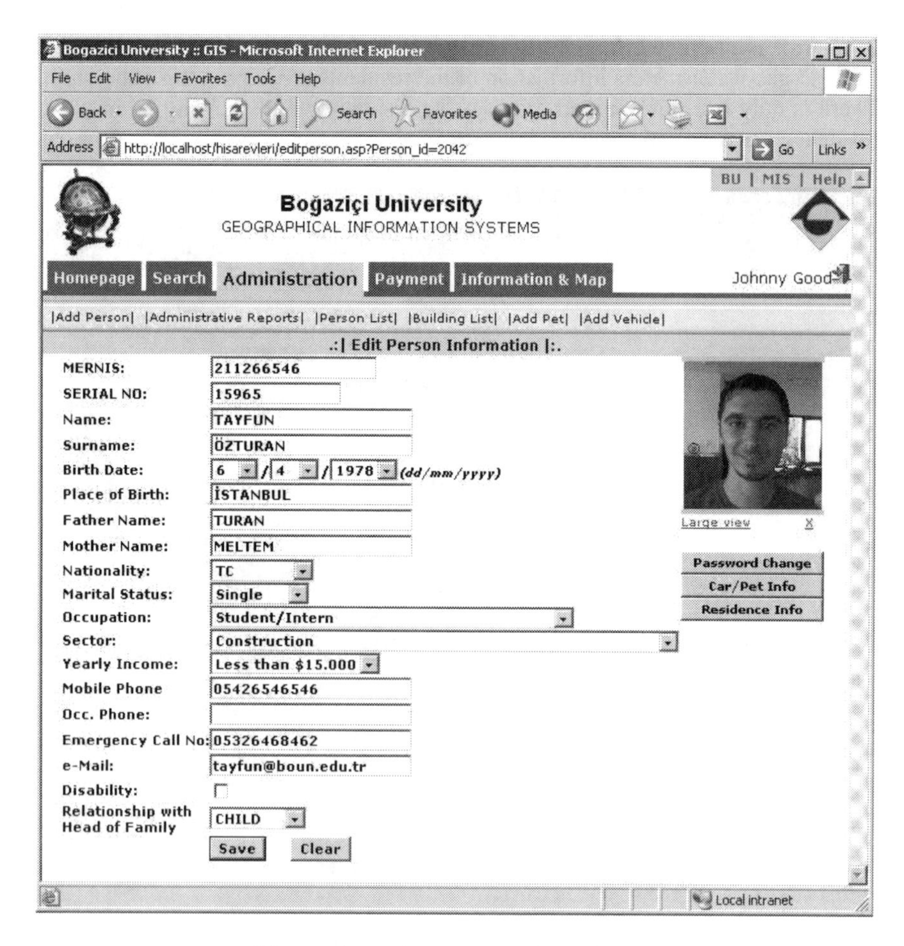

Figure 4: Person entry input form.

2.3 User interface

User interface of the system is web based and utilizes ASP technology. There are mainly five main menu items with many sub items in the user interface. All related input forms and query reports are allocated under these menu items. According to the type of the user (administrator or resident), different options are made available on the screens. Input forms for building, person, pet, car, payment requirement and payment are the main screens for data entry. Queries for residents, addresses, car/pet owners, streets, payments are all prepared as output reports and can easily be displayed on the base map. It is also possible to get related information for a map object by just clicking on it on the base map.

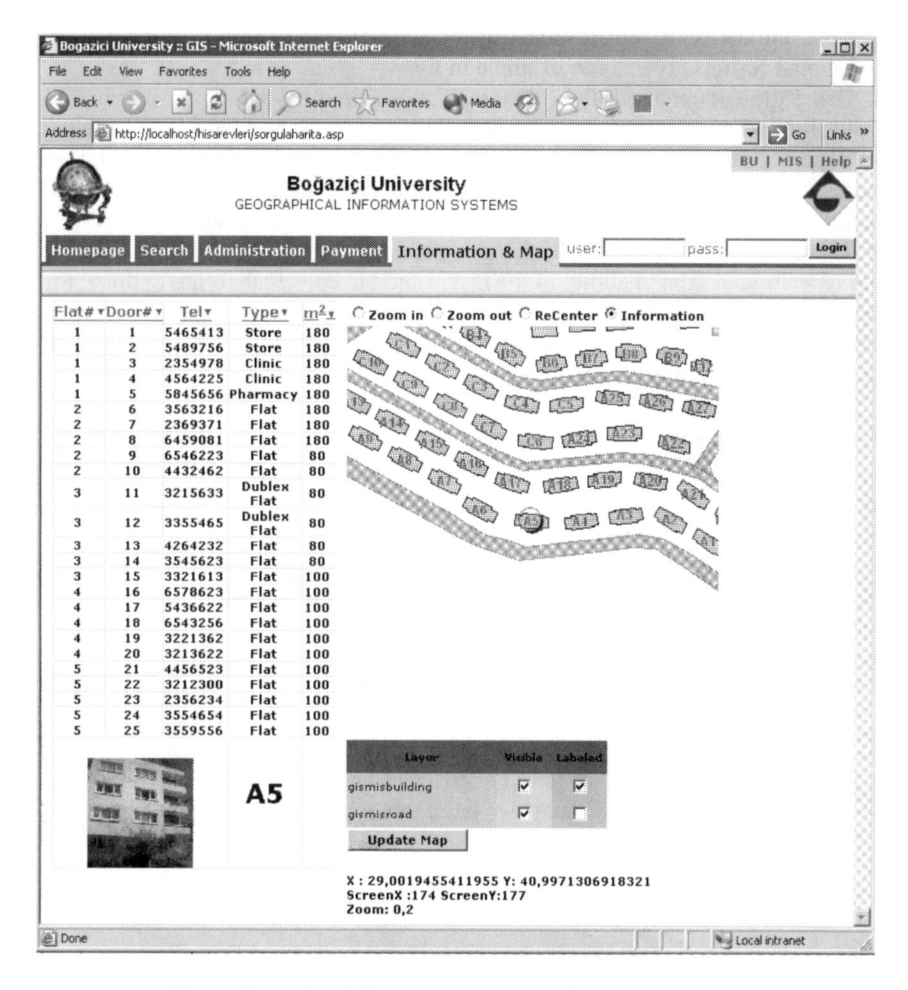

Figure 5: Building query report and its map display.

Management Information Systems, C. A. Brebbia (Editor)

3 Application

This prototype is tested for a satellite city named Hisarevleri in Beykoz region of Istanbul. The city boundary layer of Hisarevleri is formed by digitizing the map given by Hisarevleri administration using scanner. Other layers are formed by drawing the polygons representing the buildings and the roads using the scanned map. An id and a label attribute are given for each polygon and these ids are used for relating these layers to the relational database. The building layer hold 103 buildings and the roads layer includes four main streets and seven streets. The final base map with building labels is given in fig. 3.

Data related to the subsystems are also supplied by Hisarevleri administration in written format. The data are entered to the database through the web based input forms by the students of the Department of Management Information Systems and data entry errors are corrected by Hisarevleri administration. Person entry input form is given as an example in fig. 4.

Using the data entered, the query reports acquired by the administration are easily formed by the system developed. These query reports are formed using both the geodatabase and the relational database designed before. Search tab of the user interface is used to create the required query in the relational database. Using the relation formed through the objects ids to the geodatabase, the results of this query is shown on the base map. This map can be zoomed in, zoomed out and repositioned. The visibility of the layers and the object labels are optional for the users. Besides the map display, by clicking the information button, the output report for the query is also displayed on the screen. Building query report and its map display is given as an example in fig. 5.

4 Future work

In this study, a preliminary prototype web-GIS based UMIS for satellite cities is suggested. This prototype is tested for Hisarevleri satellite city in Istanbul and found to be a promising system to start with. As future work, the subsystems that are under construction should be fully completed and the integration of these subsystems must be definitely done. The authors recommend the followings as further improvements for the applied research project:

a. Integration of social, economic and educational information to cope with managerial problems
b. Integration of the system with GPS technology
c. Change of the user interface for a better subsystems integration
d. Use of SQL server as database environment due to the limitations of MSAccess

Acknowledgements

The authors would like to thank Bogazici University Research Fund for financially supporting this project and Ergin Ozturk for his help in developing the user interface of the system.

References

[1] Environmental Systems Research Institute, *ArcView Spatial Analyst: Advanced Spatial Analysis Using Raster and Vector Data*, ERIS Press: New York, 1996.

[2] Huxhold, W., *An Introduction to Urban Geographic Information Systems*, Oxford University Press: New York, 1999.

[3] Yalciner, O., Urban Information Systems for Earthquake-Resistant Cities: A Case Study on Pendik Istanbul, The Geographic Information Systems (GIS) Portal, www.gisdevelopment.net/thesis/thesis2/index.htm.

[4] Reis, M., Tematik Tabanlı Kent Bilgi Sistemi Tasarimi ve Uygulamasi, unpublished MSc thesis: Trabzon, 1996.

[5] Bolatto, G., Sozza, A., Gauna, I. & Rusconi, M., The Geographic Information Systems (GIS) of Turin Municipality, *Digital Cities: Technologies, Experiences, and Future Perspectives*, eds. T. Ishida & and K. Isbister, Spring-Verlag: Heidelberg, 2000.

Management Information Systems, C. A. Brebbia (Editor)
© 2004 WIT Press, www.witpress.com, ISBN 1-85312-728-0

GIS – application to make geomorphometric analysis in raster data

M. Moreno, M. Torres, S. Levachkine & R. Quintero
Geoprocessing Laboratory, Centre for Computing Research, National Polytechnic Institute, Mexico City, Mexico

Abstract

We present an approach to identify some geomorphometrical characteristics of raster geo-images based on a Geographic Information System (GIS). The characteristic identification involves the generation of raster layers, topographic ruggedness and drainage density. In a sense, the topographic ruggedness is used to express the amount of elevation difference between adjacent cells of the Digital Elevation Model (DEM). Moreover, the topographic ruggedness is presented by means of a Terrain Ruggedness Index (TRI). We use the Spline Interpolation Method to obtain the density layers. These layers are used to represent the amount of geographic linear objects in a specific area. The algorithm has been implemented into GIS – ArcInfo, and applied for a GIS of Tamaulipas State, Mexico.

1 Introduction

Geomorphometric analysis is the measurement of geometry of the landforms in raster images and has traditionally been applied to watersheds, drainages, hill slopes and other groups of terrain objects. In particular basin morphometric parameters attracted much attention from hydrologists and geomorphologists since watersheds have been used for analysis of different physical ecosystem processes [1]. The geomorphometry represents one set of recommended variables to analyze distribution and concentration of certain spatial objects.

Nowadays, Geographical Information Systems are powerful and useful tools as means of information, visualization and research or as decision making applications [2]. However, contrasting with the traditional topographic map methods, the GIS methods are relatively easy to apply in a consistent way on

Management Information Systems, C. A. Brebbia (Editor)
© 2004 WIT Press, www.witpress.com, ISBN 1-85312-728-0

large areas of landscape, because they allow summation of terrain characteristics for any region. They can be used to provide geomorphometric data and therefore insight the processes affected by terrain morphology for all types of mapping.

Since the mid-1980s, with increasing popularity of GIS technology and availability of Digital Elevation Models (DEM), the potential of using DEM in studies of surface processes has been widely recognized [3]. New methods and algorithms have been developed to automate the procedure of terrain characterization [4]. DEM has been used to delineate drainage networks and watershed boundaries to compute slope characteristics, and to produce flow paths [5]. In addition, DEM has been incorporated in distributed hydrological models [6].

DEM is playing an increasingly important role in many technical fields of GIS development, including earth and environmental sciences, hazard reduction, civil engineering, forestry, landscape planning, and commercial display. It is difficult to exaggerate the importance of the DEM to geomorphology, because DEM may ultimately replace printed maps as the standard means of portraying landforms. The contour maps remain an important data source for DEM, although techniques for measuring elevation directly from satellite images have been introduced in recent years. Tamaulipas State area is covered by DEM at two resolutions, 50 x 50 m and 250 x 250 m [7].

In this paper, we propose a method to make spatial analysis based on geo-image processing by means of Spatial Analyzer Module (SAM). In Section 2 we present the description of GIS Architecture. SAM is presented in Section 3 and describes its functionality. In the next sections, we describe how terrain ruggedness and drainage density have been obtained. Some results are shown in Section 6. Section 7 presents our conclusions.

2 GIS application architecture

The GIS application presents a client-server architecture. This tool counts with the following modules: an Enterprise GIS, Communication Module, Spatial Analyzer Module, Spatial Database and XML Administration Module. Fig 1 shows the GIS.

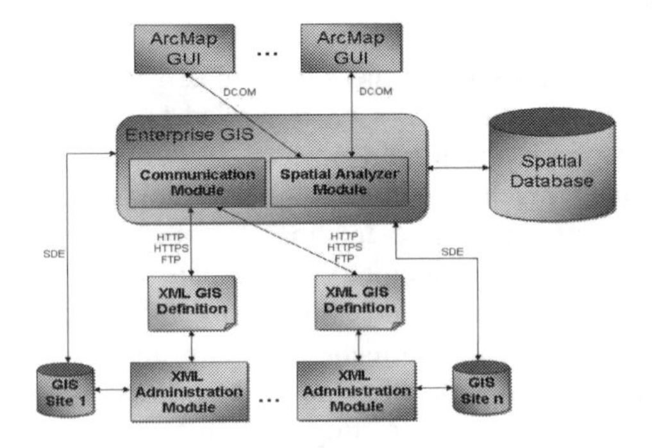

Figure 1: GIS application architecture.

Management Information Systems, C. A. Brebbia (Editor)

The functional mechanism of the GIS application is the following: ArcMap users need to make a requirement. It is sent by DCOM technology to the communication module into the Enterprise GIS. This module processes the requirement and sends the parameters via HTTP or HTTPS protocols. In the remote GIS site, the XML administration module checks the initial XML definition and queries the local XML definition, in order to locate and compare the *qualitative* and *quantitative* characteristics of the spatial data, such as scale, projection, spatial reference, representation type, thematic, DBMS type and attribute data. If the XML definition matches with the local XML definition, the spatial data are recovered and sent through SDE (Spatial Database Engine) mechanism.

The geographic objects are stored in the spatial database, in which will be analyzed by SAM.GIS application counts with a geographical database, which has been designed and implemented by a *Geodatabase*. It is a storage mechanism provided by Arc/Info, which is focused on generating independent geographical depositories. The spatial database presents an own special environment modeling, according to the characteristics of geographic phenomena. *Geodatabase* provides a topological model that is integrated by *class of elements*. This model is similar to spatial coverage. The *Geodatabase* model is supported by an object-oriented-relational database, in the other hand, it is considered as a hybrid between both techniques. With this technology users can access to spatial and descriptive attributes from different sources by means of SDE. This mechanism defines an open interface to database system, and permits to handle geographic information in an intrinsic way. In this case, the behavior of the spatial objects is defined in the system. The entities are represented as spatial objects with properties, behaviors and relationships between them. The spatial database has been designed through Arc/Info system, shapefiles and descriptive information. All these components are involved in the analysis to compute the geomorphometric analysis.

3 Spatial Analyzer Module

SAM is a special module, which has been designed to make spatial analysis procedures. SAM uses vector and raster data to make the spatial analysis. This module has been implemented using Arc Macro Language (AML) to ensure portability between computer platforms executing ArcInfo 7.0 or later.

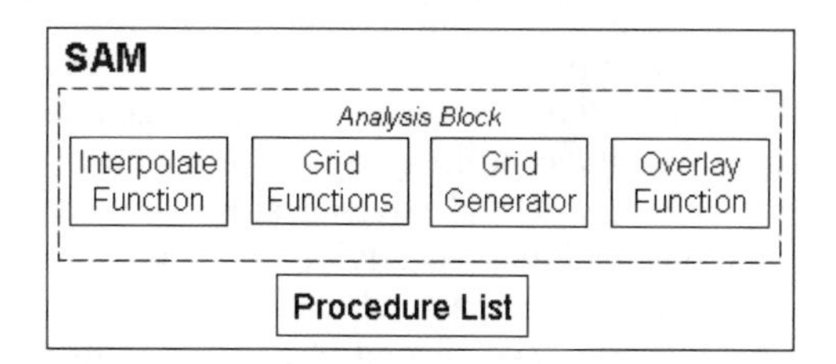

Figure 2: The Spatial Analyzer Module.

Management Information Systems, C. A. Brebbia (Editor)

The analysis is based on using different spatial data related to the case of study. SAM contains two components: Analysis Block and List of Procedures. 1) Analysis Block is composed of a set of processes to make data analysis. 2) List of Procedures stores the sequence of steps to execute the processes [8] (see Fig. 2).

3.1 Analysis block

It contains the required functions to make spatial analysis. These functions are the following:

Interpolate Function. The method used is a minimum curvature spline in two dimensions from a set of points. For computational purposes, the entire space of the output grid is divided into blocks or regions of equal size. They are represented in a rectangular shape. The equation 1 shows the spline function that has been used [9]:

$$S(x,y) = T(x,y) + \sum_{j=1}^{N} \lambda_j R(r_j) \tag{1}$$

where j= 1, 2… N; N is the number of points; λ_j are the coefficients obtained from the system of equations, which computes the point coordinates; $R(r_j)$ is the distance from the point (x,y) to the jth point.

To use this function, it is necessary to provide the set of points and tolerances, which depend on the specific case of study.

Grid Functions. They contain the set of functions for cell analysis that include operations of the map algebra, and describe how the operations are specified, the data to operate on, and the order in which operations should be processed. In this case the function is SQRT. SQRT calculates the square root of the input grid [10].

Overlay Functions. This module has been designed to make *topological overlays*, which can be used to identify areas of risk. A set of operations has been defined, and applied to the spatial analysis. This is made to establish the conditions and to combine different information layers using logical operators. These functions combine spatial and attribute data. The implemented operations for topological overlay in this application are: intersection, union and identity, which are represented by the symbols $\cap$, $\cup$ and I respectively [10].

Grid Generator. It is used to process some analyzed data, especially in density map generation. The vector grids are regular of *m* x *m* magnitude, in which *m* is the cell size. The cell magnitude in the grid is determined by the phenomenon under study characteristics (scale and covered area). Two alternatives can be used to generate the grids. First, specifying the initial and terminal grid coordinates ((x_0, y_0), (x_1, y_1)) respectively and establishing the number of required divisions for the grid. The second alternative is to specify the initial coordinate (x_0, y_0), cell size, number of columns and rows in the grid [11] (Fig. 3).

Management Information Systems, C. A. Brebbia (Editor)

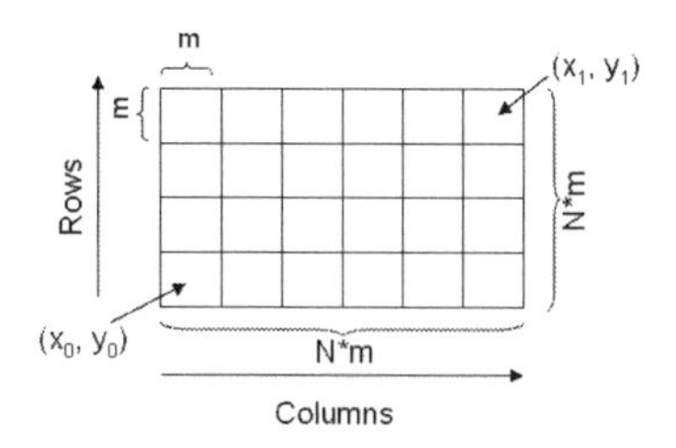

Figure 3: Specifications of the grid.

3.2 List of procedures

It stores the set of procedures for each one of the analysis processes. It has a description of the required data type and the restrictions. However, the users can change the selection criteria. This provides a list of functions as an alternative for the analysis, in which the parameters can be modified. SAM has a wide range of applications, not only to make geomorphometric analysis. It is also possible to perform the detection of *landslide* and *flooding* areas [11].

4 Generation of topographic ruggedness layer

The Terrain Ruggedness Index (TRI) is a measurement developed by Riley to represent the amount of elevation difference between adjacent cells of a digital elevation grid [12]. The process essentially computes the difference in elevation values from a center cell and the eight cells surrounding it immediately. Then it squares each of the eight elevation difference values to make them all positive and averages the squares. TRI is then derived by taking the square root of this average, and corresponds to average elevation change between any point on a grid and its surrounding area. The authors of the TRI propose the classification for the values obtained for the index (Table 1):

Table 1: Terrain Ruggedness Index classification.

TRI	Interval (m)	Represent
1	0-80	Level terrain surface
2	81-116	Nearly level surface
3	117-161	Slightly rugged surface
4	162-239	Intermediately rugged surface
5	240-497	Moderately rugged
6	498-958	Highly rugged
7	959-4367	Extremely rugged surface

The pseudo-code [12] to generate TRI layer is:

```
program TRI
 {dem   - Input Grid
  tmp1  - Grid to store the Standard elevation
             difference
  tmp2  - Grid to calculate the Topographic
             Ruggedness Index
  tmp3  - Grid to verificate tri range
     outgrid - Output grid
  /* Standard elevation difference */
  /*Execute cell by cell*/
    tmp1(X,Y)  :=((SQRT(dem(x,y)-dem(x-1,y-1))+
            (SQRT(dem(x,y)-dem(x,y-1))+
      (SQRT(dem(x,y)-
            dem(x+1,y-1))+  (SQRT(dem(x,y)-dem(x+1,y))+
            (SQRT(dem(x,y)-dem(x+1,y+1))+
      SQRT(dem(x,y)-
            dem(x,y))+   (SQRT(dem(x,y)-dem(x-1,y+1))+
            (SQRT(dem(x,y)-dem(x-1,y)))
  /* Evaluate cell-by-cell
            tmp2(X,Y)  := SQRT(tmp1(x,y))
  /* Evaluate cell-by-cell
            tmp3(X,Y)  := If  (tmp2(x,y)>=5000)
             then tmp3(x,y)  := 5000
               Else tmp3(x,y):=tmp2(x,y)
  /* Evaluate cell-by-cell
  outgrid(X,Y)  :=(if(tmp3(x,y)>=0  && tmp3(x,y)<=80)
                     then     outgrid(x,y):=1
                  if(tmp3(x,y)>=81  && tmp3(x,y)<=116)
                     then     outgrid(x,y):=2
                  if(tmp3(x,y)>=117  && tmp3(x,y)<=161)
                     then     outgrid(x,y):=3
                  if(tmp3(x,y)>=162  && tmp3(x,y)<=239)
                     then      outgrid(x,y):=4
                  if(tmp3(x,y)>= 240  &&tmp3(x,y)<=497)
                     then     outgrid(x,y):=5
                  if  (tmp3(x,y)>=498  &&tmp3(x,y)<=958)
                     then          outgrid(x,y):=6
                  if(tmp3(x,y)>=959  &&tmp3(x,y)<=5000)
                     then      outgrid(x,y):=7)}
```

5 Generation of drainage density

Drainage density is defined as the total length of channels divided by area and measured the degree to which a landscape is dissected by channels [13]. To generate the drainage density layer, it is necessary to build a regular grid of 1 km^2 per cell [14]. Using this layer, we can construct the centroid layer. Later, the drainage layer is intersected with the grid layer. For each cell of the grid the lengths by area unit are added into centroid layer. The centroid layer is

Management Information Systems, C. A. Brebbia (Editor)

interpolated and the drainage density layer is obtained. Fig. 4 shows the process to generate the drainage density layer.

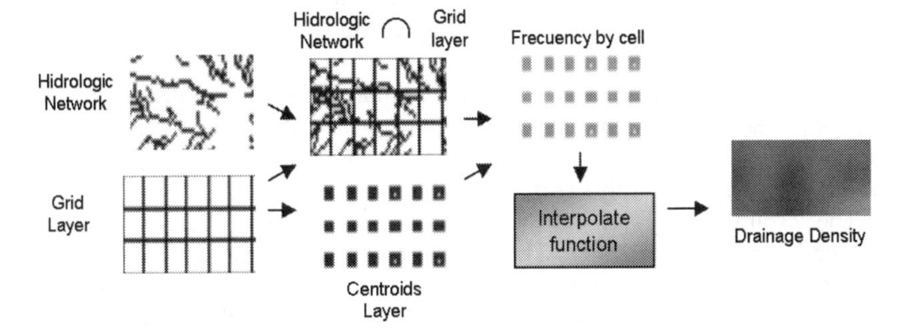

Figure 4: Process to generate the drainage density layer.

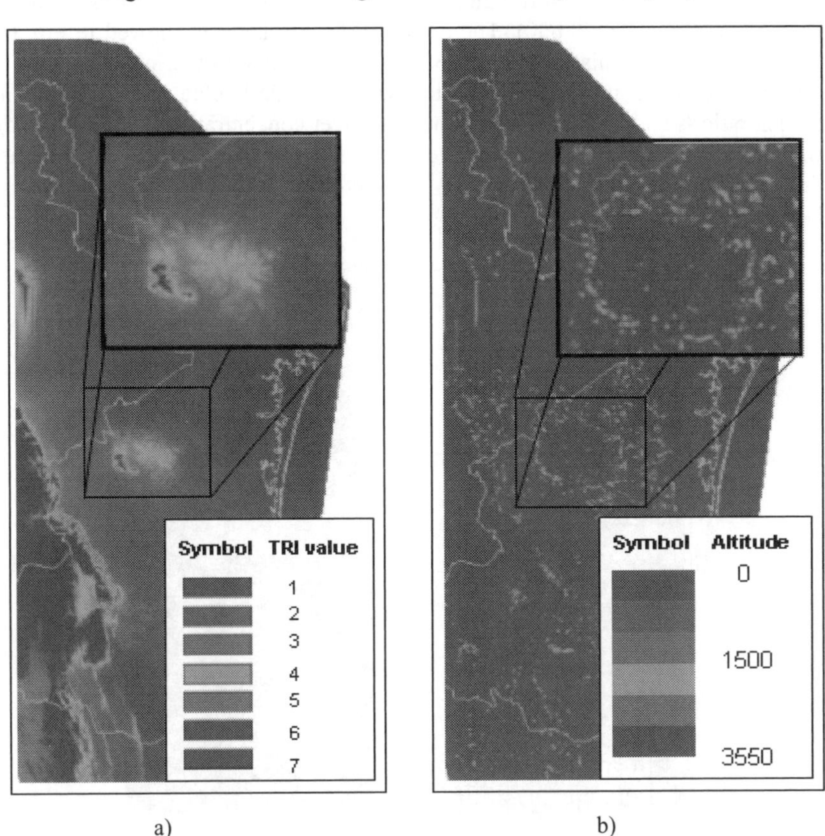

Figure 5: a) Digital Elevation Model, b) Terrain Ruggedness layer.

Management Information Systems, C. A. Brebbia (Editor)
© 2004 WIT Press, www.witpress.com, ISBN 1-85312-728-0

6 Results

Using SAM, we construct drainage density and terrain ruggedness layers. The method has been applied to the Tamaulipas State, Mexico. Some results are presented in this section.

Fig. 5a shows the original DEM. The minimum value is 0 m, maximum value is 3496, mean value of this layer is 227.40 m and the Standard Deviation is 498.469. Fig. 5b shows the Terrain Ruggedness layer constructed by SAM, and the TRI classification of this area.

The terrain index layer has the following values; mean is 2.386 m and the Standard Deviation is 2.457. This means that Tamaulipas State has slightly rugged areas in its territory. The extremely rugged areas are principally concentrated at the southwestern part of Tamaulipas State. DEM and TRI Layers are composed by 8000 rows and 2478 columns.

Fig. 6a shows the hydrological layer, this layer contains all streams of Tamaulipas State (1:200,000). The drainage density layer is showed in Fig. 6b. The mean value of this layer is 24857, which is nearly to the lower value. The concentrations are represented in blue scale, the dark blue represents higher concentrations and light blue represents the lower concentrations. We can see the highest concentrations of drainage are situated in the south coast, near Tampico City. While the lowest density is presented in the northwestern part of Tamaulipas State.

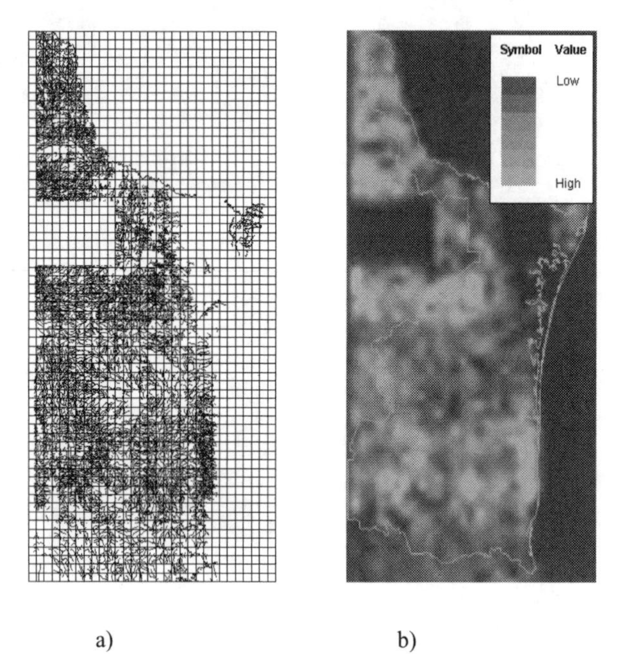

Figure 6: a) Hydrological Layer, b) Drainage density.

7 Conclusion

In this work, a GIS-application (SAM) has been developed to analyze geomorphometric characteristics of geo-images. SAM detects drainage density and terrain ruggedness using raster image data. In this method, spatial and attribute data are used to generate raster data. Using SAM, it is possible to define the *semantic* importance of the characteristics of the spatial data. Users can modify the criteria to have different scenarios to improve the decision making process.

The geomorphometric analysis is traditionally performed using the methods based on topographic map-processing in manual way. Our approach significantly decreases the amount of time and effort required to quantify selected terrain characteristics. Other methods are designed to evaluate additional characteristics, which are different to the properties proposed in our approach. However, these methods can be integrated into SAM. The generation of drainage density and terrain ruggedness layers facilitates the extraction of spatial characteristics that can be used in other cartographic processes, for instance in the generalization.

Acknowledgments

The authors of this paper wish to thank the Centre for Computing Research (CIC-IPN), General Coordination of Postgraduate Study and Research (CGEPI), National Polytechnic Institute (IPN) and the Mexican National Council for Science and Technology (CONACYT) for their support.

References

[1] Bonk, R., Scale-dependent Geomorphometric Analysis for Glacier Mapping at Nanga Parbat: GRASS GIS Approach. *Proceedings of the Open source GIS - GRASS User's conference 2002*, Italy, 2002.

[2] Goodchild, M., Perspective: Browsing metadata, where do we go from here?. *Geo Info Systems,* 10, pp. 30–31, 2000.

[3] Wharton, G., Progress in the use of drainage network indices for rainfall-runoff modelling and runoff prediction. *Progress in Physical Geography*, 18, pp. 539-557, 1994.

[4] Desmet, P.J.J. & Govers, G., Comparison of routing algorithms for digital elevation models and their implications for predicting ephemeral gullies. *International Journal of Geographical Information Systems*, 10, pp. 311-331, 1996.

[5] Wiche, G. J., Jenson, S. K., Baglio, J. V., and Dominguez, J. O., *Application of digital elevation models to delineate drainage areas and compute hydrologic characteristics for sites in the James River Basin, North Dakota*, U.S. Geological Survey Water-Supply Paper, 1992.

[6] Garrote, L., and Bras, R. L., A distributed model for real-time flood forecasting using digital elevation models, *Journal of Hydrology*, 167, pp. 279-306, 1996.

Management Information Systems, C. A. Brebbia (Editor)

[7] Instituto Nacional de Geografía Estadística e Informática (INEGI), *Modelos Digitales de Elevación - Generalidades y Especificaciones*, INEGI, 1999.

[8] Torres, M., Moreno, M., Menchaca, R. and Levachkine, S., Making Spatial Analysis with a Distributed Geographical Information System. *Proc. of the International Conference on Databases and Applications (DBA'2003),* pp. 1245-1250, 2003.

[9] Mitas, L. and Mitasova, H., General Variational Approach to the Interpolation Problem, *Journal of Computers and Mathematics with Applications*, 16, pp. 983 – 992, 1998.

[10] Molenaar, M., *An Introduction to the theory of spatial object modelling for GIS*. Taylor & Francis, U.K., 1998.

[11] Torres, M., Moreno, M. and Levachkine, S., SIGETAM: Herramienta GIS de Escritorio enfocada a la Detección de Zonas de Riesgo de Deslave e Inundación, *Proc. of the 1st International Congress of Informatics and Computing*, Guadalajara, México, pp. 156-168, 2002.

[12] Riley, S. J., DeGloria, S. D., and Elliot, R., A terrain ruggedness index that quantifies topographic heterogeneity, *Intermountain Journal of Sciences*, 5, pp 23-27, 1999.

[13] Horton, R. E., Drainage basin characteristics, *Transactions American Geophysical Union*, 13, pp. 350-36l, 1932.

Management Information Systems, C. A. Brebbia (Editor)

GIS and GPS as useful tools to determine transportation noise levels

D. Fabjan[1], D. Paliska[1] & S. Drobne[2]
[1]Faculty of Maritime Studies and Transportation, University of Ljubljana, Slovenia
[2]Faculty of Civil and Geodetic Engineering, University of Ljubljana, Slovenia

Abstract

Road traffic is one of the biggest sources of environmental problems. Traffic needs and the motorization rate are increasing due to economic growth, which in turn tends to further generate rapid increase in traffic flow on the Slovenian road network. Thus an attempt to establish the influence of traffic flow on environmental degradation has been made.

For this purpose road vehicle noise was analysed. After evaluating the mean speed of vehicles and the gradient of each road section, a model for computation of noise levels was applied. The Global Positioning System (GPS) device was used to determine the actual speed on chosen road sections which enabled the calculation of the influence of the slope and the number of crossroads on the average speed. The ArcView program was used for the Geographical Information System (GIS) approach to forecast noise levels, as well as to present the results of the analysis. The results confirmed a significant influence by traffic flow on noise.
Keywords: road traffic, traffic noise, GPS, mean vehicle speed, road gradient, GIS, noise levels forecast.

1 Introduction

Road traffic is one of the most widespread sources of noise. It affects practically all people, thus its negative effects are of great concern and a lot of attention is paid to its study. Every road vehicle emits noise – some more, others less –

Management Information Systems, C. A. Brebbia (Editor)

therefore a road with more intensive and more constant traffic flow is more aggravating to people and the enviroment than a road with less traffic.

When assessing the intensity of traffic effects on noise, noise levels caused by vehicles circulating on roads have to be calculated considering different circumstances in which the noise is generated. The process of noise level determination is a very complex one as it takes into account as much conditions and assumptions as possible. The level of traffic noise is caused by factors such as number of vehicles per hour, mean vehicle speed, percentage of heavy vehicles, road surface characteristics, road gradient, distance of measuring point from the road edge, height of the measuring point, ground cover characteristics, view angle, reflection from the objects in the vicinity, etc.

Most of the time it is not possible to consider every factor that influences the noise generation due to lack of information. In such cases the condition has to be simplified or even omitted.

1.1 Existing methods for estimation of noise levels

The best way to determine noise levels is of course to go to the site and measure them or to install a permanent measuring device that measures the noise at all times. Both of these options are still quite expensive, and so most researchers base their research on measurements of representative samples, which then enable them to create models for approximate determination of noise levels considering the affecting factors as well.

When deciding on the method to be used for measuring the noise levels caused by traffic, two strategies should be considered. The first one enables the determination of noise levels at points situated within a relatively regular grid of measuring points in the area studied. The second strategy offers the possibility of measuring noise levels by taking into account the previously classified studied area use, demographic density or the importance of the chosen streets and roads (Calixto et al. [1]).

1.2 Models FHWA and CORTN

There are several models developed for calculation and evaluation of noise levels caused by road traffic. The most frequently used are the FHWA, developed in the United States of America and the CORTN, which was developed in Great Britain. Both models are quite precise compared those which consist of most factors that affect the intensity of noise, and at the same time they are quite simply designed. (Pamanikabud and Tansatcha [2]).

The mathematical simulation model FHWA was developed for estimation, analysis and forecasting of noise levels generated on roads by a continuous traffic flow. The model divides the vehicles into three groups. For each group a separate calculation is required for basic noise levels, traffic flow correction, and shielding and barrier correction. Other adjustments such as distance correction and finite road adjustment are the same for all vehicle groups. The noise level in this model is expressed as L_{eq} in decibels (dB(A)) [2].

Management Information Systems, C. A. Brebbia (Editor)
© 2004 WIT Press, www.witpress.com, ISBN 1-85312-728-0

The model CORTN is also used for forecasting the noise levels in uninterrupted traffic flow conditions. The noise levels are expressed here in decibels (dB(A)) as L_{10} (noise level that is excessive in 10% of the time of measurement) for the loudest hour, and as L_{10} for an 18 hour time interval. The basic noise level in this model is calculated considering only passenger cars, adjusted by speed correction, percentage of heavy vehicles, gradient correction, road surface with propagation correction [2].

Most models for estimation of noise levels use the value L_{10} expressed in dB(A), which represents an average value of the 18 hourly values L_{10} for the period from 6 a.m. to midnight on working days. The Slovenian decree on noise owing to road and railway traffic defines a methodology to estimate sound pollution, which is based on the German standard RLS 90 and enables the evaluation of road vehicle noise emission percieved at the distance of 25 meters from the road axis from an approximate hight of 2.25 meters and by the mean speed of 100 kilometers per hour.

In this paper a method to calculate and forecast the noise levels that would be perceived at the distance of 10 meters from the road edge and by the mean speed of 75 kilometers per hour will be used and simplified with the assumption that the ground is flat and that there are no barriers or other objects in the road vicinity.

2 Noise levels on road sections

Data analysis and graphical presentation of the study results were effective, fast and simple due to the GIS approach using the ArcView program. This approach enables road traffic noise simulation and forecasting, thus makes the understanding of traffic effect on noise levels better. Geographical and attribute data from different databases were merged to make the analysis simplier and clearer, which increased its efficiency in changing and testing the parameters and conditions.

The reference database in this analysis was the Slovenian road network database with several attribute data and defined road sections that represented the basic unit on which noise levels were estimated. The major road network is shown in figure 1. The data on road sections enabled also the merging of other databases such as the attribute data on traffic count on the Slovenian road network. Based on traffic count data it was possible to calculate the hourly traffic flow and the percentage of heavy vehicles (vehicles >3.000 kilograms) for each road section.

2.1 Calculation of the road section mean speed

It has been proven that vehicle speed affects the noise level and that is why all noise level forecasting models contain the speed correction factor. The question arises how to determine the speed at one point or on one road section. Most of the models use the speed limit defined by regulations according to the road category. A lot of the times roads of any category are equipped with traffic signs

Management Information Systems, C. A. Brebbia (Editor)

that additionally limit the speed, or road characteristics and road conditions cause traffic flow to be slower or faster than the set speed limit.

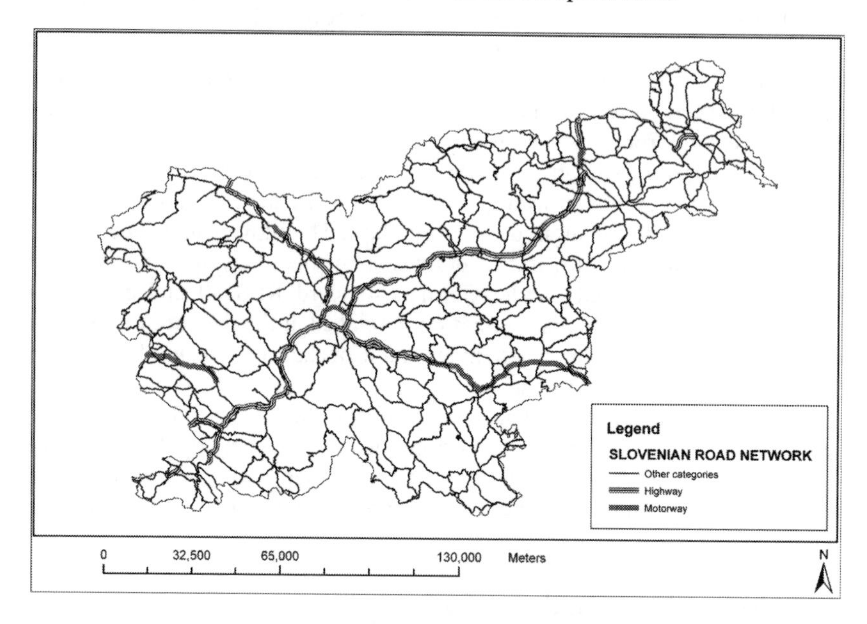

Figure 1: Slovenian major road network.

In our study we developed a model to evaluate the mean speed within road sections using the GPS device. The device enabled us to gather a sample of data for geographical coordinates and speed at different points during each trip on chosen road sections. Trips on roads of different categories and in different traffic conditions were made in the coastal region of Slovenia.

After collecting the GPS data an attempt was made to consider and estimate the influence that crossroads have on speed. Therefore the number of crossroads on each road section was defined by the query in the Slovenian road network database. Settlements were treated as crossroads. The coefficient of crossroads influence on the mean speed of road sections was calculated with the coefficient of determination 0.68 applying linear regression analysis by taking into account the regulated speed limit. The analysis showed that the coefficient (-1.4638) is statistically significant, which was confirmed by the global F test as well. The mean speed for each road section can be then defined by the following equation:

$$v = -1.4638 * C + v_l \quad (1)$$

where
v ...mean speed
C ...number of crossroads on road section
v_l ...speed limit for different road categories

Management Information Systems, C. A. Brebbia (Editor)

The negative value of the regression coefficient shows that crossroads play the role of a barrier to traffic flow, thus causing a decrease in speed.

The linear model introduced is simple and at the same time represents a good challenge to work on its improvement. The mean speed of course depends on several further factors that at this point were not considered due to lack of data.

The estimated mean speed of each road section was then used in further calculations of noise level correction.

2.2 The estimation of noise levels

The model CORTN was used to estimate noise levels (Cardiff University [3]) The formerly estimated mean speed and the road gradient were considered as well as the traffic flow structure and the immision point distance. The noise level was expressed as L_{10} in one hour intervals to better understand the intensity of the traffic influence on noise.

2.2.1 Basic noise levels

Basic noise levels for each road section were estimated using the equations provided by the CORTN model.

Matematically the basic noise level can be defined as:

$$L_{10} = 42{,}2 + 10 * \log_{10} q \tag{2}$$

where
L_{10} ...basic noise level for one hour intervals (dB(A))
q ...number of passenger cars in one hour intervals (vehicle/h)

2.2.2 Mean speed and heavy vehicles percentage corrections

In reality, noise levels tend to be much higher than basic noise levels, which is due to several factors that cause such an increase. Thus the correction due to heavy vehicles percentage was considered additionally to the previously calculated mean speeds of road sections. The equation follows:

$$L_{v,p} = 33 * \log_{10} (v + 40 + 500/v) + 10 * \log_{10} (1 + 5p/v) - 68{,}8 \tag{3}$$

where
$L_{v,p}$ ...mean speed and heavy vehicles percentage correction
v ...mean speed of vehicles (km/h)
p ...percentage of heavy vehicles (%) – number of heavy vehicles divided by number of passenger cars

2.2.3 Road gradient correction

Road gradient also affects noise levels caused by road vehicles. Vehicles use more power to move uphill, creating more noise, and brakes used to move downhill are also an additional noise generator. Thus, evaluating the road gradient for each road section was essential to show the intensity of gradient influence on noise levels.

Management Information Systems, C. A. Brebbia (Editor)

A digital model of relief (DMR) had to be made to determine the road gradient of each road section. A DMR added a third dimension to a two dimensional map of the Slovenian road network. Based on the data given a DMR could be made applying the IDW (Inverse Distance Weighing) interpolation.

The added third dimension enabled a calculation of average road gradients for all road sections. Further calculations were made under the assumption that the traffic moves uphill on each road section. This approach does not show the real condition of road slopes, but it is a good approximation to estimate the correction for the gradient factor on each road section.

It was estimated that each additional percent of the gradient increases the correction of noise level by 0.3 dB(A) [2]:

$$L_g = 0.3 * G \tag{4}$$

where
L_g ...road gradient correction
G ...gradient (%)

2.2.4 Immision point distance correction

The lack of information on ground characteristics, road surface material, and other factors made it impossible to evaluate their corrections. However, the immision point distance correction was considered for the case of the immision point 2 meters high and 4 meters away from the road. The following equation derives from the CORTN model:

$$L_d = -10 * \log_{10} (d/13.5) \tag{5}$$

where
L_d ...immision point distance correction
d ...the shortest distance between the emission and immision points

The shortest distance between the emission and immision point was calculated simply by the Pitagora rule:

$$d = ((r + 3.5)^2 + (h - 0.5)^2)^{1/2} \tag{6}$$

where
r ...horizontal distance of immision point
h ...height of immision point

Using the above formula for immision point distance correction that rests on the assumption that the surrounding ground is hard, flat and without barriers, which is quite rare in reality, it is possible to come to 7.65 meters of total shortest distance between the emission and immision points. The immision point distance correction is then 2.47 dB(A). This value was included into a final estimation of noise levels and was constant for all road sections.

Management Information Systems, C. A. Brebbia (Editor)
© 2004 WIT Press, www.witpress.com, ISBN 1-85312-728-0

2.3 Total noise levels on road sections

The final estimation of noise levels for all road sections is actually a sum of all values explained previously. First the basic noise level is calculated to which we then add the speed correction, the heavy vehicle percentage correction, the road gradient correction, and the correction of the immission point distance for the hard and flat ground without barriers.

There is a sum assigned to each road section. This sum represents the noise level (L_{10}) perceived by the 2 meters high and 4 meters distant immission point and exceeded by 10% of one hour traffic flow. These noise levels are the best indicators for traffic flow influence intensity on generating noise.

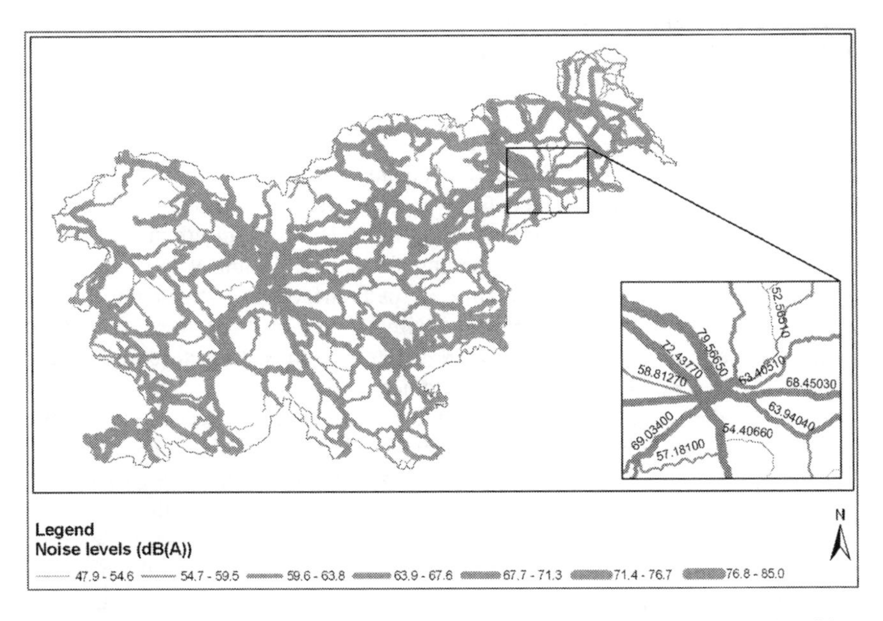

Figure 2: Noise levels on road sections of Slovenian road network with a detailed map of a chosen geographical area.

Figure 2 presents noise intensity on road sections of the Slovenian road network. Figure includes a detailed map showing the noise bands for a chosen geographical area in the north-east of Slovenia. Each road section is represented according to the noise intensity wide band using the average noise level L_{10} value for a better understanding.

3 Conclusion

The band width is subject to noise intensity generated on road section by vehicles. The pictures confirm the assumption that noise levels are higher where the traffic flow is more intensive. Thus the more vehicles using the road section,

Management Information Systems, C. A. Brebbia (Editor)

the higher the noise levels. It is possible to notice though that some road sections with a relatively lower annual traffic flow are represented with a much wider band, which in turn means that the noise levels are higher compared to other road sections with similar annual traffic flow. This leads to the conclusion that the correction factors in such cases are of greater importance.

Traffic flow speed plays an important role on some road sections. Compared to the road sections with a slower traffic flow, noise levels can be higher on road sections where vehicles can reach higher speeds.

Noise is also very dependent on the traffic flow structure. Following the rule that heavy vehicles generate more noise, noise levels on road sections with a high percentage of heavy vehicles tends to be higher.

The same situation is found when considering the road gradient factor. In fact, road sections with relatively more slope add more to a road gradient correction factor than those with less or no slope at all.

Providing that data on location of the surrounding barriers such as trees, buildings and other objects, and data on condition of the road surface were available, adjustments could be calculated and later on added to total noise levels, which would certainly increase even more. It is important to consider the barriers especially in the case of estimating noise levels in an urban area where the sound pollution is the most damaging for people (Ramiset al. [4]). Noise generation depends also on the quality of road surface. Relevant data on road condition is quite difficult to be managed with geographical coordinates and kept up-to-date due to unequal use and material characteristics. Therefore in this case the approach described in this paper would not make sense.

However, the introduction of the simple yet challenging linear model for the computation of mean speed of road sections is a step towards the implementation of valuable tools like GIS and GPS technology, and a motive for improvement. The possibility of simulation and more and more precise estimation of sound pollution caused on our roads by vehicles represents also a good tool to perceive the problem accurately, to make better decisions and to act against the negative effects of the intensity of traffic flow influence on our health and lives in general.

References

[1] Calixto, A., Diniz, F.b., Zannin, P.H.T., The statistical modelling of road traffic noise in an urban setting. *Cities*, Vol. 20, No. 1, pp. 23-29, 2003.

[2] Pamanikabud, P., Tansatcha, M., Geographical information system for traffic noise analysis and forecasting with the appearance of barriers. *Environmental Modelling & Software*, Vol. 18, Issue 10, pp. 959-973, 2003.

[3] Cardiff University. Square one research PTY Ltd. Web Site, www.squ1.com/flash-menu.html.

[4] Ramis, J., Alba, J., Garcia, D., Hernandez, F., Noise effects of reducing traffic flow through a Spanish city. Applied acoustics, No. 64, pp. 343-364, 2003.

Management Information Systems, C. A. Brebbia (Editor)

Investigation of asthmatic disorders using GIS technology

S. Jamaludin S.A, M. S. Zainol & Z. M. Saat
Universiti Teknologi MARA (UiTM), Shah Alam, Selangor, Malaysia

Abstract

With environmental health moving into a critical stage in Malaysia, environmental management becomes more relevant and challenging, There is increasing evidence that exposure to air pollutants has adverse effects on public health, in particular on cardiac and respiratory disease. The increasing number of asthmatic and other respiratory cases has affected the health and quality of life of the population. With the belief that air pollution contributes to the rise in asthmatic and respiratory cases, there is a need to explore the relationship between air quality and respiratory disease.

As yet, there is no known report of the use of a Geographical Information System (GIS) in epidemiological research in Malaysia. It is our belief that making spatial statistical techniques available for epidemiologists in GIS applications can assist epidemiological research. An attempt by the Ministry of Health, Malaysia to monitor the incidence of acute exacerbation of asthmatic cases in the year 2002 did not seem to give the required results. It did not provide an accurate picture of the effect of air pollution on public health.

This paper describes an ongoing attempt at combining the use of GIS and statistical analysis in an epidemiological related research. The problems of increasing incidence of asthmatic problem and its relationship to environmental pollutants are being investigated. The task is also to investigate whether respiratory disease incidence increases, particularly when levels of specific air pollutants are elevated.
Keywords: air pollutants, asthma, Geographical Information System, spatial statistical analysis.

Management Information Systems, C. A. Brebbia (Editor)
 ISBN 1-85312-728-0

1 Introduction

Studies conducted by the Ministry of Health, Malaysia found escalating incidence of asthma within the country and identified the state of Perlis as having the highest percentage of respiratory and asthmatic cases [5]. On the other hand, an asthma surveillance program conducted by the Disease Control Division of the Ministry of Health identified incidence of acute exacerbation of asthmatic cases in 5 locations in the country [1]. However the program did not produce the required results due to insufficient data and the health data areas being not geographically represented.

Some works involving the use of GIS in health research, linking air pollution models to GIS for the purpose of defining areas of exposure have been reported. Disease incidence were then linked to pollution. Gatrell *et al.* [2] studied the use of modern point pattern methods in exploring and modelling disease risk. Second-order methods were also included for detecting disease clustering. The result suggested that disease clusters could not be investigated unless their sizes and boundaries coincide at least roughly with the spatial units for which the data have been encoded. There were tendency for cases to cluster or aggregate more than the population at risk.

Martin [3] applied statistical methods for spatial epidemic modelling, which include use of spatial regression, test for spatial randomness and map smoothing techniques. These analyses portrayed patterns of disease rates on a choropleth map and showed rates with different statistical reliability in different areas. Rogerson [4] developed a spatial version of the *chi-square goodness-of-fit statistics* in a test for spatial clustering. This approach was able to filter disease cases and the people at risk for areas that can be controlled in both size and shape. The technique indicated the likelihood that clusters exist at particular locations.

2 Objective

The current study is a preliminary stage of an ongoing research into identifying factors which contribute to the increasing incidence of respiratory problems among the state population using a combination of GIS and statistical techniques. The association between respiratory disease and the environment will be investigated using a GIS analysis to examine disease patterns and disease rates at various levels of spatial resolution. It is complemented by statistical analyses in order to identify causal factors of respiratory problems. The current preliminary stage reported in this paper is a research into finding facts to test the hypothesis that non-environmental factors such as physical stress, emotional stress and physical surroundings are not significant causes to the incidences of asthma among the population. It is an attempt to try to isolate these factors from the set which are thought to be causal to the problems, i.e., environmental factors.

Management Information Systems, C. A. Brebbia (Editor)

In the current study data on asthma patients who seek treatment at the local hospital from January 2003 through March 2004 were collected. Patients' addresses were geo-coded and added to a GIS layer of census tract. Incidence rates were calculated for each census tract. Maps were created using Mapinfo and ArcView. They will later be used for identifying areas at risk of respiratory disease. Smoothed rate maps will be produced to identify spatial patterns of respiratory problems. The results will then be used to produce GIS maps for the Health Department of the state of Perlis which will be useful in respiratory disease prevention programs in specific areas and help community groups understand the impact of air pollution on respiratory disease.

3 Geographical area of study

The study area, the state of Perlis is the smallest and northern most state in Peninsular Malaysia which covers an area of 810 square kilometres. A large portion of the state is low lying and well under 61 meters. The state capital is Kangar. Arau, the *Royal Town* is 10 km away. There are 22 districts (or mukim) in Perlis. The state is relatively poor with a fairly high rate of unemployment and inadequately developed infrastructure and water supply. The economic activity of the state is predominantly agriculture which made up about 65.3% of the land use, with an insignificant industrial sector.

There are several reasons for choosing the area. Firstly, the state of Perlis has been identified to have the highest percentage of respiratory and asthmatic cases in Malaysia. Secondly, the required map and information on asthmatic patients are available and accessible. Thirdly, it is an area of environmental interest since the area, although small, is concentrated in its industrial activities. A large cement factory, a fairly vast sugar cane plantation complete with a refinery, as well as a vast padi plantation. The fairly well known post-harvest open burning activities of both the sugar cane and the padi plantations has invited numerous speculations on their roles in the aggravation of respiratory problems among the population. However, to date there have been no study to support the speculation.

The climate of Perlis is tropical monsoon. Temperature is relatively uniform within the range of 21°C to 32 °C throughout the year. Humidity is consistently high on the low lands ranging between 82% to 86% per annum. The mean rainfall is between 2,032 mm to 2,540 mm with the wettest months from May to December. During the months of January to April the weather is generally dry and hot.

4 Data acquisition methodology

The implementation of the study involves the process of data acquisition among a sample which consists of (a) a sample of persons registered as patients in clinics plus (b) a sample of non-patients selected randomly from among the neighbors of the sampled patients. The data acquisition process involves the planning and execution of a *sample survey* to capture data pertaining to factors

Management Information Systems, C. A. Brebbia (Editor)
 ISBN 1-85312-728-0

and dimensions of asthmatic problem, profiles of respondents (category of respiratory problems, demographic profile) as well as geographical and environmental profiles of areas. Data acquisition also includes (i) the dissipation of information about the impending survey to the respondents and (ii) getting the permission and cooperation of the guardian of other data sources (in this case the Perlis Health Department.

4.1 Measured variables

Variables measured for the study up to the current stage include, among others, category of respiratory problems, demographic, geographical and environmental profiles. The variables measured in the survey for the preliminary study are as follows:

Table 1: Variables in the survey.

Variable Group	Section in Questionnaire
1. Demographic Profile of Respondents	A
2. Geographical and Environmental Profiles of Areas of Residence	B
3. Standard of Living Profile of Respondents	C
4. Health Profiles (Related to Respiratory Problems)	D1 – D3
5. Causal/Trigger Factors of Respiratory Problems	D4 – D6
6. Stress Factors	D7

4.2 Survey sample size

The size of the sample for the survey was based on the chosen level of precision (B) of the estimate of the proportion (p) of respondents falling into a particular category or having a certain perception of an issue investigated by the survey, and the actual proportion of respondents agreeing to an issue or having a certain perception of an issue raised in the survey. The size of the sample n is calculated using the following expression $n = p(1-p)\left(\frac{z_{0.025}}{B}\right)^2$. For example if the *precision level B is chosen to be 0.03* and the *value of p obtained from the pilot survey is 0.6* (i.e., 60%), then $n = 0.6(1-0.6)\left(\frac{1.96}{0.03}\right)^2 = 1024$. Thus if the proportion obtained in the survey is 0.60, one is 95% confident that the true proportion in the overall population is between 0.57 (0.60-0.03) and 0.63 (0.60+0.03), i.e., between 57% and 63%. In this preliminary stage a total of 599 respondents were selected.

4.3 The asthmatic database of the Perlis Health Department

An important data source for this research is the filled questionnaires which form part of the records of more than 1000 patients between ages of 1 to 80 years kept

in the Perlis Health Department. The Perlis Health Department is involved in the study in the following manner:

1. The collection/compilation of data about frequency and seriousness of asthma and allergies in the population from various age categories under different living conditions.
2. The collection of basic epidemiological data, in order to make predictions about variations in frequency and seriousness of these illnesses in future years.
3. The development of a framework for future research, examining links with genetics, lifestyle, environmental factors and medical care.

5 GIS analysis

Visualization in GIS is important for better understanding, while statistical tests provide new understanding of associations between epidemiological and environmental phenomena. The currently available health data is not yet complete, and do not cover the whole of the state of Perlis.

In the next stage of this ongoing study a GIS database will be built by collecting and converting topographic maps, land use maps and other related map data into a GIS system. A GIS spatial analysis will be used to examine disease patterns and disease rates at different levels of spatial resolution. The detection of clusters will also be carried out in order to investigate the likelihood their existence at particular locations. A logistic regression analysis will be applied to estimate the probability of occurrence of respiratory problems at a particular location.

A frequency map of the distribution of asthmatic incidence within the study area has been constructed using data obtained in this preliminary stage. Figures 1 shows a spatial distribution of the asthmatic incidence in the area under study while Figure 2 shows the location of factories of the area. On closer scrutiny it becomes evident that asthmatic incidence tends to cluster in the Western and in the middle of the region. The North-East is the location of the cement factory and the sugar refinery, conjectured to be the sources of chemical emission and burning activities. Other activities such as quarrying, rice milling and other smaller industries are distributed over the region. The extreme western part of the high asthmatic incidence area is centred at the relatively high populated area of the capital town of Kangar.

5.1 Spatial analysis at the address level

The preliminary analysis of the data takes place at address-based levels since the asthma patients can be located at exact geographical locations. The first step is to find any spatial disease patterns. The results of this analysis show that the number of cases with diagnosis of asthma, asthma symptoms during the last 12 months is more than expected. In these cases a further analysis is definitely worthwhile.

Management Information Systems, C. A. Brebbia (Editor)

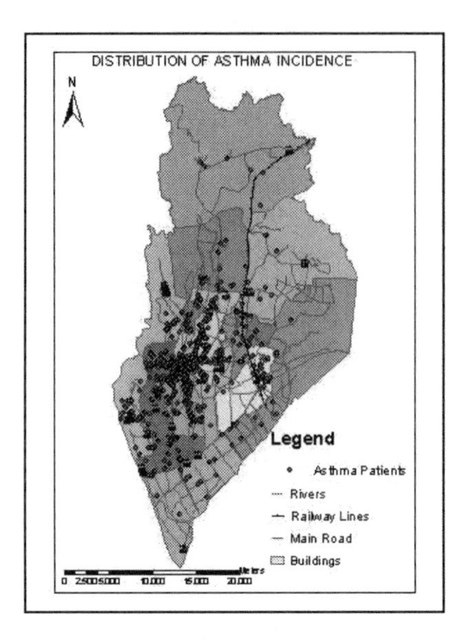

Figure 1: Distribution of asthmatic incidence in the study area.

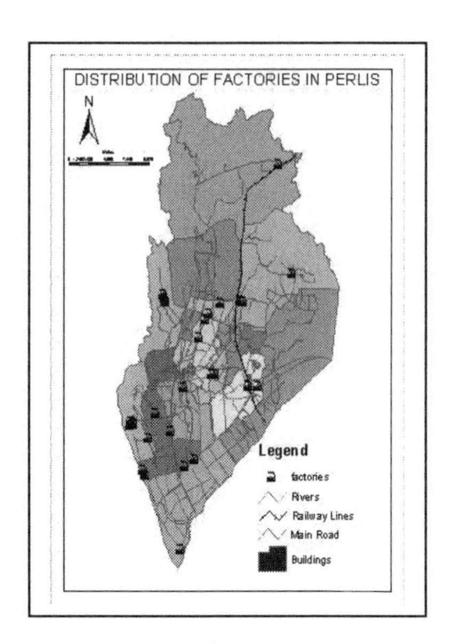

Figure 2: Location of factories in the study area.

The point data at address level have the following properties: Since their complete address is known, a pair of xy-coordinates is attached to every patient. There more than 100 attribute values known at each point, information about disease symptoms, the environment, lifestyle and related attributes, including stress factors. The information about disease symptoms is mainly bivariate: the individuals have the symptom or they do not have [2].

Because of confidentiality, raw point data cannot be visualized as they stand but has to be aggregated to small areas. The research region will be divided into small administrative areas called Mukim.

The epidemiologist of the Health Department has constructed two broad hypotheses as a starting point for the spatial analysis on this level. First, that there is an obvious relationship between the air pollution (conjectured to originate from the cement factory, sugar refinery/plantation and the padi plantation), and different allergy/asthmatic symptoms. Second, that there could be a relationship between social status and different allergy/asthmatic symptoms.

The study aims to investigate if there is any clustering in the 21 symptoms. A variety of methods exist to detect clusters and clustering in a point map [2]. Wartenberg and Greenberg [6] describe a strategy to select an appropriate method of cluster detection. First, the selection of the data type: the location of an event, the distance between all pairs of events, the nearest-neighbour distance between events, or the distance to a fixed point.

After finding an eventual spatial pattern in the data, this pattern will be compared with the spatial pattern of possible causal factors. Correlation,

Management Information Systems, C. A. Brebbia (Editor)

covariance and regression methods are often used to detect relations between variables. Since disease data often have a binomial character, logistic regression will be used.

6 Statistical analysis

The results of the analysis on whether non-environmental factors (non-environmental factors investigated in this study include emotional and physical stress, physical surrounding, and socio-economic and habitual factors) contribute to asthmatic problems among the population yield the following results:

a. there were significant differences between the patients and the control group in the prevalence of 21 symptoms related to asthma. Hence it is concluded that respondents representing the control group has no significant asthmatic problems.

b. Analysis on whether stress factors contribute to the asthmatic problems among the population showed no significant difference in the effect of each of the 39 emotional and physical stress factors on the patients and those in the control group. One-Way ANOVA analysis between each of the 39 stress factors and respondents' category (whether they are patients or control) yielded p-values greater than 0.05 in all cases. This means that in all these cases the *null hypothesis of no difference in the effect of stress* between the patients and the control group are accepted. Hence it is concluded that these factors has no causal effect on the respiratory problems.

In the next stage of the research, measurements of particulate matter (air pollutant index) will be obtained. These will then be used to obtain census tracts which in turn will be used to examine the spatial distribution of respiratory illness and its relationship to areas of elevated particulate matter.

7 Conclusion

The use of spatial and statistical analysis in health research projects is increasing, but there are still important issues to be resolved. Spatial statistical analysis is also found to be quite complex, but produces significant statements about spatial patterns in epidemiological data. In the current study is part of an ongoing project, we have digitised basic map coverages, georefenced patients' data, visualized queries on different aggregation levels and made a fundamental statistical analysis.

Analysis on the contribution of stress factors to asthmatic problem showed no significant difference in their effect between the patients and the control group. It is thus concluded that these factors has no causal effect on the respiratory problems. The investigation into whether environmental factors have significant effects on asthmatic problems in the area will be carried out in the

Management Information Systems, C. A. Brebbia (Editor)

next stage of the study. It will involve, among others, investigations and into the conjectured sources of pollution (the cement factory, the sugar refinery, the sugar cane and padi plantations), the contents of pollutants in the ambient air and their role in asthma morbidity in Perlis.

References

[1] Daud, A.R., Rozlan, I., Doraisingam & Ikhsan, M.S., *Environmental Health: Malaysian's Perspective*, Disease Control Division (NCD), Ministry of Health Malaysia Publications, **1(2)**, pp. 4-8, 2002.

[2] Gatrell, A.C., Bailey, T.C., Diggle, P.J. & Rowlingson, B.S., Spatial point pattern analysis and its application in geographical epidemiology. *Transactions*, Institute of British Geographers, No. 21, pp. 256-274, 1996.

[3] Martin Kulldorff, Statistical methods for spatial epidemiology: Tests for randomness. *GIS and Health GISDATA 6*, eds. A. Gatrell & M. Loytonen, Taylor & Francis, Series Editors: Ian Masser & Francois Salge, 1998.

[4] Rogerson, P.A., The detection of clusters using a spatial version of the chi-square goodness-of-fit statistics. *Geographical Analysis*, No. 31, pp. 130-147, 1999.

[5] Rozlan, I., The Study on Asthma Admissions in Malaysia. Disease Control Division (NCD), Ministry of Health Malaysia, 1(1), pp. 10-17, 2002.

[6] Wartenberg, D. & Greenberg M., Space-time models for the detection of clusters of diseases. Spatial Epidermiology, ed. R.W. Thomas R.W. Pion: London, pp. 17-34, 1990.

Management Information Systems, C. A. Brebbia (Editor)

GIS/data mining applied for identification of environmental risk factors for diseases

S. Anno
School of Engineering, Shibaura Institute of Technology, Japan

Abstract

This study has tried to identify environmental risk factors for respiratory system diseases in Tokyo metropolitan area using GIS and data mining. GIS were applied to the analysis of spatial relationships between the distribution of the diseases and SPM exposure levels and distances from the national roads as environmental factors. A CART model as a data mining tool was applied for assessment of habitat types where there was a potential risk of the incidence of the diseases with the databases obtained from GIS analysis while identifying environmental risk factor levels concerned with the incidence of the diseases. Another purpose of this study was to provide a comprehensive outline of the methods of data mining, which have wide applications in a GIS and other areas where spatial data are used, whilst indicating its strengths and weaknesses, how and when to apply the methods and their role when used.
Keywords: geographic information systems (GIS), spatial analysis, data mining, classification and regression trees (CART), decision tree.

1 Introduction

Human respiratory effects of environmental factors are well documented; numbers of epidemiological studies have examined affect of environmental factors on respiratory health [1, 2, 3, 4, 5]. However, those studies have not analyzed spatial associations between respiratory system diseases and environmental exposures; moreover, those studies have not identified environmental risk factor levels concerned with the incidence of the diseases. Then, our study tried to analyze relationships between spatial pattern of the diseases and environmental factors; and identify environmental risk factor levels concerned with the incidence of the diseases using spatial analysis in geographic

Management Information Systems, C. A. Brebbia (Editor)

information systems (GIS) and data mining. Another purpose of this study was to provide a comprehensive outline of the methods of data mining, which have wide applications in a GIS and other areas where spatial data are used while indicating its strengths and weaknesses, how and when to apply and its role to be used.

2 Material and methods

2.1 Spatial and attribute data

The study site was focused on Tokyo metropolitan area. The digital map for the area with a scale of 1:2,500, which contains the national roads, was used as spatial data. As for attribute data, data on a hundred of the subjects diagnosed with typical symptoms of respiratory system diseases, which were collected from a health survey by the agency of Aozora, were used for the study. Also, data on the average SPM concentrations for 22 years (i.e., from 1978 through 1999) measured along the major roadways and residential districts of Tokyo consisting of 70 observation sites in total were used as attribute data [6]. Those spatial and attribute data were integrated within a GIS database of the study area and used for spatial analysis.

2.2 Spatial analysis

Spatial analysis in a GIS was conducted in the following procedures. First of all, SPM data plotted as points on a layer in a GIS were interpolated using the method of Inverse Distance Weighted (IDW) to estimate spatial distribution of SPM concentrations throughout the study area and converted into raster data, which divides space into a mesh unit. There are some intervening factors at the individual scale that may influence normal process of environment and diseases phenomena. Within the raster data, the spatial resolutions of the data were set to 100 m^2 and 500 m^2 respectively for addressing health events and environmental exposures at different spatial scales. The data sets consisted of 97,500 meshes for 100 m^2 and 5,246 meshes for 500 m^2 respectively. Secondary, the patients' addresses were converted into longitude and latitude by using geocoding service for CSV formatted file on WWW [7]; and geocoded to the digital map, which represented by points on a layer. Each data set including SPM concentrations, the national roads and the patients was represented by a separate layer. Those layers were overlaid and outputted as the raster databases. Each mesh of the raster databases represents the number of the patients, the value for SPM concentrations and the attribute value for distances from the national roads. Patient data within a mesh were maintained at the individual level and the data were aggregated by each mesh. In other words, the meshes that represent a patient/patients were assigned an attribute value of present; and the meshes that did not represent a patient/patients were assigned an attribute value of absent. The raster databases obtained from the spatial analysis in a GIS were exported as databases and used for further processes of data mining.

Management Information Systems, C. A. Brebbia (Editor)

2.3 Data mining

Data mining methodologies have become quite popular in recent years as computing speed and storage capabilities have increased. There are a number of classification and prediction algorithms that fall under the rubric of data mining. These include neural networks, clustering programs, association rules and classification and regression trees (CART). The ultimate objective of data mining is knowledge discovery within a large database. In other words, using a combination of machine learning, statistical analysis, modeling techniques and database technology, data mining finds patterns and subtle relationships in a large database and infers rules that allow the prediction of future results. This usually involves the prediction of some criterion variable value (i.e., outcome or dependent variable value) from a number of predictor or independent variable values. Data mining techniques have wide applications in a GIS and other areas where spatial data are used [8, 9, 10]. In our study, a CART approach [11] was applied for analyzing the databases, which were exported from the raster databases.

According to Breiman et al [11], a CART is an algorithm that learns binary decision tree representations. Decision tree models classify data using a series of if-then rules depicted in a tree representation. The basis of decision tree algorithms is the recursive partitioning of the data into more homogenous subsets. A CART model was trained in a round-robin fashion (i.e., leave-one-out or k-fold cross-validation with k=N) in order to assess the model performance [11, 12].

A CART model was implemented for assessment of habitat types where there is a potential risk of the incidence of respiratory system diseases using the databases in terms of the mesh size of 100 m^2 and 500 m^2 respectively as the estimated suitable area for population at risk. The following variables were considered as possible predictors of the presence/absence of patients in the modeling. The outcome variable has two categorical variable values: present and absent. The two predictor variables are measurements of distances from the national roads and SPM concentrations. A CART model was built using the databases in order to interpret the differences in habitat types between the patients' presence and the patients' absence; and developed to predict the presence/absence of patients allowing the ranking of variables. The CART approach produced a decision tree that could form the basis of simple if-then rules that could be used to predict the outcome of interest. The tree can be viewed as providing a probability model, with partition defined by the overall probability of misclassification. More, a chi-square test on a two-dimensional contingency table was used to examine strength of relations between variables in order to evaluate the CART model. Based on the threshold on the output decision variable obtained from the results of data mining, two groups were established: under and over the threshold; and a chi-square test was performed to examine if there were strength of relationships between patients' presence and the predictor variables.

Management Information Systems, C. A. Brebbia (Editor)
© 2004 WIT Press, www.witpress.com, ISBN 1-85312-728-0

3 Results

The decision tree in case of mesh size of 100 m^2 is shown in figure 1. The hierarchical structure of the tree can be directly used to evidence the predictor variables, which appear to have effects on the distribution of patients in the study area. The tree can be interpreted that habitat types and areas at high risk of the diseases are locations of: 1) less than 1509.25m from the roads and 2) 1509.25m and over away from the roads with SPM concentrations of 0.19255mg/m^3 and over. Of particular interest is the predictor variable value of SPM concentrations selected by the model exceeds the threshold (0.10mg/m^3) that is defined by the Ministry of Environment as a critical value for the notification of the citizens based on long-term evaluation. The results go for the case of mesh size of 500 m^2. The result obtained from a qui-square test suggests that, in case of mesh size of 100 m^2, there were significant differences ($p < 0.01$) between presence of the patients and distances from the roads; and between presence of the patients and SPM concentrations. In case of mesh size of 500 m^2, there were significant differences ($p < 0.05$) between presence of the patients and distances from the roads; and between presence of the patients and SPM concentrations. Especially, the diseases were more significantly associated with distances from the roads compared to SPM concentrations.

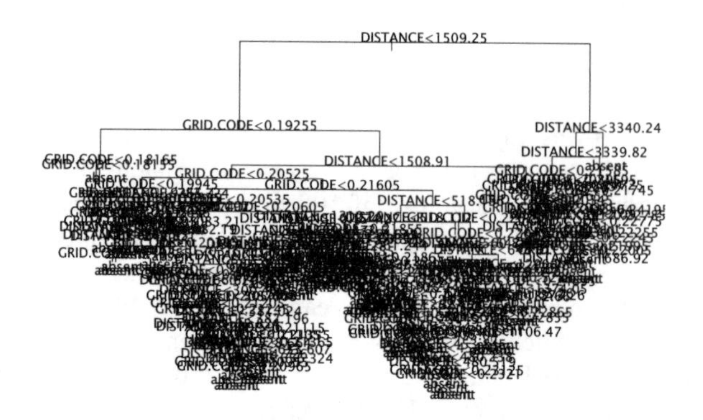

Figure 1: The decision tree in case of mesh size of 100 m^2.

4 Discussion

The resulting decision tree was clearly illustrated that the most important predictor was a distance from the roads; in other words, the incidence of the diseases in the study area were strongly influenced by a distance from the roads. It suggested that automobile exhaust could be considered as the environmental risk factor and affect the distribution of patients of the diseases in the area. Mesh size may affect results of analysis and should be based on the minimum mapping unit because using too large a cell size will cause some information to be lost. For addressing health and environmental problems in this study, mesh size of

Management Information Systems, C. A. Brebbia (Editor)

100 m^2 was a suitable spatial scale to represent and analyze the spatial variation of the diseases and the environmental factors while taking the results of a qui-square test into consideration. Methods of data mining are useful for not only finding patterns and subtle relationships in a large database but inferring rules that allow the prediction of future results. A CART methodology as a data mining tool, which yields a set of decision trees, allows the accurate prediction of outcome for future patients based on the values of their predictor variables and can be used as a new method of analyzing health data within the context of public health concerns. Furthermore, the methodology of data mining used in this study can be expected to make an approach to an effective application of GIS.

Acknowledgements

The author would like to thank Dr. Sagara at Research Center for Information Sciences, University of Tokyo for help with providing the geocoding service for this study.

References

[1] Bartonova, A., Clench-Aas, J., Gram, F., Gronskei, K.E., Guerreiro, C., Larssen, S., Tonnesen, D.A. & Walker, S.E., Air pollution exposure monitoring and estimation. Part V. Traffic exposure in adults. *Journal of environmental monitoring*, 1(4), pp. 337-40, 1999.

[2] Guo, Y.L., Lin, Y.C., Sung, F.C., Huang, S.L., Ko, Y.C., Lai, J.S., Su, H.J., Shaw, C.K., Lin, R.S. & Dockery, D.W., Climate, traffic-related air pollutants, and asthma prevalence in middle-school children in Taiwan. *Environmental health perspectives*, 107(12), pp. 1001-6, 1999.

[3] Revitt, D.M., Muncaster, G.M. & Hamilton, R.S., Trends in hydrocarbon fleet emissions at four UK highway sites. *The Science of the total environment*, 235, pp. 91-99, 1999.

[4] Roorda-Knape, M.C., Janssen, N.A., de Hartog, J., Van, Vliet, P.H., Harssema, H. & Brunekreef, B., Traffic related air pollution in city districts near motorways. *The Science of the total environment*, 235(1-3), pp. 339-41, 1999.

[5] Sauerbeck, G., Parkin, C.S., Houston, S., Whall, C., Newton, A. & Carlyle, J., Nitrogen dioxide and particle pollution near trunk roads and in towns of the south Midlands in England. Journal of the Royal Society of Health, 120(3), pp. 183-7, 2000.

[6] Report on air pollution, 2001; Bureau of Environment, Tokyo government, Tokyo, Japan.

[7] CSV address matching service; the Center for Spatial Information Science, the University of Tokyo, Online. http://fujieda.csis.u-tokyo.ac.jp/cgi-bin/geocode.cgi.

[8] Piatetsky-Shapiro, G. & Frawley, W. J., (eds). *Knowledge Discovery in Databases*, AAAI/MIT Press: Menlo Park, CA, 1991.

[9] Shaw, G. & Wheeler, D., (eds). *Statistical Techniques in Geographical Analysis*, David Fulton: London, 1994.
[10] Fayyad, U.M., Piatetsky-Shapiro, G. Smyth, P. & Uthurusamy, R., (eds). *Advances in Knowledge Discovery and Data Mining*, AAAI/MIT Press: Menlo Park, CA, 1996.
[11] Breiman, L., Friedman, J.H., Olshen, R. & Stone, C.J., *Classification And Regression Trees*, Wadsworth and Brooks/Cole: Pacific Grove CA, 1984.
[12] Clark, L. & Pregibon, D., Tree-based models. *Statistical Models in S*, eds. J. Chambers & T. Hastie, Champman and Hall: New York, 1992.

Management Information Systems, C. A. Brebbia (Editor)

Database design and GIS application in brownfields management: the case of Torviscosa site, Italy

P. L. Fantozzi[1], R. Falaschi[1], A. Lordo[1], A. Callerio[2], L. Giannetti[2], V. Pastore[2], N. Nosari[2] & P. Gardini[3]
[1]*Center of Geotechnology, University of Siena, Italy*
[2] *Studio Geotecnico Italiano S.r.L., Milan, Italy*
[3]*CAFFARO S.p.A., Udine, Italy*

Abstract

This study concern the preparation of a spatial database serviceable to the management of brownfields remediation, particularly for planning of activities, collection of subsurface data and for detailing the site conditions to a technical consulting and administrative authority. The study site, owned by Caffaro SpA, is located in Torviscosa (Udine, Italy); the industrial plants of Torviscosa have been established by the company SNIA in 1938. The large extension, importance and complexity of the site demanded a set of tools to analyse geographical data coming from various sources, in an efficient way. The database includes auxiliary data coming from geotechnical and field tests performed during the investigation phase, in order to define a detailed soil profile of the site. The GIS system used to manage the data is ARCGIS™ by ESRI™ with the extension required for geostatistical, vector and raster analyses. The designed geodatabase (in Microsoft™ Access format) contains different feature classes and tables relative to the modelled entities. All the system has been conceived and designed in order to have a reliable, comprehensive and user friendly tool to manage all the assessment, characterisation, decommissioning and clean-up operations of a huge industrial area, and to allow the full control and information exchange at any time of a presumably long duration project.
Keywords: GIS, database, DBMS, archive, brownfields, Arcgis™, Oracle™.

Management Information Systems, C. A. Brebbia (Editor)

1 Introduction and scope of the work

This study concern the preparation of a spatial database (Geographic Information System, GIS) [1], [2], [3] serviceable to the management of brownfields remediation. The aim of the work is particularly focused to the establishment of a tool for planning of activities, collection of subsurface data and for detailing the site condition to technical consulting and administrative authority. The study site, owned by Caffaro SpA, is located in Torviscosa (Udine, Italy). The industrial plants of Torviscosa has been established by the company SNIA in 1938 an initially devoted to the production cellulose as primary component for synthetic fiber (see Figure1 for an aerial view of the industrial site). During the following decades the plants have been revamped in order to produce primary base and specialized chemicals. Currently the site can be seen as a cluster of different chemicals plants which products span from the base chemical field to intermediate products for fine chemistry. The large extension, importance and complexity of the site clearly demanded for a set of tools to analyse geographical data coming from various sources, in an efficient way. In order to establish a clean-up plan for the area, an extensive site characterisation plan has been prepared by the Management of the Caffaro Group with the co-operation of Studio Geotecnico Italiano. This latter site investigation plan has been issued according to the guideline of the Italian law in force (with particular reference to the D.M. 471/99), collecting all the available information regarding the industrial processes, industrial plants characteristics (working or not working) and the localisation of chemicals produced during the entire life of the factory. In order to accomplish the design stages described below, the characterisation plans have been analysed in details. The flows of activities has been organized according the following steps [4]: A) Database design; B) conceptual, logical and physical design of data base; C) Software development; D) Input data; E) Editing data and definition of standard procedure for analysis and mapping.

Figure 1: General view of the Torviscosa industrial site.

Management Information Systems, C. A. Brebbia (Editor)
 ISBN 1-85312-728-0

1.1 Data base design: conceptual design

The conceptual modelling allowed for a definition of an entities set which represent the most important feature of the data base. Throughout the development of an entities-relationship model, the entities have been traduced to geographical features with appropriate informative attributes. The most important entities are investigations and samples, which record all the data coming from subsurface coring, and chemicals analysis carried out on the water and soils samples obtained. The database include auxiliary data coming from field and geotechnical laboratory tests performed during the investigation phase, in order to define a detailed soil profile of the site. The reference documents which represent the sources for database developments are listed in the following:1) Characterization Plan relative to the factory site; 2) Characterization Plan to the waste deposits. 3) Characterization Plan relative to the filling areas. 4) Topographic surveys of the factory site. 5)Maps, images and aerial photographs. 6) Laws and rules (D.M. 471/99). Essentially, a Characterization Plan (CP) preparation is articulated in three main phases/parts: a) Gathering and organization of the existing data, also defined as historical data; b) Site Characterization and preliminary formulation of the general conceptual model relative to the potential contamination sources; c) Plan of further investigations. The characterization plan gathers the most important information relative to the chemical and geotechnical analysis already realized in the site. The designed database will not contain data referred to production or naturalistic features of the site. The efforts of the analysis conducted has been focused on the geographical feature which will allow for a spatial analysis between chemicals encountered in the samples and physical elements of the industrial plants as equipments, waste areas, tanks, deposits, etc.

1.1.1 Peculiar of requisitions

The requisitions analysis of the data base is a result of a detailed analysis of the existing documentation, i.e. the cited characterization plan (in the following denoted as CP). The CP allows one to define the entities of the data base and their functional and/or spatial relationships. This entity represents in other words the "characters" of the database with their specific geometry, position, behaviour and informative and spatial-temporal attributes. For instances, the spatial extension of the site is defined not exactly coincident to the physical extent of the industrial plant, but it is enlarged up to consider the whole polluted area. For every site the data relative to the carried out investigations are collected and integrated to the topographic survey. Every investigation can be seen as composed by: a) sampling of soils and fluids to be analysed in terms of chemicals and geotechnical characters. In the chemicals analysis the verification respect to the limit concentration (admitted according to the current laws) is regularly performed; b) daily or monthly measurements of groundwater; c) on site measurements of the permeability and soil tests; d) GPS positioning and detailed topographic survey; e)stratigraphy of subsurface geology; f) Collection, description and cataloguing of the soils coring results. In order to establish a Geographic Information System, the positional information (i.e. georeferencing aspects) of the entities has been deeply analyzed and defined. In the following the most important entities of the data base are listed as a selection of the most peculiar features among more than one hundred entities on which the data base is focused on (the complete list of such entities is supplied in database reports).

SITE, defined as the spatial extension of environmental interest on which characteristics and properties are known and defined in the PC. Every site own its geographic location within an optional administrative boundary (a county, for instance). For every site, a set of aerial photograph or simple photo are included in the database with topographic surveys eventually accomplished with on site GPS measurements. Every site contains investigations (chemical, geo-chemical, geotechnical) realized or not. Every site can be considered as a set of geographical features (polygon) within a geographical database, with appropriate shape attribute as area and coordinates.

INVESTIGATION, as any analysis, sampling o borehole conducted within the PC. All the investigations belonging to the site area catalogued according to PC prescription and sub-plane indication (for instance, preliminary plane investigations -PIP investigations -). For any investigations, a location in an official map is contemplated; generally every investigations will be modeled as a simple geographical feature (point) with given x,y,z coordinates. Some investigations will be used for measurement of the water table; other investigations will be employed in site permeability or geotechnical tests. When the investigations surround a collection of soil cores, this latter are catalogued in cataloguer boxes. The samples are carried out from the investigations, and chemical and geotechnical analysis carried out on them. Each investigation is stored in the database accomplished with a set of information reported in the so called "Investigation Card".

SAMPLE, intended as any substance collected during an investigation, despite of its form, dimension, volume and the type of collected material. General information regarding the samples are collected in the so called "Sampling Card". A sample can be submitted to the laboratory for chemical and geotechnical test2. In the case of chemical analysis, the results of test are inserted by the laboratory in a file (so called "Chemicals Card") which is directly imported in the geodatabase. The "Chemicals Card" report general parameters like normative concentration or borderline limits. In case the of geotechnical test conducted on the sample a so called "Geotechnical Card" is filled with the information contained in the report of the geotechnical tests performed on the sample.

CHEMICAL BIBLIOGRAPHY, which collects all the chemical encountered in the samples analysis, classified with their complete names and optional C.A.S. number (if available).

HORIZON, as the lithologic horizons encountered in the sampling and coring of the subsurface geology.

HYDROGEOLOGICAL UNIT, defined within the site area and correlable to the permeability value measured in the permeability tests.

POLLUTER, as substances o compounds which can pollute a defined area.

Management Information Systems, C. A. Brebbia (Editor)

POLLUTING PONTENTIALLY AREA, defined according to the historical information available a set of polluting potentially area are defined. These areas area characterized by the presence of the chemicals produced, stored or wasted. All the entities (site, site area, Characterization Plan, Locality etc) constitute the geodatabase™ in terms of geographic features that (see below). After the requisition of the entities definition, and relative specifications, the design of the database has been carried out following the entity relationship schema. The Entity-Relationship schema (in the follows E-R schema), has been designed through CASE (Computer Aided Software Engineering) instruments as Oracle Designer™. Being the most important feature of the database Site, Investigations, Samples, Analysis, the first rough E-R schema will consist only in a simple schema regarding this entities (see Fig. 2). The definition of a complete scheme with the full list of attribute and values has been reached gradually extending the initial simple schema. In the follows an example of the over mentioned procedure is described by means of the working flow for the section of the data base relative to the SITE, INVESTIGATION, SAMPLES and PC. These entities are tied each other according to spatial, informative and logical operation; the analysis of the connections between the mentioned entities leads to the elaboration of a complete data base conceptual schema. In Figure 2 is represented the ER Skeleton Schema relative to the relationships between SITE-INVESTIGATION-SAMPLES-PC.

Each SITE is defined by the following informative attribute (database fields; the original name in Italian language is preserved): a) COD_SITO, alphanumeric code specific for the site; b) RAG_SOC, alphanumeric value related to the administrative asset and trademark of the site owner; c) ATTIVITA', alphanumeric value related to the company mission .

Each AMMIN will be defined by the following informative attributes: a) COD_ISTAT, alphanumeric code registering the information of Region-County-Municipality where the Site is placed; b) N_REGIONE, name of the Region where the Site is Placed (for instance, Friuli Venezia Giulia; c) N_PROVINCIA, name of the County where the Site is Placed (for instance, Udine); d) N_COMUNE, name of the Municipality where the Site is Placed (for instance, Udine).

The CHARACTERIZATION PLANE (PC) will be defined by the following informative attributes: a) ZONA_PC, alphanumeric value indicating the zone of the site linked to the PC; b) PROC_PC, alphanumeric value indicating the procedure indicated by the law in force for the redaction of PC; c) RELATORE_PC, alphanumeric value indicating the reference designer of the PC (company or person); d) DATA_PUB_PC, the date of publication of PC; e) COMMESSA, an alphanumeric value indicating the order number; f) COD_EMISS, alphanumeric value indicating the internal code given to the document; g) N_ARCHIVE, alphanumeric value indicating the archive number of the document.

Each PLACEMENT will be defined by the informative attributeNOME_LOC, an alphanumeric value indicating the geographical name of the site.

Management Information Systems, C. A. Brebbia (Editor)

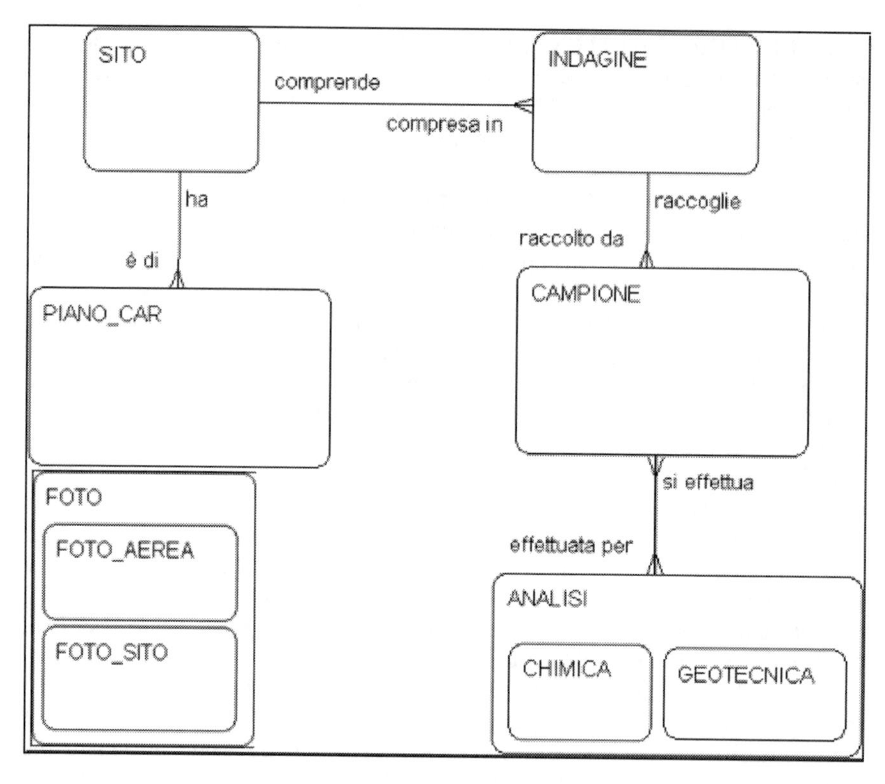

Figure 2: Skeleton schema relative to SITE. Key Ita.-Engl. sito= site;indagine=investigation;campione=sample; piano_car = characterization plan; analisi=analysis; chimica=chemical; geotecnica=geotechinical.

The definition of SITE stated it as the site of interest which is defined by the PC (characterization plan as defined before). For each site one can have different PC's according to the nature of the site and the different sector or zone which compose it. From a database rules point of view, this type of relationship is one (SITE) to many (PC's). Every SITE is extended onto several parcels that belong to a LOCATION, inside the administrative boundary. From this point of view the relationship between SITE and LOCATION is many (SITE) to many, (LOCATION) In other words, many SITE can include many LOCATIONS. The same relationship (one to many) exists between ADMINISTRATIVE BOUNDARY (for instances County) and LOCATION.

Other graphical documents can be related to each SITE (Figure 3): AERIAL_PHOTO, SITE_PHOTO, SKETCHES etc.

The relationship occurring between SITE and AERIAL_PHOTO will be many-to-many type, with optional participation of the SITE (the SITE can be depicted bay many photographs and one photo can depict many sites).

Management Information Systems, C. A. Brebbia (Editor)

The relationships occurring between CHARACTERIZATION PLANE (PC) and SITE_PHOTO, CHARACTERIZATION PLANE (PC) and SKETCH, CHARACTERIZATION PLANE (PC) and OTHER_DOCUMENTS will be one-to-many type with optional participation of the CHARACTERIZATION PLANE (PC). In this way the CHARACTERIZATION PLANE (PC) can include several pictures, tables, figures and sketches, etc, but all this documents belongs to only one PC.

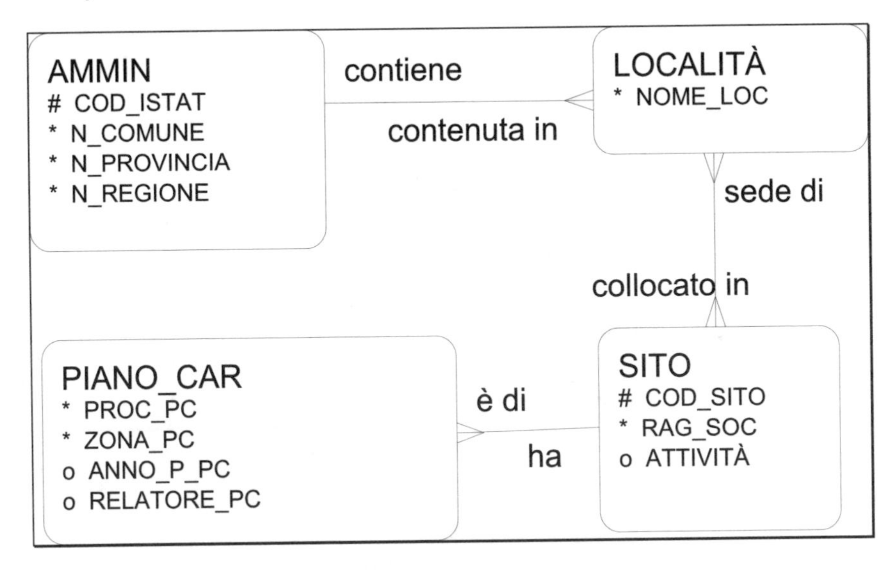

Figure 3: Preliminary enrichment of the SITE entity.

In Figure 4 a first version of the definitive conceptual scheme relative to SITE-INVESTIGATION, PC, ,LOCATION, AERIAL_PHOTO, , PHOTO_SITE, SKETCH, OTHER_DOCUMENTS is presented. In the above example relative to SITE and other important entities as investigation, has been explained the realization of the conceptual schema for the brownfields of Torviscosa. A complete view of the Conceptual model is synthetically reported in Fig. 6. The other steps relative to logical and physical design of data base are a close consequence of the conceptual modeling and are driven by the chosen CASE tool (Oracle Designer™, [5]), and due to this reason will not be here described.

2 Software development

The software component of the whole informative system is split in two main parts: a) the application of commercial GIS system (ArcGis™,[6]), and it is relative to the managing spatial (georeferenced data) contained in a geodatabase ™; b) a software for managing field data (investigations, samples, analyses, etc).

Management Information Systems, C. A. Brebbia (Editor)

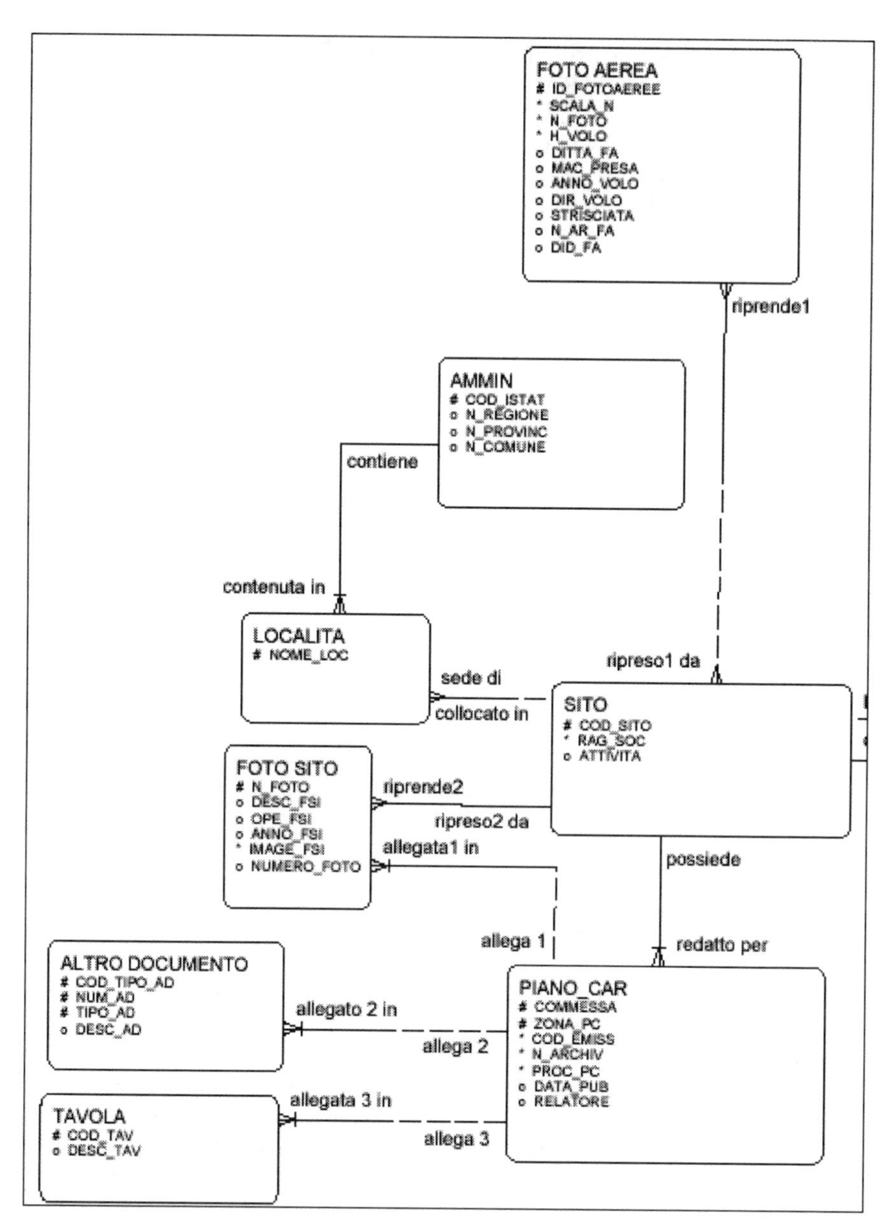

Figure 4: First sketch of definitive conceptual schema for the site.

2.1 The Geodatabase and the use of ARCGIS™

The database design phases described in the previous paragraphs allowed for a definition of a set of tables, values domains and procedures for loading , editing and elaborate data. The data management tables (either for geographic and not geographic data) have been defined inside the database, bye means of the

ARCGIS software tools. The geodatabase, on the personal version, is in the format of a Microsoft Access™_file where all the geographic data (with the geographic and informative component) and the database alphanumeric data are managed. This approach allows one to manage all data of a geographic and alphanumeric archive in a single file through the common and well known database rules. Besides the normal space analysis functions of the data, through the Geostatistical analysis tools one can calculate the maps of Probability distribution of the pollutants elements, applying the Rigging functions to the data resulting from the analyses on samples. Besides the Geostatistical analysis, some samples of 3D visualizations have been executed, allowing for the rebuilding of the three-dimensional space relationships between sampled zones and technical structures (plants, production zones, waste deposit, etc, see Figure 5).

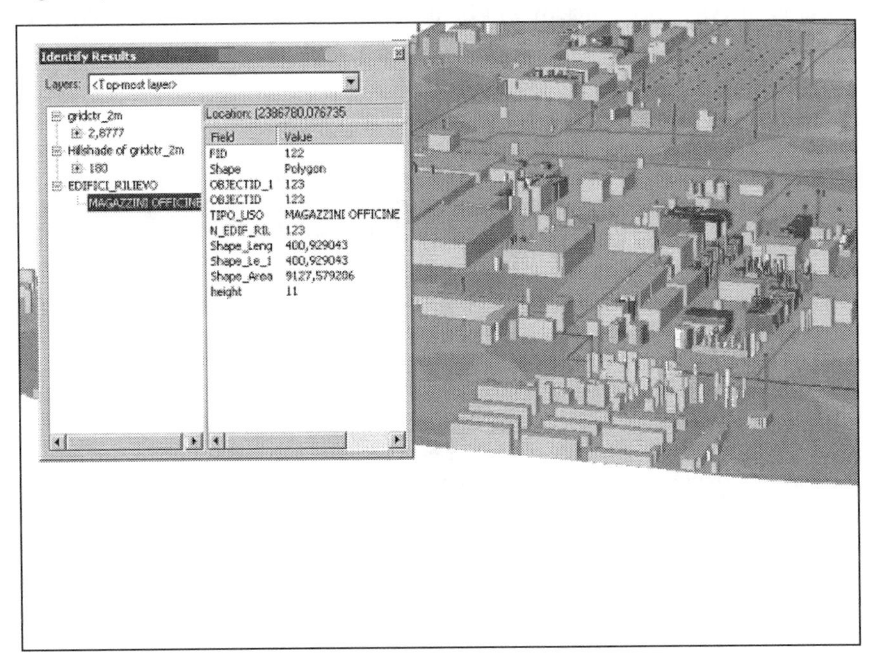

Figure 5: Distribution map of reservoirs, drillings and plan.

2.2 System devote to field data management

In order to simplify the collection, input, editing and recovery of data gathered from the chemical and geotechnical analyses, dedicated software has been developed. One of the requisites that drove the development of the software was the easy management of investigations, samples, and the analysis obtained from chemical laboratory. In particular, was required to the software to import directly the analyses data obtained from the laboratory by means of a substances and materials database, within a written protocol for the electronic transmission of data, allowing always a direct control by the user. Then, the database obtained

Management Information Systems, C. A. Brebbia (Editor)

can be exported on the intranet permitting in that way the interaction from non-GIS workplaces, by means of a standard Windows™ interface for viewing and importing data. Stratigraphies of soil, obtained from sampling, are saved as image files in a pre-defined directory tree, directly accessible within explorer interface. As an example of software functionality, in Figure 6 and Figure 7 a set of screen captures obtained during the software utilization is presented. In particular, in Fig. 7 the main form show the fields for the management of site, investigation (either geographic or not) and sampling data. A view of the chemical analysis results is also shown, allowing for a fast exploring of such data. The investigation form allows, for the definition of the geographic localization of the entity, the method adopted and the design phase in which the investigation is enclosed (as an example in the characterization or remediation phase).

A set of ARCGIS™ macro will be developed in the near future in order to simplify the user effort during the ritual activities of data representation on screen and printing.

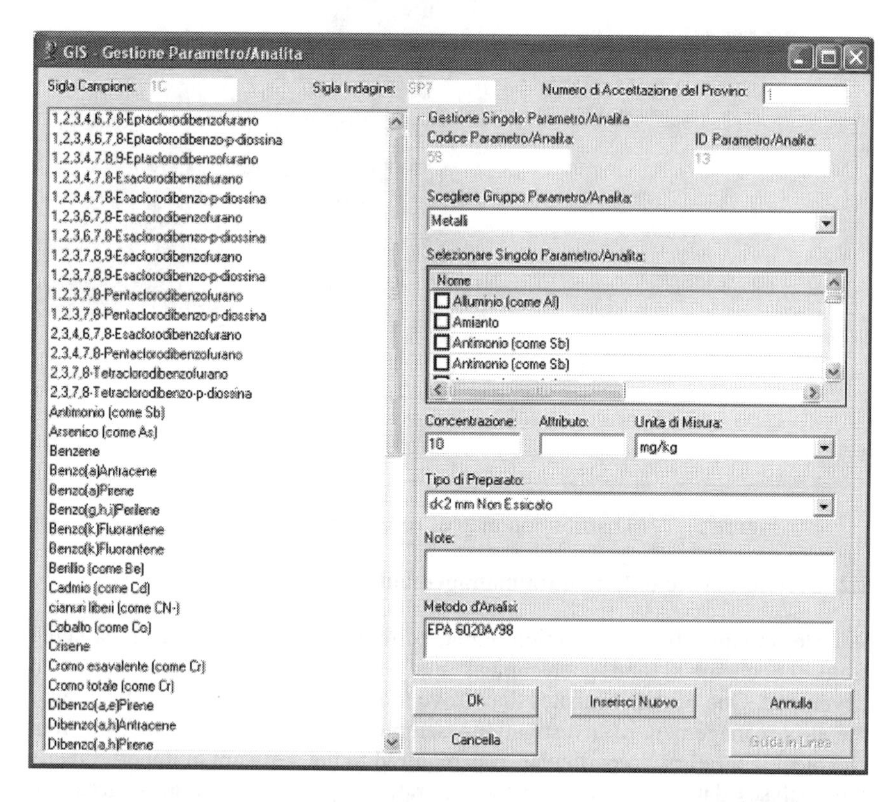

Figure 6: Form devoted to the exploring of chemical analysis results.

Management Information Systems, C. A. Brebbia (Editor)

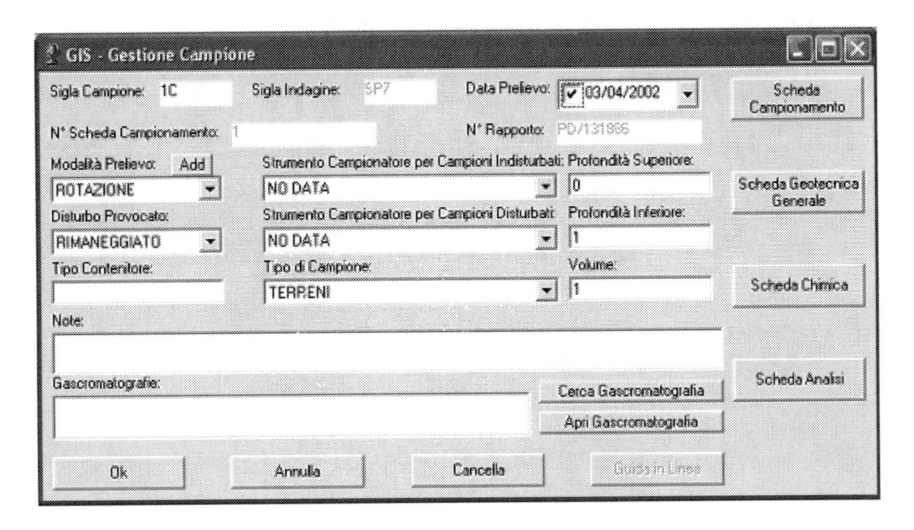

Figure 7: Sampling management form.

3 Conclusion

The use of a geographic informative system for the management of the brownfields represents an excellent system of record, analysis and reporting of the countless amount of data necessary to the editing of the plans for the placing in safety of the sites. All the system has been conceived and designed in order to have a reliable, comprehensive and user friendly tool for managing all assessment, characterisation, decommissioning and clean-up operation of a huge industrial area, and to allow the full control and information exchange at the any time of a presumably long duration project.

Acknowledgment

We wish to thank F. Chiereghin (Etruria Telematica, Siena) who have collaborates to the develops of dedicated Software.

References

[1] G. F. Boham-Carter *Geographic Information System*. Pergamon, 1994.

[2] PressT. Bernhardsen. *Geographic Information System*. Viak IT, 1992.

[3] Burrough and R. McDonnel. *Principles of Geographical Information System*. Oxford University Press, 1998.

[4] P. Atzeni, S. Ceri., S. Paraboschi, R. Turlone. *Basi di Dati*. McGraw Hill, pp. 1-561, 1996.

[5] M. Abbey, M.J. Corey, I. Abramson. *Guida a Oracle8i*. Oracle Press™, pp. 1-678.

[6] B. Boot; S. Crosier, J. Clark, A. Mc Donald (2003) ARCGIS™ rel. 8.3. *Building Geodatabase*. ESRI, pp. 1-460, 2003.

Management Information Systems, C. A. Brebbia (Editor)

Section 2
Applications of MIS

Will the real MIS please stand up?

I. Caddy
School of Management, University of Western Sydney, Australia

Abstract

One problem for information and communication technology (ICT) is its lack of commonly agreed upon classification frameworks and definitions, due mainly to the ICT field not possessing a sound theoretical foundation. It is considered that viable classification frameworks are important. In particular this paper considers issues of how to best classify different information types comparing a 'Yes/No' or digital approach with an analogue approach that allows different information system types to overlap. For example, in defining 'true' management information systems (MISs) it is considered that these systems should have all of the following characteristics: (a) focus is on information retrieval rather than input of data; (b) use of the system is voluntary not mandatory; (c) user population is predominantly at management level across the organization; (d) data contained in the MIS repository comes from both internal sources as well as external sources; and (e) major interactions by end users with the system is for structured decision making. Other information systems types may also possess one or more of these features but not all. Finally, as a test of the analogue approach to classifying information system types comparisons are made of decision support by transaction processing systems (TPSs) and MISs.
Keywords: information systems types, classifying information systems, management information systems, MIS, defining MIS, using MIS.

1 Introduction

After more than 30 years of information system use by commercial, public sector, and third sector organizations we still appear to have problems in defining the types of information systems that organizations use (Leek [1]). One reason for this is the fact that the ICT field lacks a sound theoretical foundation, both overall as well as for some of its component parts. The lack of a theoretical foundation leads to problems which were very eloquently stated by

Management Information Systems, C. A. Brebbia (Editor)

Kirs *et al.*, [2:183]:

> One disadvantage of frameworks in IS stems from their relative ease of construction. Not bound to a single theoretical approach, they are both easily generated and readily discarded and forgotten. Since they are not necessarily required to meet any minimum standard of theoretical consistency, trivial conceptual modifications are often heralded as novel and definitive. The consequences are failure to build cumulative research findings, failure to establish an enduring scientific presence for IS in the academic community, and promotion of disciplinary alienation for IS researchers.

Often in the ICT field there seems to be an implicit assumption that validation of generally accepted models or frameworks is not necessary. Furthermore, many ICT terms are used with a presumption by the author that their definition, a priori, is correct and that the reader knows what the author is talking about. This lack of discipline in terminology can be seen in how the term 'management information system' (MIS) has been applied to all sorts of information systems that more often display features atypical rather than characteristics in common (Small and Yasin [3]; Faber, *et al.* [4]; Mandal and Baliga [5]; Smith [6]; Wang [7]). For example, Smith [6:327] states that a key issue for a small business is:

> to install, as soon as practicably possible, a cohesive method for handling the various information pertinent to the running of the business. Such a system has variously been called a 'management information system' (MIS), an 'accounting information system', or a 'management accounting system'

Even as late as 1999 there appears to be difficulty in differentiating between what should be seen as transaction processing systems ('accounting information system', 'management accounting system' - although Smith [6:335-336] is really referring to financial accounting information such as profit and loss statements and balance sheets, rather than management accounting information) and a MIS, which is pertinent to the management of the business rather than to the operation or "running of the business". These two activities should be seen as distinct and separate, though obviously inter-connected. Such would be the case in differentiating the core objectives between TPSs and a MIS. Murphy and Daley [8:63] make the same mistake as Smith [6] in confusing what is a transaction processing system with a management information system, viz.:

> International Freight Forwarders (IFFs) face an increasingly volatile business environment. One way for IFFs to differentiate themselves in today's competitive marketplace is through [the] development of sophisticated management information systems, particularly electronic data interchange (EDI).

Management Information Systems, C. A. Brebbia (Editor)

In addition, use of the terms MIS, decision support system (DSS), or executive information system (EIS) have suffered a similar fate as that discussed above for TPSs and MISs. For example, Liang and Hung [9:303] state that: "a typical DSS must meet three criteria:

(1) support but not replace decision makers;
(2) tackle semi-structured decision problems; and
(3) focus on decision effectiveness, not efficiency (Keen and Scott Morton, 1978; Sprague and Carlson, 1982; Turban, 1995)."

These three criteria obviously define in part what a DSS should do (although there is no mention of either structured decision making or unstructured decision making), but could equally apply to a DSS as well as a MIS. Indeed Liang and Hung [9] go on to refer to a study by Alter [10] in which he classified "DSSs" into two sub-categories: data-oriented and model-oriented. Data-oriented DSSs are focused on data manipulation through simple analysis procedures. On the other hand model-oriented DSSs are more interactive and provide facilities such as simulation and model building.

Liang and Hung [9:303] also allude to the development of EISs as a "special type of data-oriented system". Again there appears to be a problem in classification: whilst Liang and Hung [9] have identified two entirely different modes of interaction by users between data-oriented DSSs and model-oriented DSSs they still group these systems together under the one category of DSS. However, the distinctiveness of the end user interaction should argue for data-oriented DSSs to be separately classified, namely as MISs. This would then be more consistent with these authors use of the category EIS to identify a separate "special" category of a "data-oriented" DSS. Would it not be more logical to state that an EIS is a special type of MIS in that the core objective of an EIS is information provision to the executive rather than the provision of simulation or model building facilities?

Finally there is a problem in the use of the label "EIS", or "executive information system by Liang and Hung [9:307]. Although the authors quite rightly identify an "executive" as an upper-level manager, they then claim that the EIS deployed by China Steel is used by 5,000 of the 10,000 employees in this organization. Surely China Steel does not have 5,000 executives (indeed not even 5,000 managers); and so this system should not be classified as an EIS but possibly a MIS. From the above, it would also seem that within an organization there may be a high level of agreement about what is or what is not a MIS, or indeed various other types of information systems. For example, it would appear that China Steel is quite comfortable classifying their system an EIS. But this level of agreement in terms of classification does not allow comparisons of information systems types between organizations, and could lead to confusing or erroneous comparisons. For instance the operating costs for the China Steel EIS would be an order of magnitude greater than the costs of an EIS deployed by another organization in which the user population is restricted to senior executives and so would be numbered in tens rather than thousands.

Management Information Systems, C. A. Brebbia (Editor)
© 2004 WIT Press, www.witpress.com, ISBN 1-85312-728-0

More recently Alavi and Liedner [11] have discussed the emergence of knowledge management systems (KMS), although the definition provided for this type of information system (Alavi and Liedner [11:114]) is somewhat tautological. Is this yet another IS category? If so, what are the distinguishing features or characteristics of KMSs compared to other information systems in an organization's ICT portfolio? Alavi and Liedner [11] do provide insight into some of the core functions of a KMS (for example, see Table 1, p. 111) but do not go on to define boundaries for KMSs. The authors also do not consider the sorts of relationships that should exist between KMSs and other information systems such as MISs which also have an important role to play in organizational decision making and knowledge management. It seems that whilst we are often very concerned with issues of boundary setting during requirements analysis of an information systems development, it would appear that we are not as concerned with developing boundaries and unambiguously classifying information systems as they emerge from development into production. Lacking commonly agreed upon boundaries for information systems means that the classification process is all the harder – or maybe not as we will see later.

This paper supports the view that organizations, researchers, and teachers should endeavour to better define the boundaries between the different types of information systems that comprise their ICT portfolio. A principal benefit of doing this would be to provide to organizations, particularly senior management, a greater degree of clarity and understanding about the sorts of information systems the organization should possess, i.e. a superior ability to conduct appropriate ICT portfolio analysis. Understanding the core objectives of information systems would mean that a more informed and robust analysis could be conducted by senior managers to answer questions such as:

a. Do we really need a MIS; do the benefits outweigh the costs?

b. If we build a MIS what purpose will it serve? For example, to what degree does the organization's current information systems, such as its transaction processing systems, support decision making within the organization?

c. If we implemented a MIS: what level of usage would this system get; what sorts of problems could it address; who are the potential users of this system?

Surely these are significant and substantial questions to ask if there is to be effective management of the organization's ICT service delivery.

2 Is classification of information systems important?

Some may wish to argue that it is not important to differentiate between different types of information systems. They all receive, process, and store data and/or information: so the various labels developed over time can be equally applied to any type of information system. There is no real value in developing a classification framework of information systems types. The response to this

Management Information Systems, C. A. Brebbia (Editor)

argument is why were these labels developed in the first place? We still find in most recent editions of ICT textbooks used in higher education institutions in Australia (e.g. Turban, *et al.* [12], and many others), sections or even whole chapters devoted to developing a classification of information systems types. The implication is that organizations use information systems which have different core structures and objectives, and classification leads to greater clarity rather than confusion in understanding why these different types of information systems are developed and used by organizations.

Indeed, many organizations apparently do find value in attempting to determine the core objectives of their various information systems. That is, the organization possesses a portfolio of distinct information systems that need to be managed both as an individual information system in their own right, as well as part of an overall, integrated and inter-connected ICT service delivered to that organization. Analysis of Information systems portfolios then allows important questions, *inter alia*, such as the following to be asked and hopefully answered:

a. How much of the total ICT budget should the organization invest in its transaction processing systems as compared to its other information systems?

b. What sort of return on investment should the organization expect by developing a management information system?

c. Having deployed a management information system, does the current usage level of this information system justify continuing ICT budget support?

d. What types of information systems do the organization's competitors use?

Another issue that needs to be addressed with respect to a classification framework is how to treat exceptions. Often, difficulty arises with exceptions due to the method adopted in classifying the things of interest. Classification can adopt a binary or 'Yes/No' approach or it can adopt an analogue or continuous approach. If the latter approach is adopted then classification of an information system as a MIS can then be approached on the basis of the information system either possessing or not possessing a number of MIS characteristics, rather than requiring that an information system should have a complete set of MIS characteristics, and if the system does not then supposedly it will satisfy all the characteristics of another category. But information systems, like organizations, are dynamic rather than static artefacts and so one information system may become more like a MIS (or less like a MIS) as these systems are enhanced, merged, or even re-developed over time. It is considered that analogue-based classification schemes have less problems dealing with exceptions than is the case if a digital approach was used. In graphical terms, adopting an analogue approach would mean that information system types necessarily overlap as indicated in Figure 1 shown below, which essentially provides a generic map of the information systems portfolio possessed by organizations:

Management Information Systems, C. A. Brebbia (Editor)

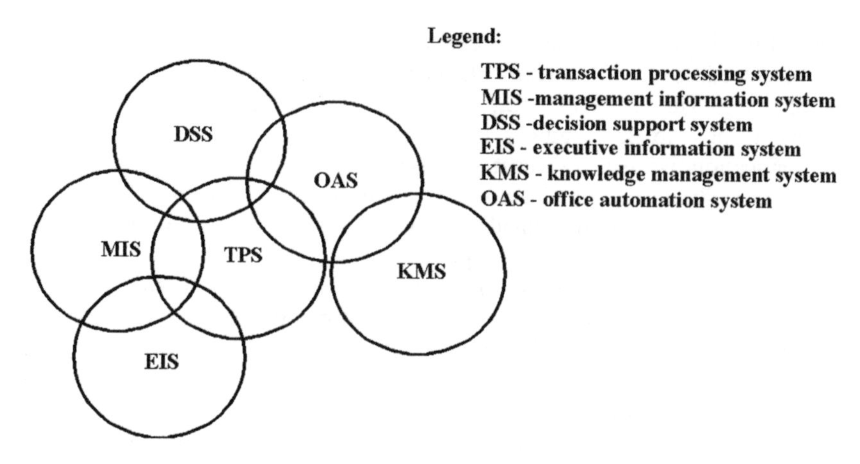

Figure 1: Generic information systems portfolio.

(Note that Figure 1 above does not attempt to include an exhaustive list of all information system types used by organizations but at least shows the major categories.) Accordingly, for these reasons, the discussion outlined below will presume an analogue approach rather than a digital approach with respect to classifying information systems types.

3 Generic MIS characteristics

The first task in terms of an analogue approach should be to define comprehensively the features that define an information system type. So what features should an information system possess in order to be considered a MIS? To be a MIS it is considered that an information system should possess all of the following characteristics:

- focus of the information system is on data retrieval rather than data input,
- use of the information system is voluntary rather than mandatory,
- user population is composed predominantly of middle or senior management levels, and spans most areas or functions of the organization,
- data contained in the MIS repository comes from both internal sources as well as external sources,
- 'internal' data is predominantly time series data that are not available from other information systems such as TPSs, and
- major interactions by end users with the information system are to assist in structured decision making.

Furthermore in adopting an analogue approach, one major change is that information systems lacking these characteristics should not be considered a 'true' MIS but rather an information system that has features of a MIS: some

Management Information Systems, C. A. Brebbia (Editor)
 ISBN 1-85312-728-0

may have more features than others, and so be closer to a 'true' MIS. As such it is contended that very few organizations possess what could be labelled a 'true' MIS; most would have other types of information systems that have some but not all of the above characteristics. For example, a transaction processing system that produces a large number of exception reports and other summary reports used in routine decision making is essentially a transaction processing system that has features of a MIS. On the other hand there has been the emergence recently of data warehouses as a new type of information system. Data warehouses have been considered to assist in decision making (Ma, *et al.* [13]) – and so may be a MIS; facilitate access to and use of corporate knowledge (Hurley and Harris [14]), or support the creation of new information and/or knowledge (Gargano and Raggad [15]) – and so may be a KMS. Using the analogue approach to classification we may find that data warehouse systems share features of both MISs and KMSs, indicating again the superiority of the analogue approach to classifying information systems when compared to the more simplistic digital approach. Using the analogue approach should lead to better and more informed comparisons being made about how ICT service delivery assists decision making or other activities by end users between organizations as well as within organizations.

4 Decisions and information system types – TPSs and MISs

What sort of new insights does the discussion above provide in terms of either how information systems work or how users interact with information systems? The first insight is that in terms of decision making or decision support, these activities and their support by information systems are not the preserve of MISs, DSSs, or KMSs. Consider the following two decision scenarios regarding inventory replenishment. The first scenario, used by Organization A, adopts a very simplistic and structured approach, as indicated in this organization's inventory replenishment business rule:

> When stock on hand of an inventory item falls to a particular level and no purchase order exists to acquire further stock, initiate a purchase order for a certain quantity of this inventory item.

In fact this business rule, and the information required to make a decision, is so simple that it could be incorporated into the inventory management transaction processing system itself. Yet it is a decision no less, and firms that operate in very stable markets may find that this business rule or a variant that only needs access to data stored in the inventory management transaction processing system is appropriate for its purposes. Given these circumstances to develop a MIS to perform the same decision task would be redundant. Furthermore, referring back to Figure 1, Organization A's inventory management transaction processing system would be located somewhere in the overlap area between a 'true' transaction processing system and a 'true' MIS. Obviously there are many other decision making activities within Organisation A

Management Information Systems, C. A. Brebbia (Editor)

that can be made based on solely on the use of the organization's transaction processing systems - transaction processing systems that assist decision making but are still considered to be predominantly transaction processing systems rather than another information system type. In fact Organization A may find, after analysing all its decision making scenarios, that there is no real need to develop and implement a MIS, DSS, or KMS!

On the other hand, Organization B operates in a more unstable business environment and so inventory replenishment is driven by a more complicated process which requires more careful analysis. In this situation, Organization B. may find that the algorithms required or the underlying data needed are beyond the capacity of its current inventory management TPS to provide. For instance, presume that Organization's B inventory replenishment business rule is:

> When inventory falls to 50% of the average (seasonally adjusted) level for the same quarter of the past three years, re-order an appropriate quantity of inventory to satisfy the perceived level of demand forecasted for the next three months and adjust by the mean production time required to produce the finished goods for which the inventory item is needed either as raw material or as a finished component.

There are two important points to note about the above decision scenario: (a) transaction processing systems would not normally provide all the information processing functionality required to support the decision making process; and (b) one transaction processing system would not normally store all the data required by this business rule to complete the decision process. First, in the case of '(a)' above, the data needs to be seasonally adjusted, an algorithm that would not normally be required in terms of any data integrity or validation checks performed during either input or storage of inventory receipts, issues, or other adjustments say due to the end of month stocktakes. Second, in the case of '(b)' above, transaction processing systems normally would store data for an accounting period, typically a year. That is, their core objective is to record the history of the organization (in the form of its various transaction streams) rather than assist managers make decisions. The quantity and size of the individual transaction records normally dictate that only current year data are stored and these data would be archived out of the information system at the end of the accounting period. As such the four years of time series required for this business rule would not normally be found in a transaction processing system but rather a system that possesses one of the key requirements for a MIS. However, if the algorithms necessary to process this decision rule and four years time series data are part of the inventory management transaction processing system, then this information system would be closer to a MIS than is the case for Organization A. Adapting Figure 1 above, these systems would be plotted as:

Accordingly, one conclusion to be drawn here is that identifying solid boundaries between information systems types may be an exercise in futility and ultimately not be reflective of real world practice. However, adopting the analogue approach allows greater flexibility and better captures the complexity

Management Information Systems, C. A. Brebbia (Editor)

that is always evident for ICT in the real world. Finally, Figure 2 provides the basis for developing information systems portfolio maps of organizations.

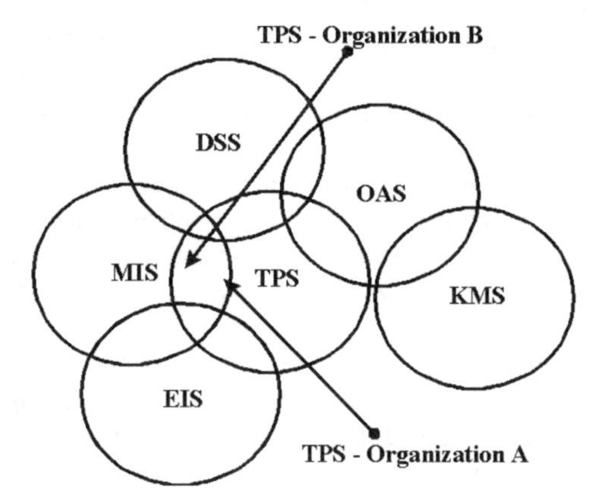

Figure 2: Comparing information systems, organizations A and B.

5 Further research

It is hoped that this paper will initiate further discussion on the major issues outlined above. In addition, further research is intended both from a theoretical and an empirical perspective. From a theoretical perspective validation of the framework across all information system types should be conducted. Further work will also occur on refining the generic information systems portfolio map shown in Figure 2 above. There are also other classification issues to consider: classification may be relative to the user of the information system. That is, an information system may be a TPS for operational personnel, but the same system may provide a decision making capability to managers using that system and so be perceived as a MIS by these users. What impact would this have on the development of organization information systems portfolio maps? From the applied perspective work will be conducted in the form of case study analysis of information systems for the financial services sector, the manufacturing sector, and the public sector to gain more insight into what sort of information systems are deployed by organizations and to generate and compare information systems portfolio maps for these organizations. Finally, additional work will focus on data warehouse systems to determine where these systems would be located on the information systems portfolio map.

References

[1] Leek, C., Information systems frameworks and strategy, *Industrial Management & Data Systems*, **97(3)**, pp. 86-89, 1997.

Management Information Systems, C. A. Brebbia (Editor)

[2] Kirs, P.J., Sanders, G.L., Cerveny, R.P., and Robey, D., An Experimental Validation of the Gorry and Scott Morton Framework, *MIS Quarterly*, June, **13(2)**, pp. 183-197, 1989.
[3] Small, M.H. and Yasin, M., Advanced manufacturing technology adoption and performance: the role of management information systems departments, *Integrated Manufacturing Systems*, **14(5)**, pp. 409-422, 2003.
[4] Faber, N., de Koster, B.M. and van de Velde, S.L., Linking warehouse complexity to warehouse planning and control structure: An exploratory study of the use of warehouse management information systems, *International Journal of Physical Distribution & Logistics Management*, **32(5)**, pp. 381-395, 2002.
[5] Mandal, P. and Baliga, B., MIS-user interface design for job shop manufacturing environment, *International Journal of Operations & Production Management*, **20(4)**, pp. 468-480, 2000.
[6] Smith, J., Information technology in the small business: establishing the basis for a management information system, *Journal of Small Business and Enterprise Development*, **6(4)**, pp. 326-340, 1999.
[7] Wang, S-H., An object-oriented approach to plant configuration management information systems analysis, *Industrial Management & Data Systems*, **99(4)**, pp. 159-167, 1999.
[8] Murphy, P.R., and Daley, J.M., International Freight Forwarder: Perspectives on electronic data interchange and information management issues, *Journal of Business Logistics*, **17(1)**, pp. 63-84, 1996.
[9] Liang T-P. and Hung S-Y., DSS and EIS applications in Taiwan, *Information Technology & People*, **10(4)**, pp. 303-315, 1997.
[10] Alter, S., A taxonomy of decision support systems, *Sloan Management Review*, **19(1)**, pp. 39-56, 1977.
[11] Alavi, M. and Leidner, D.E., Review: Knowledge management and knowledge management systems: Conceptual foundations and research issues, *MIS Quarterly*, **25(1)**, pp. 107-136, 2001.
[12] Turban, E. McLean, E. and Wetherbe, J., *Information Technology for Management: Transforming Business in the Digital Economy*, John Wiley and Sons, Ltd., New York, 2002.
[13] Ma, C., Chou, D.C. and Yen, D.C., Data warehousing, technology assessment and management, *Industrial Management & Data Systems*, **100(3)**, pp. 125-135, 2000.
[14] Hurley, M.A. and Harris, R., Facilitating corporate knowledge: building the data warehouse, *Information Management & Computer Security*, **5(5)**, pp. 170-174, 1997.
[15] Gargano, M.L. and Raggad, B.G., Data mining - a powerful information creating tool, *OCLC Systems & Services*, **15(2)**, pp. 81-90, 1999.

Management Information Systems, C. A. Brebbia (Editor)

Insula: urban maintenance system

R. Todaro
Insula spa, Italia

Abstract

Insula is a public limited company set up by the Venice City Council in 1997. The Council has 52% of the share capital, the four underground utility companies having equal 12% shares. Insula's task is maintenance of this city on the lagoon: it has to dredge canals so that the waterways remain navigable, restore bridges and raise pavements to allow pedestrian passage even at high tide, repair embankments to ensure the stability of buildings, keep the sewer system running to provide the best possible sanitary conditions and renew and complete the underground service grids (water, electricity, gas and telephones). Insula has equipped itself with a geographical management system to run and control the maintenance process: a set-up based on the city features (canals, embankments, pavements, buildings, etc.) identifies the various elements, roles, rules and actions, applying them to the geographical information system.

The heart of the system is the organisation of geographical information levels in a database engine, which not only finds references in accordance with the principles that have been established for it, but ensures simultaneous accessibility and editing for users with different roles in various parts of the area covered, thus ensuring the solidity of the database. In accordance with the same principle, application instruments have been designed according to a web-orientated logic. This allows the information and instruments to be shared among different categories of operator, such as the authorities, citizens, the company and third parties.

The planning, design and execution processes of the maintenance works are managed by dedicated applications. Taken from the basic map, the geographical features involved in the work are designed, and thus the general characteristics of the job and the details of each single element in the planning of the area as a whole are identified.

In the planning and design of underground works, the management and geographical system provides a working level for cooperation among the different bodies concerned, making detailed information available regarding underground service systems and the planning of the work to be done on an area basis.

The execution of maintenance works is supervised by on-line collaboration among the site engineer, the job manager and the authorities.

Keywords: maintenance, to dredge canals, waterways, bridges, high tide, embankments, sewer system.

Management Information Systems, C. A. Brebbia (Editor)

1 Introduction

Insula spa is a Venice City Council company. Its task is the extraordinary maintenance of the city and the other built-up areas in the lagoon. Maintenance takes the form of dredging canals, restoring embankments, repairing pavements, renewing and modernising sewers and rationalising underground systems and their technological grids.

In order to run and control the maintenance process as well as possible Insula has decided to set up its information structure in a geographical information system. This system, called SMU (Urban Maintenance System) provides methods and procedures for the processing of geographical data through management information tools and also supports management through tools that make forecasts and support decision-making. In this way Insula is able to work out an all-round intervention strategy following urban maintenance principles.

Figure 1.

For the conduct of its work Insula has developed an architecture based on five fundamental geographical elements: canals, bridges, embankments, buildings and pavements. All geographical elements have been subjected to detailed technical surveys. Geometrical and structural data and information sheets regarding the state of repair of the different elements that make up their structures have been catalogued. Each level of information has been organised into univocal codes, in order to create a system of relationships between distinct elements. The geographical elements are thus set out with relational links that

Management Information Systems, C. A. Brebbia (Editor)
 ISBN 1-85312-728-0

allow correlation and complex analysis. The universe of the canals of Venice and the larger islands in the lagoon have been laid out in a network of segments and intersections following water flow principles.

Insula also has an information system to support design. The codes according to which the area has been divided up are used for both planning and designing works and for accounts management. This type of organisation allows Insula to run a work control process that, for each geographical element that needs to be maintained, identifies a data sheet regarding the work involved that summarises technical and financial information.

Insula manages its activities through an Urban Maintenance System that is not a mere graphic description, but a thoroughgoing process control tool. In order to achieve this, a series of management applications have been developed that use the geographical information system as a base from which they provide a complete process control system. The process is not confined to Insula's sphere, but, its objective being the maintenance of a city, has as fundamental actors the public administration and all the bodies that operate in the area. For this reason some of the elements of the system are based on web technology.

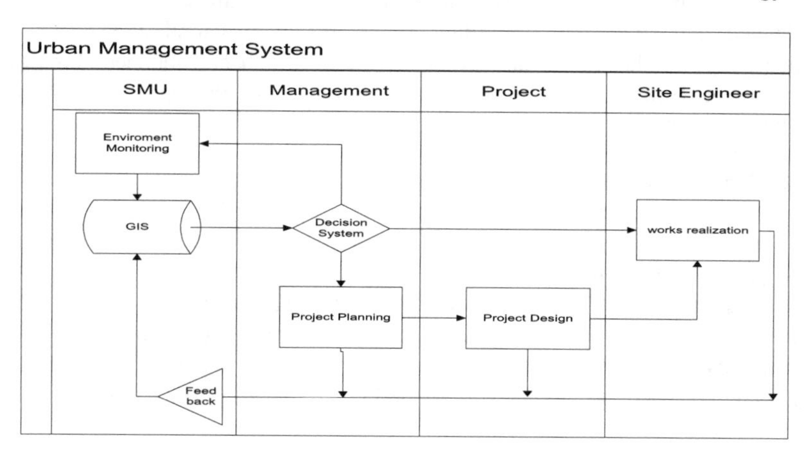

Figure 2.

2 The elements that make up a system

Insula puts the geographical elements in the city of Venice at the disposal of users at various levels. Databases are added to and updated by means of investigations and surveys necessary for or consequent on integrated works. A complex system of data has been created for each element, such as geometrical characteristics, type and a series of identification details.

2.1 The fundamental elements

The fundamental elements into which Venice has been catalogued, and which constitute the basic "units of measurement" are five information levels:

Management Information Systems, C. A. Brebbia (Editor)

pavements, bridges, embankments, canal segments and buildings. These elements are defined as SMU elements.

Each element has characteristics such as type, state of repair, area in square metres and the presence of particular superstructures. When these elements are handled each variation is recorded whenever necessary so that in time the variations in structure or changes in the state of repair can be retrieved.

2.1.1 Pavements

Pavements are divided into their fundamental elements; their types, properties and various identification particulars have also been recorded.

2.1.2 Bridges

All bridges have been subjected to detailed technical survey. Geometrical and structural data have been recorded and data sheets have been compiled regarding the state of repair of their various constituent parts.

2.1.3 Embankments

Canal embankments, whether waterfronts, quaysides or outside walls, have been catalogued and subdivided according to criteria of structural and ownership homogeneity. A large amount of information has been obtained for each embankment element, such as the number of mooring rings, the type of material and the state of repair.

2.1.4 Canal segments

The physical information about canal segments includes length, average width, area, average depth and a series of bathymetric measurements describing the beds. Their description also includes a range of embankment data regarding the structures looking on to them: ownership, material, state of repair, sewer outlets, types of construction, etc. Each canal segment is connected to the adjacent segments by means of hydraulic intersections. The canal network defined in this way can be used as input for modelling tools (hydrodynamics, traffic, inconvenience).

2.1.5 Buildings

Buildings are part of the structural elements looking on to the Venetian canals and therefore detailed knowledge is necessary of the data regarding constructions or properties from both structural and other points of view.

3 The elements in the works

3.1 The works

Urban maintenance works are geographically defined through outlining the confines of the area involved in a "job". Maintenance works are identified by means of several geographical information levels: the first basic level is that of the job to which the area involved is related.

Management Information Systems, C. A. Brebbia (Editor)

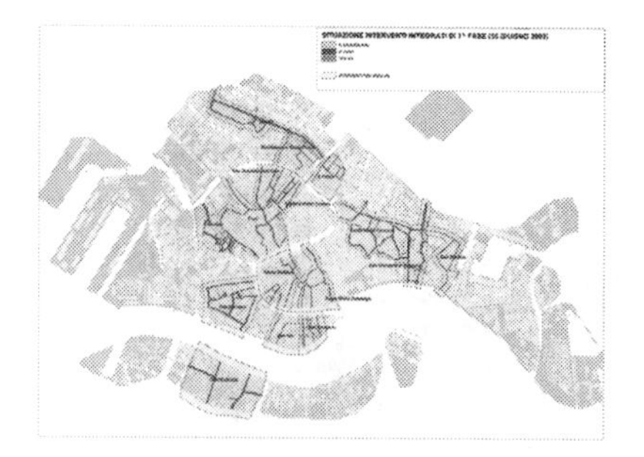

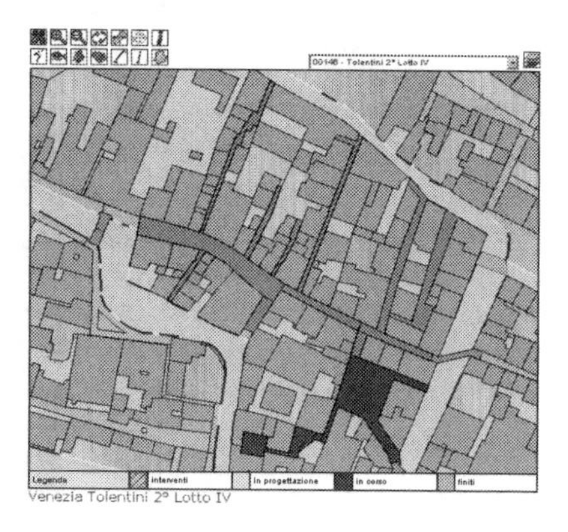

Figure 3.

3.1.1 UTP

UTP (Geographical Design Units) are associated with the job object: these are geographical objects that identify the areas on which homogeneous types of work are carried out. They derive direct from the SMU elements and at first inherit both their physical and alphanumeric characteristics, but can subsequently be modified from a geographical point of view in order to make them more relevant to the work done. Among the main features of the UTP is the indication of status, which shows whether the site in the area concerned is still under design, is active or has closed.

Organisation of the area by sub-dividing it by objects and codes is useful in the design phase both in planning works and setting up their accounting. In the same way a works control process is operated that identifies a technical and financial data sheet for each geographical element subject to maintenance; when the works are completed the data sheet is compiled by the site engineer, who thus contributes to keeping the system up to date.

3.2 Supporting elements

3.2.1 Basic maps

This is the combination of all the fundamental levels of information that are the base for the supporting and complementary mapping allowing generation and consultation of the levels of work. Here there are more than 300 information levels, which are the concern of various organisations operating in the area of Venice.

3.2.2 Benchmarks

A high-precision geodetic network has been created. Global positioning system tools were used to materialise 51 network vertices, allowing the setting up of planimetric and altimetric benchmarks valid for reference, re-attachment and linking to the Punta Salute mareographic system, the IGM95 (Italian Military

Management Information Systems, C. A. Brebbia (Editor)
 ISBN 1-85312-728-0

Geographical Institute) vertices in the surrounding area and the existing IGM geometrical levelling lines.

Figure 4.

3.2.3 Underground services

These elements, out of the majority's sight but none the less vital for the city for this reason, unravel through lanes and bridges following routes that are mostly unknown. The system intends to show the routes that are known and achieve a complete route map by continuous updates as works by Insula or cooperation with the utility companies make it possible to plot the course of sections not previously surveyed.

Figure 5.

4 Organisation of data

The technological solution adopted to develop the urban maintenance system rests on two fundamental principles:
- centralised management of data;
- decentralisation of working tools.

4.1 Centralised management of data

Data have to be organised in a univocal, solid and compact way in order to provide information that is always correct and up-to-date. Hence data, both

Management Information Systems, C. A. Brebbia (Editor)
© 2004 WIT Press, www.witpress.com, ISBN 1-85312-728-0

spatial and non-spatial, are managed in a spatial database. This type of solution allows geographical data to be memorised, accessed, checked and handled by means of the solid tools of relational databases and also ensures the typical procedures of a GIS environment. All the properties of the database as a relational instrument are extended to spatial data, so that normalisation, lock, mirror, commit or rollback can be applied. Data therefore correspond completely both in the various applications and through distinct access interfaces.

Another important feature of this technology is the possibility of having all the normalised tables on the database, connected with definition or coding tables, which avoids typing errors when data are entered and makes searches and joins quicker; at the same time de-normalised displays can be downloaded for users so that the database structure is completely transparent for its end-users. For example, this allows simultaneous editing of the same record by more than one user, with correct and solid management of conflicts. Several operators can thus work at the same time without compromising the solidity of the information.

This type of solution also allows the adoption of typical management system techniques or the activation of geographical object historicisation mechanisms.

A spatial database provides:

- complete integration of attributes with spatial data;
- protection of data;
- a wide range of spatial functions directly available at database level;
- ease of use;
- access to the database through standard program interfaces (ODBC).

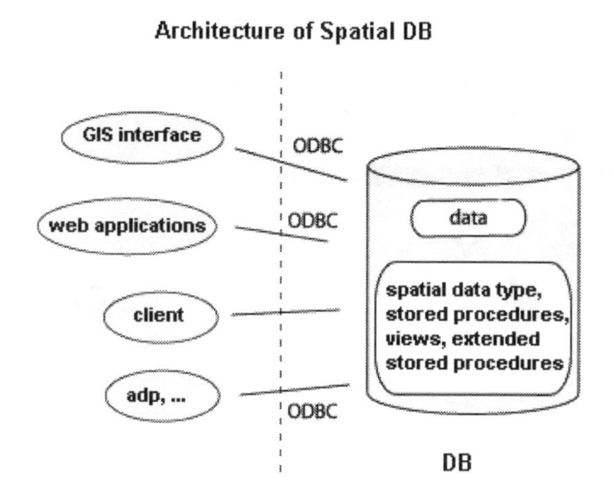

Figure 6.

4.2 Interfunctional sharing

Just as it is important to have a univocal database, it is essential for this base to be shared by all the organisations operating in the area. These operators (City Council offices, *Magistrato alle Acque* – the local Ministry of Public Building

Management Information Systems, C. A. Brebbia (Editor)

and Works office – and the underground utility companies) have diverse roles, points of view and requirements, but they are all engaged in the same area and therefore need to manage information in a coordinated manner.

The solution lies in developing different applications that share the same base while complying with the system's procedures. The univocal codes identifying geographical objects thus also become a unit of measurement that is shared in dialogue among different operators.

Furthermore, the operators are in different parts of the area, so they all have to be reached at their various locations. The way the system is organised allows physical decentralisation of data management maintaining univocality.

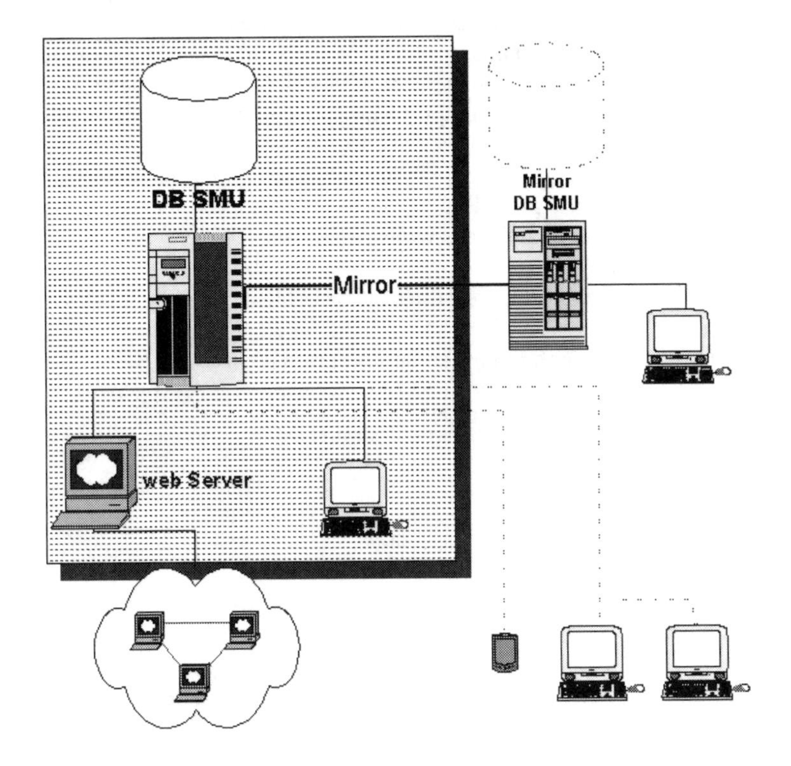

Figure 7.

Technically, there is more than one way to achieve this type of management, and this allows operators to adapt to the various environmental problems (both technological and human) that need to be solved. Among the diverse solutions are database mirror, direct connection, remote console, web applications or asynchronous server client applications. For this reason the applications that have been created have been developed for use via web.

Management Information Systems, C. A. Brebbia (Editor)

5 Application tools

5.1 Management tools

5.1.1 Puma and Marte

These are two web applications that have been developed to manage urban maintenance works in the city of Venice, an activity that is divided into individual works identified by a single job, each with its number and a description that sets out the type of work and its feature.

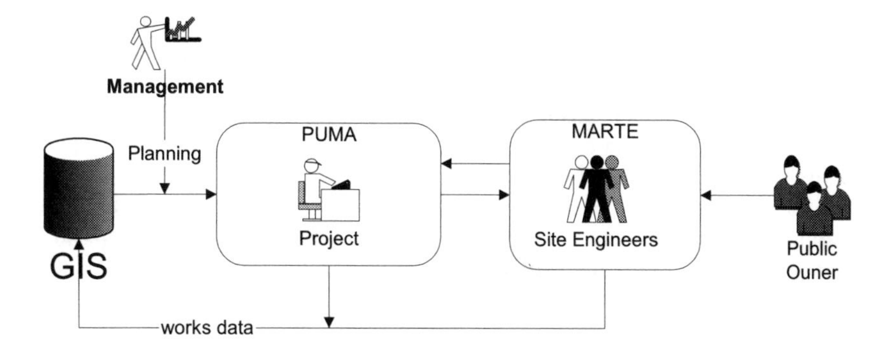

Figure 8.

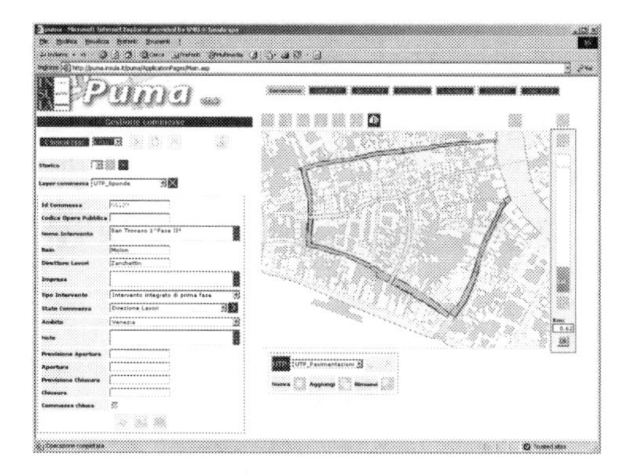

Figure 9.

One of the applications (PUMA) is devoted to design, the other (MARTE) to execution. They are therefore intended for two different kinds of user.

PUMA (Planning Urban Maintenance) is devoted to the design phases of urban maintenance works. It therefore provides very sophisticated design and planning tools. Alphanumeric and graphic functions regard the various kinds of geographical element involved in the works: segments and intersections (canals),

embankments, pavements and bridges. The application is based on the totality of geographical elements.

These elements can be broken down into basic elements, called SMU elements (segments and intersections, embankments, waterfronts, pavements and bridges) and other accessory elements (information layers such as structures, lagoon, street furniture, etc.). Essentially, Puma has three aims:

- design of the "job" elements;
- design and definition of the geographical design units for each job;
- publication (for consultation) of the SMU;
- modification of the attributes (not graphic modification) of the SMU elements.

A job can be historicised to trace its status at a certain date: for example on the date of approval or rejection of a preliminary design, on the date it was stored in the archive when an executive or final design project site was closed or after significant changes were made to key fields or areas of interest. The job and its geographical design units are either historicised or put away in a separate archive, no longer modifiable.

The attributes of the basic elements (SMU elements) may be modified by the designer after surveys and investigation campaigns, and a record is kept of all the changes made, with the name of the person responsible for them and the date of the variation (historicisation).

MARTE (Maintenance and Restoration by Geographical Element) is used in the monitoring and management of the physical progress of the works carried out. Access can be obtained through internet to a physical description of the works and information about the different types of basic element: segments, intersections, pavements and bridges. The MARTE application allows photographs to be taken of the physical progress of the works carried out by a certain date and provides information about the state of repair of bridges, embankments and pavements monitored on the day concerned.

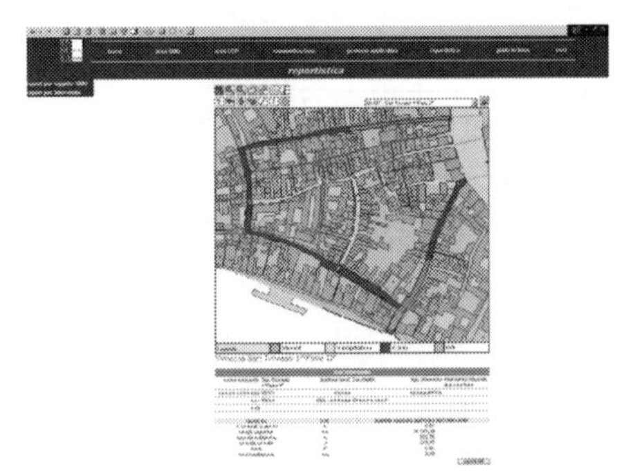

Figure 10.

The data regarding each object can be updated by using special geographical data update sheets and to each basic element may be added information of other kinds, such as images, photos and drawings.

Management Information Systems, C. A. Brebbia (Editor)

Conceived to monitor and manage work site progress, this application puts Insula managers in touch with designers and site engineers and provides the company management with a control tool. The use of the application ensures constant updating of the state of repair of the geographical objects in the Urban Maintenance System.

Its characteristics are:

- designers, job managers and site engineers are put into contact on internet;
- each of these operates on the same geographical data and objects;
- geographical objects are updated whenever they undergo any change;
- each object is also the container for all the documents that concern it;
- each geographical object bears with it the history of the works and monitoring that has involved it;
- there is always information regarding the actual progress of the works;
- the use of the "time" variable allows reconstruction of the stages in the system at any time, and also assessment of the dynamics of change (evolution of deterioration, accumulation of sediment, etc.).

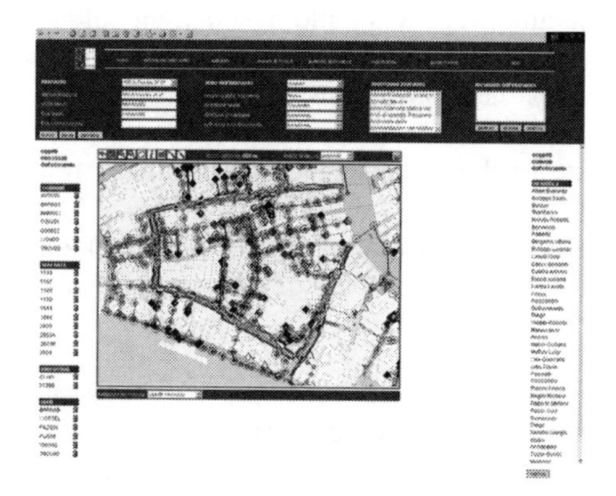

Figure 11.

5.1.2 Venere (Venice Network Restauration)

This application was prepared to manage the process of definition of joint works underground and puts Insula (which is delegated by the City Council to manage this activity) into communication with project partners (Telecom, Italgas, Enel, the Council, etc.) enabling Insula to conduct overall management of the process of coordination of underground works. The use of the application meets the requirements of the Micheli rulings of 3 March 1999 concerning coordination of works and management of data and documents. It also ensures that the state of repair of the underground utility systems is constantly updated within the Urban Maintenance System.

Its characteristics are:

- the members of the inter-company working group taking part in the definition of an underground project are put into contact on intranet;
- all the partners operate on the same geographical objects and have access to the same information;

- definition and planning of the works take place in a transparent manner;
- the database is always up-to-date with the details of the work that has been carried out;
- each object is also the container for all the documents that concern it;
- documentation is always accessible to all the partners.

5.2 Publication tools

These are applications that have been conceived for publication of the Urban Maintenance System on internet and to provide the city's inhabitants with useful information.

5.2.1 GISPORTAL

Venice GIS Portal collects the information regarding the urban elements that undergo maintenance by the Venice City Council.
Derive: analyses the state of deterioration and the structural conditions of embankments, foundation walls and buildings looking onto the canals.
Underground services: a map of the known underground service routes.
Benchmarks: provides the monographs of the network of Venice benchmarks.
Bridges: a database catalogue that collects information regarding geometry, architectural type, accessibility and materials of all the bridges in the historic centre of Venice and the islands.
Altimetry: contains the entire database of all the detailed topographical surveys.
Bathymetry: contains the entire database of the measurements taken by Insula.

5.2.2 WEBGIS

Street map of the city and basic map, which provides access to the map by address or important place and gives the main information about canals, bridges, etc.
Council ordinances are an application for input, display and download of ordinances concerning closures of sections of pedestrian thoroughfares and/or of waterways.
Navigability of canals is an application allowing users to check canal navigability according to the type of craft to be used and the tide level (the latter directly taken from the Venice City Council tide level forecasting office), also taking account of active worksites.
Pedestrian routes is an application that calculates and displays the best route connecting two places in the city, taking account of active worksites (set up through the council ordinances application) and, in this case too, of the tide level.

Management Information Systems, C. A. Brebbia (Editor)
© 2004 WIT Press, www.witpress.com, ISBN 1-85312-728-0

MIS systems in Armenian water resources management: a case study

M. Corkran[1], R. Hoffpauir[1], B. Britton[1], M. Vardanyan[2], V. Narimanyan[3], L. Harutunyan[4] & M. W. Hall[1]
[1]*Consultant, USA*
[2]*USAID, Yerevan, Armenia*
[3]*Armenian Water Resources Management Agency, Yerevan, Armenia*
[4]*ARD, Inc., Yerevan, Armenia*

Abstract

The *Water Code of the Republic of Armenia* mandates the collection and storage of a broad range of data related to water resources and requires that these data be stored in a MIS designated as the *State Water Cadastre*. By regulation, the Cadastre will consist of hydrological, meteorological and water quality data, ground water data, and data on water extraction, water use and water systems. The Cadastre will be the legal repository for water use permits and registration of water control structures. The Cadastre will include a spatial database and GIS.

Current work is proceeding with review and analysis of the data available for inclusion in the State Water Cadastre and will result in recommendations for digitizing those data. A review of the computer network systems planned for managing the Water Cadastre is underway, and policy issues affecting development and proper operation of the Cadastre are being identified and addressed. Specific recommendations for priorities in development of the Cadastre will be made. It is estimated that a functional Water Cadastre framework could be created within a period of approximately three to five months.
Keywords: water resources management, Armenian public policy, water cadastre, spatial database, GIS.

Management Information Systems, C. A. Brebbia (Editor)

1 Introduction

The Republic of Armenia, an independent nation in the south Caucasus Mountains is not well known outside the region. Armenia once stretched from the Caspian Sea to the Mediterranean, included significant parts of modern day Turkey, Syria, Iraq and Iran, and part of today's Georgia. Totally landlocked today, it is a smallish country in area, about the size of Belgium, with a population of only some 3.5 million. Its long and world famous history includes becoming, in 301 A.D., the first nation to formally adopt Christianity. The people known ethnically as Armenian, have inhabited the area for something like 6000 years. Its location between East and West makes it important to the clash of cultures currently afflicting the world.

Armenia was the smallest republic of the Soviet Union from 1920 until 1991. Upon independence, the new country found that its inheritance included extreme environmental degradation and overexploitation of natural resources, with water showing up in both categories. Water is an important resource everywhere, but in Armenia, it is of special importance, to the national economy, to the environment and to its people's distinctive culture. Understanding this, in 1999, Armenia began development of an *Integrated Water Resources Management Program.*

Armenia today, however, is an economically poor country; half its citizens existing on $1 per day, or less. It has not always been so. As recently as 1987, its economy was among the strongest, and its living standard one of the best in the Soviet Republics. Within a few days, in December, 1988, its industrial heartland was destroyed by earthquakes that killed more than 25,000 of its citizens and left a half million homeless. This was followed by the break up of the Soviet Union, which closed much of the market and flow of raw materials, that Armenia had depended on for decades. It has yet to recover, economically, from either of these events.

Seeking assistance for rebuilding its economy and its environment, Armenia turned to international donors. Among others, the United States Agency for International Development (USAID) answered this call. In late 2000, with the assistance of Armenia's government, it initiated a project entitled *Strengthening Sustainable Water Management for Enhanced Environmental Quality in Armenia.* This project was designed to enhance Armenia's environmental quality through promoting means for mitigating adverse effects of surface water over-exploitation and the resulting consequences for sector development in energy, tourism, health, and other areas.

This endeavour led, in early 2001, to approval for the reformation of the nation's water management system. This eventually meant the creation of a resourceful, contemporary, national law by which the state's water could be managed. This new law, the *Water Code of the Republic of Armenia* (Water Code), with its implementing regulations, mandated the creation of several new water resources management agencies, as well as the collection of a wide range of information and data related to water resources and their management.

Management Information Systems, C. A. Brebbia (Editor)
© 2004 WIT Press, www.witpress.com, ISBN 1-85312-728-0

2 Project background

The main features of the Armenian Water Code are (1) the maintenance and enhancement of a *National Water Reserve*, (2) that water resources are state property, thus their every use is predicated on a state issued permit, and (3) integrated basin planning and management. Each of these features is information intensive, a problem in a place which, for the past fifteen years, has been ravaged by earthquakes, war and political upheaval; where for the 70 preceding years, public use of all strategic information was extremely limited.

The Water Code recognizes these problems and addresses them by requiring that the information needed for nationwide sustainable water resources management be assembled and stored in a *Management Information System* (MIS) which it refers to as *The State Water Cadastre* (Cadastre). Such information includes that which pre-existed the Water Code, that which has been generated since its passage and that which will be generated in the future. When completely operational, the Cadastre will consist of hydrological, meteorological and water quality data; ground water data; data on water extraction, water use, water use and water systems; data on Hydrotechnical structures; environmental and biological data; and additional related information.

The Cadastre also will be the legal repository for water use permits and for registration of Hydrotechnical structures. As a part of the nation's move towards transparency in government, information in the Cadastre will be available to the general public; hopefully, online.

USAID funded two separate but connected projects related to Armenia's sustainable water resources management information needs. One of these, the *Sustainable Water Resources Management Project* (*SWRMP*) was implemented by ARD, Inc. (ARD), Burlington, Vermont. A complementary project, part of the initiative, *Water Management in the South Caucasus* (*Water Management*), ran concurrently with the ARD project and was implemented by Development Alternatives Inc. (DAI), Bethesda, Maryland.

The two projects have different areas of emphasis but are closely related; each used resources developed by the other, and the training and assistance provided under both projects were carefully coordinated to take maximum advantage of project resources and funding.

3 Cadastre (MIS) development

The first task assigned to the creators of the Cadastre was to find the needed information, and to make recommendations for getting it assembled into an MIS. The *Water Code* specifies an extensive list of data types that are to be included in the State Water Cadastre. These are:

- *Quantitative indices of water resources, including data on snow cover;*
- *Qualitative indices of water resources;*
- *Water resources use efficiency;*

Management Information Systems, C. A. Brebbia (Editor)
© 2004 WIT Press, www.witpress.com, ISBN 1-85312-728-0

- *Results of anthropogenic impact on water resources, including pollution;*
- *Wastewaters and composition and quantities of substances in them;*
- *Rehabilitation of water resources;*
- *Forecasts on floods, mudflows, droughts;*
- *Protection zones of aquatic ecosystems and their status;*
- *Atmospheric precipitation and atmospheric temperature regime;*
- *Water use permits and water systems use permits;*
- *Implementation of the National Water Program.*

In addition to these data types the Cadastre is designated as the official, legal repository for registration of Hydrotechnical structures (HTS), including such things as dams, canals, and wells; and it is the official, legal repository for resource and permit data on ground water.

The Water Code clearly states also, that "... data of the State Water Cadastre shall be considered official...." and that "... information of the State Water Cadastre shall be freely accessible to the public...." Subsequent chapters of the Water Code further establish the crucial official legal position of the Cadastre.

Being both a comprehensive national MIS for water resources and a legal repository for water permit and water control information, the Cadastre clearly must fulfill a number of functions. As these separate functions require different types of information and fall under the purview of different agencies, the Cadastre is best conceptualized as a network of separate but inter-related databases, housed in separate agencies.

3.1 The current water Cadastre

In Armenia, as in most nations, a host of agencies are involved with the creation, collection, storage and analysis of water resources information. In the majority of cases information was collected during the Soviet period, but neither collection nor archiving of data has been systematically maintained since that time. The primary exception to this trend is in the area of hydrological and meteorological data.

The three principal such agencies in Armenia are:

- ArmStateHydromet (ASH) a recently created Non-Commercial Entity (NCE; more about this later) which actively maintains extensive archives on both hydrological and meteorological data, dating from 1929 to the present day, some in electronic format;
- The Environmental Impact and Monitoring Commission (EIMC) another NCE, collects and stores, on paper, environmental (water quality, among other things) data;
- The Water Resources Management Agency (WRMA) collects and maintains, on paper, information needed for issuing and monitoring water use permits.

Management Information Systems, C. A. Brebbia (Editor)
© 2004 WIT Press, www.witpress.com, ISBN 1-85312-728-0

All these agencies are part of the Ministry of Nature Protection (MoNP.) Other water resources management projects now underway in Armenia will eventually contribute information to the Cadastre and will thus also have a part in its overall development.

In July and August of 2003, ARD and DAI, working in consultation with the MoNP, developed the overall MIS requirements for ASH, EIMC and WRMA. Specifications for technical equipment and training were completed, and plans made to install networked database management systems at each of these agencies. Some $500,000 was committed to the installation of data servers and networks at ASH, EIMC and WRMA, to house the State Water Cadastre. In addition, some $250,000 was spent on a hydrologic monitoring system and a water quality laboratory. Hydrological and meteorological information will be kept at ASH, and water quality information at EIMC. Water use and permit information will be kept at WRMA. According to Armenian water law and policy, information will be shared among these and other government agencies as needed, but each portion of the Cadastre will have a designated "owner," responsible for the generation and validation of the their portion of the information.

It was not deemed feasible within the water resources management reforms currently underway in Armenia to build the complete Cadastre. In numerous cases the required data do not exist, and it will require a number of years of observations to collect usable sets of data. Instead, the requirements for the Cadastre were reviewed and prioritized, and the highest priority elements were developed as a pilot project. It is anticipated that additional elements of the Cadastre will be developed in subsequent international assistance projects, beginning as soon as late 2004.

Research into the data available for the Cadastre showed that there were detailed and reliable records for meteorology (water entering the system) and for surface water hydrology (water flowing through the system) for the period 1929 to present. These data are collected country-wide, through an extensive network of observation stations. The practice of routine data collection was established by the Soviets and was continued almost without interruption after the collapse of the Soviet Union. The only times such data were not collected were those periods in the early 1990's, when there was no electricity to run instruments and no fuel available to get observers to their posts.

Very limited water quality data are available, beginning from about 1965. These data are not in a standardized format, are available for limited areas only, and are generally of unverifiable accuracy. Limited data on water use and withdrawals also are available, but there is no central repository for them. Major water users and several state agencies and committees keep their own withdrawal and use records. These data are of varying quality, and each dataset must be examined individually to determine its suitability for use in the Cadastre.

The minimum requirement for any water rights allocation program is information on water entering the system and water leaving the system. Therefore, the first element of the Cadastre to be developed was the hydrological database. Flow data were the first to be populated, with withdrawal data being

Management Information Systems, C. A. Brebbia (Editor)

added as the data sources are discovered. Prior to the current water resources projects in Armenia, the hydrological data were kept in hard copy format. As part of its regional project, DAI developed a hydrological database in Microsoft Access software and installed it at ASH, the agency responsible for generation and maintenance of these data. ARD is currently modifying this database so that it will include water use and withdrawals. The modified database will be installed at WRMA, the agency responsible for water rights allocation and permits.

As of this writing, a water quality database has been developed by DAI, but has not been populated. As Armenia's water resource management agencies re-initiate water quality observations, these data will be added to the system. This database is not currently being used in the water allocation permitting process. An extensive meteorological database was developed prior to the ARD and DAI projects. The database runs on proprietary software and is maintained at ASH. This database will not be altered for the Cadastre; the data will be available to state agencies as needed.

The hydrological and water quality databases at ASH are already linked to a spatial database running in on ArcView software, allowing for easy searching from, or output to, map media. At WRMA the hydrological and permit tracking databases as well as the modelling system, will be linked to ArcView.

Wherever possible data are recorded automatically at the point of observation and saved to unattended devices or sent by telemetry to ASH. Data that must be collected manually are handled according to established quality control protocols and are sent via Excel spreadsheets as e-mail attachments to ASH. Data are sent from the generating to the using agency via CD.

3.2 Ongoing Cadastre development

With the computer and network systems in place or being installed, the present work now is focused on the information to be managed, and on the manner in which it will be used. The following priorities have been recommended for rapid development of the Cadastre:

1. Establish Quality Control and Database Security Protocols. Before any further data are converted to digital format or loaded into the databases, quality control and security protocols must be in place. The DAI technical staff may be in a position to provide this service for the coming year at ASH and EIMC. At WRMA this service should be obtained through outside contracting, which could be managed by ARD.
2. Digitize the Hydrological Data Archive. The hydrological data set is available at ASH and should be converted to digital format immediately. The digital database already created by DAI, in consultation with the Hydrometeorological agencies of the Trans-Caucasus nations, should be used.
3. Develop a Database Structure for Water Use and Permit Registration. At present there is no database for storing data on water use or for registering permits, as required by the Water Code. The capability to develop this database does not exist within the WRMA, the agency that will issue and

Management Information Systems, C. A. Brebbia (Editor)

manage permits. Therefore, database development should be contracted to outside technical sources. The recommended approach is to make limited use of international, short-term consulting to direct the work and to contract with local technical staff to program the database. The developers must work in close cooperation with WRMA staff to guarantee that the product will meet the needs of the agency. The database should be developed in Access to ensure compatibility with the databases developed by DAI.

4 Finalize Water Permit Application Forms. WRMA has prepared draft application forms in accordance with the new Water Code, and has circulated the drafts to other agencies of the Armenian government for comment. Completing the forms and development of the database for WRMA should proceed interactively. There should be close correlation between the information items requested on the application forms and the individual fields in the database. The services of outside technical consultants will be required for this activity.

5 Collect, Evaluate and Enter Data on Water Use and Permits. Some data on water use and existing permits are available but are scattered among several agencies and water users. Recommendations have also been given above for a method of gathering these data. While the database is being developed, these data can be collected and copies of the data can be kept at WRMA.

6 Load Current Water Quality Data. DAI has developed a water quality database, and both ARD and DAI have upgraded selected facilities of EIMC for collecting and analyzing water quality data. As soon as collection of water quality data starts for the pilot project areas (Sevan-Hrazdan and Krami-Debed), the data should be loaded into the Cadastre.

7 Digitize the Meteorological Data Archive. DAI is currently developing a database to hold meteorological data. The structure of the database will be determined through consultation between DAI and the Hydrometeorological agencies of the three Trans-Caucasus nations.

8 Collect Spatial Data for Water Use, Permit Registration, et al. The model established by DAI can be followed for this task. Global Positioning System (GPS) equipment should be provided to WRMA. WRMA staff should be trained to use GPS to collect information for major water users, permit applicants, and the like.

This list is not all-inclusive. As the Cadastre develops, the needs of the users will change, so that it is not practical to set up a complete list of priorities for Cadastre development at this early stage.

ARD is now developing an Access-based system to track and manage water use permit applications at WRMA. The system provides user-friendly templates for each page of the permit application form as well as a large array of pre-scripted, push button queries. In addition to providing administrative tracking of permit applications and ancillary actions such as environmental statements, the system will capture the water use and withdrawal information from the permits

Management Information Systems, C. A. Brebbia (Editor)
© 2004 WIT Press, www.witpress.com, ISBN 1-85312-728-0

and pass that information to the hydrological database. In turn, the hydrological database will provide input to basin models.

All of the major stakeholders in the projects realize the limitations of databases based on Access and are fully aware of the need to migrate to a more robust relational database management system in the foreseeable future. However, the consensus among the project sponsor and implementing agencies was that priority should be given to making the initial MIS user-friendly and to shortening the start-up time as much as possible. The use of Access shortened the personnel training and system development times, and reduced initial software costs.

3.3 Modelling

The developing State Water Cadastre will create a rich source of information for constructing water resources management models. The WRMA is in charge of modelling for water use permitting and basin management planning. As the Cadastre is not entirely functional and populated with data, individual water availability models for only a select few river basins are being prepared by ARD, for presentation to the WRMA. The limited amount of available data are being collected and applied to these basins. Eventually, the WRMA will have at its disposal the Cadastre as the official data source in Armenia. The individual basin models being prepared then can then be integrated into a single modelling package that best suits the needs of the WRMA for aiding in permit allocation and the preparation of basin management plans.

A specific use of water availability models will come in the water use permitting process. As new applications are submitted to the WRMA or old applications are due for renewal, the technical staff at the WRMA can evaluate the permit's impact on the system. Questions relating to the probable effect on existing permit holders by granting a new permit, or changing an existing permit, can be answered through modelling.

3.3.1 Modelling of the Hrazdan River Basin

The computer modelling package being used by ARD, Inc. to create a water availability model of in the Hrazdan basin is the Water Rights Analysis Package (WRAP.) WRAP is an open source code model that is freely distributed via the internet (http://ceprofs.tamu.edu/rwurbs/wrap.htm) or through contact with the Texas Commission on Environmental Quality (TCEQ.) WRAP was created in the state of Texas, through public funding, during the mid-1990s to meet the water availability modelling needs of the state's comprehensive river basin management project. WRAP was selected for use in Armenia by ARD, Inc. because of its applicability to Armenian needs, as well as the open access to the model and manuals.

The current modelling effort at ARD, Inc. is a sub basin of the Hrazdan River basin; the smaller Marmarik River basin. The data availability for the entire Hrazdan River basin is not at a stage for feasible modelling. However, it is judged that the Marmarik River system has sufficient available data to begin the modelling process. As more data become available for the remainder of the

Management Information Systems, C. A. Brebbia (Editor)
© 2004 WIT Press, www.witpress.com, ISBN 1-85312-728-0

Hrazdan Basin, the Marmarik sub basin model will be integrated to form a complete basin model.

4 The intersection of MIS and policy

It has become clear that there are questions of policy that must be resolved before the Cadastre can become a reality. A clear policy on ownership and control of the data needed by the Cadastre is essential. To be effective, the policy will have to be issued at the ministerial level, or higher.

The problem arises from differences in perception of the Water Cadastre by the various agencies that have a role to play in it. Two basic perceptions of the Water Cadastre were identified: first, there are agencies that see the Cadastre as an important tool to assist them in carrying out their mission; second, there are agencies that see the Cadastre as a means to ensure their survival through the control and commercial use of data.

Broadly this division follows the lines of data generators versus data users, or NCEs versus other agencies. The NCEs clearly regard the information they generate as their own property and see the large-scale release of those data to others as a threat to their continued survival.

In fact, each of the information generating agencies should ultimately have its own database for storing and managing all the information it produces. However, at least a subset of that information must be provided to the Cadastre on a regular basis, and the Cadastre itself must be accessible by all the agencies and individuals involved in managing Armenia's water resources.

Armenian water policy and law are clear that information collected and maintained at state expense will be available, freely, to all state agencies, and to others not using them for commercial purposes. However, the word has not gotten around to all agency heads. Neither has the issue of what constitutes a "commercial purpose" been settled.

Presently, transfer of ownership of equipment already purchased by USAID, and on site in Armenia is being delayed until this issue is resolved.

5 The larger issues

The Water Code mandates the creation and maintenance of a State Water Cadastre containing information on a very wide range of environmental and water resources-related topics, as has been noted above. Furthermore, the Water Code stipulates that all such information will be made available to the public. However, the Water Code does not furnish a blueprint for the Cadastre; it does not specify the sources from which this information will be obtained, the format in which it will be recorded, the mechanisms by which it will be provided to the public, or what agencies will maintain manage and use it; it does provide funding sources for creation or management of the Cadastre. Thus, it is clear that the authors of the Water Code envisioned the need for extensive information to support development and implementation of a national water resources management policy. It is not clear whether those authors appreciated the

Management Information Systems, C. A. Brebbia (Editor)

complexity of the system they mandated or the amounts of time and effort that would be required to create such a system.

Yet, those Armenian leaders who saw to the passage of the Water Code into Law are owed a debt of thanks by Armenians everywhere. At a time of utter darkness in its economy, in its national politics and in the quality of its environment, at a time when they could have been excused for cursing this darkness, they chose instead to light a candle.

This Water Code, in its entirety, is a statement of hope; of faith that things can be made better. After three generations of lives lived under central planning concepts, flexible minds could still be found; people stilling willing to take charge of their own destinies.

An example: the Water Code requires that the information in the State Cadastre be made freely available to state agencies and to the public. This requirement alone is a significant departure from the standard operating procedure in use since the beginning of the Soviet period. Under the Soviet system, official data were closely guarded and were not considered to be in the public domain. For a country that was part of the former Soviet Union, making state data available to the public, requiring public participation in decision making, and having those decisions based on information instead of dogma, requires a paradigm shift in thinking. Nevertheless, there appears to be strong commitment on the part of a group of key individuals at the ministerial and agency management level, to create such a situation.

The Water Code also requires that the concepts of transparency and sustainability be applied in Armenian water resources management. One of the most effective ways of doing this–is through use of the worldwide web. ARD and DAI have provided training as well as technical and financial assistance to Armenia's water resources management agencies, to develop websites. ASH is already providing hydrological data to the public and to other countries in the region through its website. EIMC's website is under development, but does not have water quality data available at this time; it is not certain what data will be posted as the site develops. WRMA's website is still in its initial stages of development, but the agency plans for it ultimately to contain water use permit application forms and instructions, copies of recently issued water use permits, the water use permit fee structure, and other data important to stakeholders. The availability of such information, to a public whose participation must be sought by water resources managers, will, in turn, ensure that the common sense concepts of sustainability are more likely to be followed by these managers.

6 Conclusion

The Water Resources Management and Protection Agency has been required to submit a budget and technical proposal for establishing their part of the Cadastre. The required software, computers and network have already been committed to WRMA through the United States Agency for International Development Project.

Management Information Systems, C. A. Brebbia (Editor)
© 2004 WIT Press, www.witpress.com, ISBN 1-85312-728-0

This commitment to the Agency permits the establishment of the recommended *Data Management and Modeling Unit* within the WRMA. However, this unit will be an unstable, ever changing entity for years to come. It must be allowed to develop in accordance with both the needs of the users and the requirements of the Water Code.

This Data Management and Modeling Unit will require continuing, rigorous outside technical support in database and network management, and in modeling, for at least one to two years of operation, before it can begin to stand alone. In addition, advanced training in database management, network administration, GIS and water availability modeling will be needed for a number of individual Agency employees.

Management Information Systems, C. A. Brebbia (Editor)
© 2004 WIT Press, www.witpress.com, ISBN 1-85312-728-0

A distributed system for mobile information communications

C. García & F. Alayón
Departamento de Informática y Sistemas,
Universidad de Las Palmas de Gran Canaria,
Edificio de Informática y Matemáticas,
35017 Las Palmas, Spain

Abstract

In this paper, we present a distributed system for mobile agent communication that integrates non interactive mobile systems and is based on a mailbox scheme. To describe the system, the authors have used a structured model that permits the evaluation of the meeting level of the system's requirements. This scheme characterised the mobile protocols and applications using 3D space, the properties used are: mailbox migrating frequency, operation executed for message delivering and kind of synchronization made to avoid the message loss. Nowadays, the system is working in a context of an intelligent system for public transport by road and it plays a main role in the administration and maintenance of its mobile stations.

1 Introduction

This work belongs to the mobile distributed systems filed, in which users and systems, in various locations, can work using common geographically dispersed resources. In this context, new research areas and technologies of distributed computing arise such as: peer-to-peer, pervasive and nomadic computing [1]. Nowadays, it is very common to meet systems that use mobile agents. Fields such as e-commerce, telecommunications, public information systems, network managements, etc, use this type of agents. In general we can, affirm that the mobile agents began to play a key role in networking and distributed systems. We define a mobile agent ass an autonomous element that moves between locations in an information system. This information system must provide all the

Management Information Systems, C. A. Brebbia (Editor)

necessary resources for mobility and communications. Some authors identify these resources as a distributed abstraction layer named mobile agent system [2]. The communication protocols are an important element in this kind of systems because they ensure the communication between agents. In recent years, researches have provide a wide range of schemes for mobile communications, but in general, these schemes have particular assumptions and methodology, as a consequence, there is no uniform or structured paradigm to analyse the effectiveness and performance of protocols and applications in the context of the mobile agents system [3], [4], [5]. Therefore we have used an structured scheme to design our distributed system for mobile agents administration. This scheme is based on use of a mailbox to buffer messages in each mobile agent, permitting that agent and mailbox reside in different host migrating separately. The system that we have developed permits us transferring data and commands between mobile agent systems in order to administrate and maintain a corporative information system that integrates a set of mobile information platforms in an automatic and remote way.

2 The problem to solve: objectives and requirements

The general context of our problem can be described as follows: we have a corporative information system that integrates mobile and no mobile platforms. The mobile platforms move around a large geographic area. They are systems that have no human interactivity and they are not permanently connected to the corporative network. In our case, the corporative information system is the information system of a public transport company of passengers that has 350 buses and a mobile platform is installed on each bus and it achieves the following functionalities: the control of the time tables fulfilment, the hardware and software exceptions produced in on board devices, the registration and automatic transferring of data which represent the activity produced on board. These functionalities must be achieved without a periodic communication with any control centre, using the communications resources in a efficient way. To fulfil this requirement, it is necessary that the mobile platforms have all the hardware and software components needed to work in an autonomous way. To guarantee the software resources availability, it is necessary to use a tool that permits us the transference of these software elements to the mobile systems installed on the buses. The availability of this tool facilitates the administration and maintenance of this type of systems, moreover developing this tool in a suitable way the administration and maintenance can be performed in an automatic and no supervised way, an important practical aspect of this kind of corporation information system based on mobile information system.

To achieve this main objective, additionally, it is necessary that the communication protocol used by the system fulfils the following requirements:

Location transparency. Mobile systems (mobile agent) are sporadically integrated in the corporative network using mobile stations, so the system must

Management Information Systems, C. A. Brebbia (Editor)
© 2004 WIT Press, www.witpress.com, ISBN 1-85312-728-0

support the mobile agent location, allowing to send and receive messages not taking into account the physical location.

- **Asynchrony.** To guarantee reliable message delivery, the messages forwarding and the agent migration must be coordinated by the system. This coordination must not constrain the agent mobility
- **Reliability.** The system must ensure that all messages are routed to the target agent, independently how it migrates.
- **Efficiency.** The cost of the communication must be minimized performing two operations: agent migration to new sites and messages delivery. The cost includes aspects such us: distance, number of messages, sizes of messages, etc.

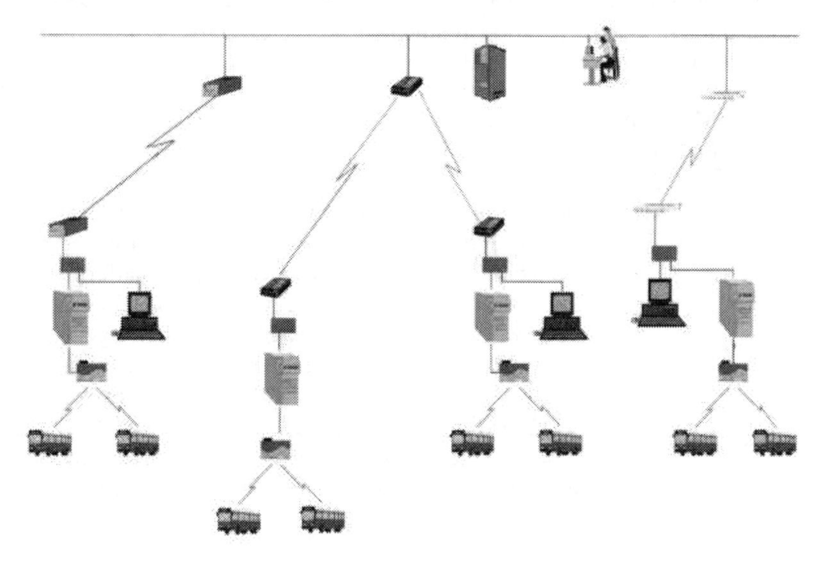

Figure 1: General vision of the system.

3 Scheme based on mailbox

In that scheme, there are three components: the mobile agents, the mailboxes and the mobile agent platform. We have assumed that each mobile agent has a mailbox that buffers the messages received. The agents are autonomous and mobile and the mailbox are mobile but not autonomous because they can not determine migration paths. The agent can send messages to the mailbox of another agent. To obtain the messages from its mailbox, the agent can execute the pull operation or the push operation, so the communication between agents has two steps: first, sending the messages to the receiver mailbox, and, second, delivering the message from the mailbox to its agent.

To apply this general scheme to our problem, we have the following settings: the mobile information systems of the buses play the role of the mobile agent platform, the applications that can send or receive messages are the agents. When the agents are being executed in a mobile platform, then they are mobile agents.

Finally, the mailboxes are objects that are not in mobile agent platforms. In order to guarantee that messages are not lost during the transmission and deliverance of messages, our system uses a high level of reliable network communication. Figure 1 represents a general vision of the system, the mobile agents execute in mobile platform installed on buses, and these are sporadically integrated by no mobile agents. Mailboxes of the mobile agents are installed in the home server

4 System description

In order to explain in a systematic way the more relevant aspects of the system, our description will be base on the general model proposed by Cao et al. [6]. In this model three key parameters are defined in every mobile communication application: mailbox migration frequency, messages delivery from mailbox to the agent and synchronization between migration and delivery.

- **Mail box migration frequency**. It consists of the number of migrations that occur during the mobile agent working. In general we can find three basic options:

  No migration (NM). The mailbox permanently stays in the same agent platform generally named home. It means that the mobile agent can migrate to different agent platforms. All the messages must be sent to the same home but these messages must be forwarded from the home to the mobile agent.

  Full migration (FM). The mailbox is considered part of the mobile agent and so it migrates with the mobile agent.

  Jump Migration (JP). The mailbox and the agent can migrate dynamically. The mobile agent determines where to establish its mailbox considering factors such as distance, number of messages, etc.

Our system uses NM mode because the tracking of mailbox produces no cost and it is simpler from the point of view of design. Moreover, generally, NM mode works well with small and medium systems with a few mobile agents and this is our case.

- **Message delivery**. The messages to the mobile agent must be sent first to the mailbox and next the messages must be delivered from the mailbox to the mobile agent. To achieve the second communication step, we can execute two alternative operations: push operation (PS) or pull operation (PL).

  Push operation (PS). By this operation, the messages are forwarded from the mailbox to the mobile agent and so the mailbox must know the mobile agent location every time. This means that every migration of the mobile agent must be communicated to the mailbox.

  Pull operation (PL). By this operation, the messages are retrieved by the mobile agent only when the messages are

Management Information Systems, C. A. Brebbia (Editor)
© 2004 WIT Press, www.witpress.com, ISBN 1-85312-728-0

needed. This means that the mailbox does not know the location of the mobile agent, so the mobile agent must query its mailbox for messages.

In our system, we have selected the push operation for two reasons: first, in some situations we need that the information arrives to the mobile information system on the buses as soon as possible (real-time delivery), and second, the pull operation would produce an increment of the message delivery cost.

- **Synchronization between migration and delivery.** This aspect refers to the reliability of the system. In order to avoid the message loss, the system can act as follows:

  No synchronization actions **(NS).**

  Synchronization between host's message forwarding and mailbox's migration **(SMH).**

  Synchronization between mailbox's messages forwarding and agent's migration **(SMA).**

  Full synchronization **(FS)** that covers the two previous synchronization actions.

In our system we use SMA synchronization; the no mobile agents are synchronized with the mobile agents. When a mobile agent contacts with a no mobile agent, it sends register message and waits for an acknowledge message. Then the messages can be forwarded and for each message forwarded to the mobile agent, it sends an acknowledge message indicating that the message has been received. In the opposite case, before each mobile agent migrating send a deregister message to the agent server that can forward pending messages or not.

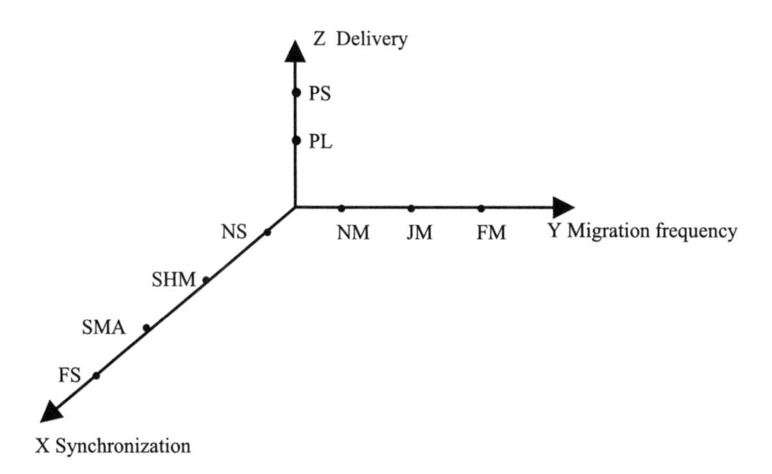

Figure 2: 3D dimensional model.

Management Information Systems, C. A. Brebbia (Editor)

Resuming, this scheme can be expressed as a three dimensional model that can be used to develop and evaluate protocols and applications in the field of the mobile communications. Using this vision of the scheme, the protocol used by our system can be characterized as a (SMA, NM, PS) protocol. For example: mobile IP is represented by (NS, NM, PS), so our (SMA, NM, PS) protocol is similar to mobile IP with synchronization and Forwarding-pointer-based protocols are represented by (NS, * , FM).

5 Conclusions

In this paper we have explained in a structured way a distributed system for non interactive mobile information system communication. With this tool we can solve a critic problem in mobile information systems that integrate a fleet of non interactive mobile systems: to administrate and to maintain the non interactive mobile systems in an automatic and none supervised way. The method used in the explanation can help to design new protocols and applications in the mobile communications and it can be also used to evaluate existing protocols and applications in this communication field. Our future work will be to reduce the protocol communication cost, using jump migration (JM), with this new characteristic our system will be more flexible and it should be used in other communication mobile contexts. Another important reason to improve the system using jump migration is that the integration of the mobile platform in the corporative network is not permanent and that connectivity between two mobile platforms could be possible.

References

[1] Steven J. Vaughan-Nichols, "Developing the Distributed-Computing OS", Computer, vol. 35, no. 9, 2002, pp.19-21.

[2] V.A. Pham and A. Karmouch, "Mobile Software Agents: An Overview", IEEE Comm. Magazine, vol. 36, no. 7, 1998, pp. 26-37.

[3] E. Pitoura and G. Samaras, "Locating Objects in Mobile Computing", IEEE Trans. Knowledge and Data Eng., vol. 13, no. 4, 2001, pp. 571-592.

[4] D. Milojicic et al., "MASIF: The OMG Mobile Agent System Interoperability Facility", Proc. 2nd International Workshop Mobile Agents (MA 98), Springer Verlag, 1998, pp. 50-67.

[5] M. Ranganathan, M. Bednarek, and D. Montgomery, "A Reliable Message Delivery Protocol for Mobile Agent", Proc. Joint 2nd International Symposium Agent Systems and Applications and 4th International Symposium Mobile Agents (ASA/MA 00), Springer Verlag, 2000 pp. 206-220.

[6] Jiannong Cao, Xinyu Feng, Jian Lu and Sajal K. Das, "Mailbox-Based Scheme for Mobile Agent Communications", Computer, vol. 35, no. 9, 2002, pp.54-60.

Management Information Systems, C. A. Brebbia (Editor)

Visibility and viewshed algorithms in an information system for environmental management

G. Achilleos[1] & A. Tsouchlaraki[2]
[1]*National Technical University of Athens, Greece*
[2]*Hellenic Open University, Greece*

Abstract

Viewshed calculation is one of the basic procedures existing in an Information System, which has the ability to manage elevation data. Visibility calculation used to be a very time consuming procedure. The rapid development of computer technology propagated the demand for faster and better algorithms, of higher quality and of a more sophisticated logic. Viewshed depends on the height of the viewing point. The alteration of this height brings alteration of the viewshed. Different techniques have been developed concerning the different ways according to which: (a) data are collected from the Digital Elevation Model (DEM) for the viewshed calculation, (b) Visual Point and Target Point is defined horizontally and vertically, (c) the decision VISIBLE or NOT VISIBLE is taken for a point. An answer "VISIBLE" or "NOT" is not enough information to define if a point is really visible or not. Visibility depends on a series of factors and conditions that must be taken into account. This article attempts to collect all necessary information about visibility and viewshed analysis concerning conventional approaches. Basic viewshed algorithms are described, alternative methods of viewshed analysis are examined and different techniques used in viewshed calculation are referred. All this analysis of algorithms and techniques is accompanied by certain environmental application examples. Further on, factors that influence the accuracy of a viewshed and also possible control procedures are referred.
Keywords: visibility, viewshed, GIS, environment, accuracy.

Management Information Systems, C. A. Brebbia (Editor)

1 Introduction

Viewshed calculation from a given observation post is nowadays one of the basic processes accomplished with the use of GIS [1]. The concept of visibility was initially examined with the appearance of Digital Terrain Models (DTM) and GIS. The development of DTMs and GIS has solved many problems related to difficult and repetitive calculations, giving the scientists the opportunity to better manage and utilise their time. A definition of visibility is [2]:

"Two points in a DTM (P, P') are intervisible, only when there is a straight line which connects point P to point P' without intersecting the DTM at a point between points P, P'."

With regard to the viewshed, V table is formulated, whose elements are a) Vij=1 when Pij is visible from P; and b) Vij=0 in the opposite case. The viewshed contains the areas of table V, where Vij=1. Fisher in 1994 [1, 3, 4, 5] and De Floriani and Mangillo in 1993 [6, 7] have proposed other visibility analysis types, which are horizons (primary and secondary) and deviations from the line of sight. Horizons are lines of viewshed that separate the visible relief from the sky (skyline). The viewshed does not extend beyond the primary horizon [1]. Deviations are defined as the elevation differences between the line of sight and the points of the relief.

Alternative visibility analysis types may be useful in landscape analyses and in monitoring any visual impact due to the spatial arrangement of activities, which must or must not be visible. Of particular use is the notion of deviation, which defines the maximum or minimum permissible elevation limits of the activities.

Viewshed depends directly on the observation elevation. Usually an increase of the observation elevation results in an increase of the surface area of the part that is visible. Calculating a viewshed is usually a time-consuming procedure. Developments in IT favour only a partial solution to the problem, which still remains unsolved. Various algorithms have been developed in order to address the problem of time-consuming calculations by using "intelligent" processing paths.

Various types of viewsheds have been devised and developed, in order to meet different needs. Therefore, the typical viewshed, where each point / pixel / triangle is determined as visible or not visible, is a binary representation of the problem. In other viewsheds, each point is codified in proportion to the number of times it is seen from the observation posts [4], or in proportion to the surface area of the wider region where it has visibility within a particular radius. One could easily identify from such maps the observation posts that provide a long range of visibility and towards a particular direction.

New types of viewsheds are the probabilistic viewsheds [3, 5, 8, 9, 10] and the fuzzy viewsheds [3] which do not determine their viewpoints in the conventional manner, as "Visible" or "Not Visible", but provide either the possibility for a viewpoint to be probably seen, or the value of the membership function in the viewshed (fuzzy).

Management Information Systems, C. A. Brebbia (Editor)

This paper presents the notions of visibility and viewsheds with references from the literature, trying to cover the main aspects of this matter. The usefulness of these algorithms in a management information system is obvious, especially when researching and addressing environmental issues and problems.

Visibility, viewsheds and the various calculation algorithms are presented within the framework of conventional approaches. This paper does not address non-conventional approaches, such as probabilistic approaches that have been developed and are researched during the last years [3, 5, 8, 9, 10].

2 Viewshed calculation algorithm

A general conventional algorithm for calculating visibility between two points is the following [11]:

- Start from a view point A
- Specify the line of sight which connects view point A to a certain target point B
- "Move along" the line of sight from view point A to target point B
- If one of the points of the relief, along the line of sight, is higher than the line of sight, then define target B as "Not Visible"
- If upon reaching target B, no point of the relief is encountered along the line of sight, which is higher than the line of sight, then define target B as "Visible"

In its generic form, this conventional algorithm is the basic logic in all visibility calculation, viewshed and visibility analysis algorithms [12, 13, 14]. What differentiates the various algorithms is the technique for obtaining the data (viewpoint location, target point location, elevations) and the decision-making technique for "Visible" or "Not Visible" targets.

In the case of viewshed calculation, the algorithms are differentiated as to the process flow, the techniques used for reducing the calculation time and the techniques applied for optimizing the quality of the results.

2.1 Traditional viewshed calculation algorithms

This section refers to traditional algorithms used for calculating viewsheds. The term "traditional" distinguishes these algorithms from other algorithms that are also conventional, but more "intelligent", more flexible and certainly more appropriate in several and, perhaps, more specialized cases. In addition to the standard algorithm, two known "traditional" algorithms are the following:

- Algorithm Weighted on the Visible Surface Area from the Observation Post
- Algorithm Weighted on the Distance from the Observation Post

In the standard algorithm, which was described in section 2, lines of sight are implemented from the viewpoint to each target point in the area of interest (usually in an area of specific radius). Each target point is examined as "Visible" or "Not Visible" and is codified accordingly using binary representation (1 or 0) [1, 4, 6, 8, 9, 11, 15]. This algorithm is the most time-consuming one.

Management Information Systems, C. A. Brebbia (Editor)

Another form of recording visibility, which does not examine whether the pixels' location is "Visible" or "Not Visible", but it examines the percentage of their surface area, which is visible from the observation post, is developed in the algorithm weighted on the visible surface area from the observation post. Pixels are not viewed as horizontal, flat or square forms, but as an inclined plane with specific slope and aspect. The observer therefore deals with the possibility not to see the pixel's surface at all or to see a part of it or its entire surface. What should be examined is the angle formed by the line of sight with the vertical line in the center of the pixel [15]. This algorithm is very useful in cases where binary information (Visible", "Not Visible") is not adequate.

It is normal for the visibility range in a viewshed to decrease when the target point is displaced away from the viewpoint. Therefore, the most distant target points, although characterized as "Visible", like the proximal target points, provide lower viewshed due to the distance. The third traditional algorithm makes use of this property and codifies the pixels, not with a binary value (1, 0), but with a weight that varies according to the distance [15]. This is accomplished, on the condition that the pixel is visible on the basis of the standard algorithm.

2.2 New enhanced viewshed calculation algorithms

Certain new algorithms are based on the basic logic of the previous methods and are differentiated as to the technique, which they try to improve and accelerate. A technique for accelerating the visibility analysis process is the main ray technique. The Main Ray Technique makes use of the property of a "Visible" point to present a wider vertical angle calculated from the Nadir, from the not visible points that follow it in the track of the line of sight over the relief (Figure 1) [16]. This technique includes the following procedure:

- Point 1 is not examined, because it is always regarded as "Visible". The ray angle of point 1 is measured and determined as the main ray
- The angle of point 2 is measured and compared to the angle of the main ray
- If it is narrower, then point 2 is "Not Visible" and the procedure moves to point 3
- If it is wider, then point 2 is "Visible" and its angle is determined to be the angle of the main ray. The procedure moves to point 3
- All points are examined along the track of the line of sight, without the need to examine each point separately with its own line of sight. Therefore, the examination of e.g. Pixel 5 does not require the examination of all the previous pixels, but only of pixel 4 (i.e. the first pixel upstream)

The three known algorithms that are presented below are the "Spiral" Examination Algorithm, the Perimetric Examination Algorithm and the "Pseudo-target" Algorithm.

Based on the "Spiral" Examination Algorithm, the visibility analysis and the viewshed calculation develop spirally outwards from the viewpoint to the

Management Information Systems, C. A. Brebbia (Editor)

perimeter of the ray defined as the visibility barrier (Figure 2) [16]. Each point of the spiral is examined with the use of the main ray technique. After examining the point, the elements of its main ray are registered, and are therefore available for use in the next spiral cycle in the corresponding direction. The elements of the main directions are then interpolated at the intermediate points beyond the 8 main rays (i.e. the majority of the points), in order to proceed with the examination of the points' visibility (Figure 2). For example, in order to examine point B03, points A01 and A02 are interpolated at the elements of their main rays.

One of the advantages of this algorithm is that each point, in order to be examined in terms of its "Visible" or "Not Visible" property, needs only the elements of the main ray (or of the main rays, if interpolation is required) of the point (or the points) of the immediately preceding spiral cycle. This accelerates significantly the algorithm in comparison to the speed of traditional algorithms mentioned above.

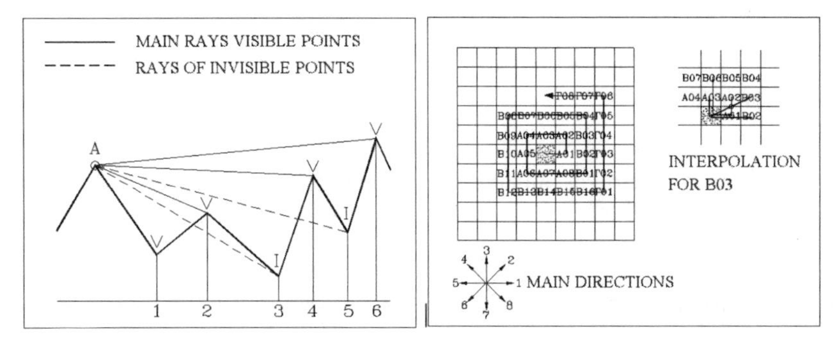

Figure 1: Main ray technique. Figure 2: Spiral examination algorithm.

The perimetric examination algorithm makes also use of the main ray technique. Its difference with the "spiral" examination algorithm is the logic on which the viewshed calculation develops [16]. The pixels that constitute the perimeter of the area whose viewshed is required are identified based on the radius within which the viewshed is calculated. The lines of sight towards each of these, resolved based on the main ray technique, are implemented from the viewpoint. The result of this resolution is that each pixel in contact with the track of the line of sight on the relief is codified as "Visible" or "Not Visible" (Figure 3). This algorithm is definitely faster than the traditional ones already mentioned. However, we couldn't say whether it is faster or slower than the first algorithm, that of the "spiral" examination. This is due to the fact that speed difference depends on various parameters in each case (pixel size, hypsometric data range, etc.).

In order to determine whether a target point is visible, the "pseudo-target" algorithm makes use of the property: "a target point is definitely visible from the intersection point of its line of sight with the horizontal plane at the highest elevation of the track of the line of sight on the relief" (Figure 4) [11].

Management Information Systems, C. A. Brebbia (Editor)
© 2004 WIT Press, www.witpress.com, ISBN 1-85312-728-0

Based on this property, the algorithm substitutes this target point with a "pseudo-target" point, which is placed at the intersection point. Therefore, the visibility examination between the viewpoint and the target point is a matter of examining the existence of visibility between the viewpoint and the "pseudo-target" point. The examination is limited within spacing D', instead of D, which under suitable conditions accelerates significantly the calculation of visibility. This algorithm may prove useful in the visibility calculation, in cases of forest and fire protection network monitoring, scheduling of flights of identification aircraft, specification of air-defence locations (anti-aircraft weapon systems), etc.

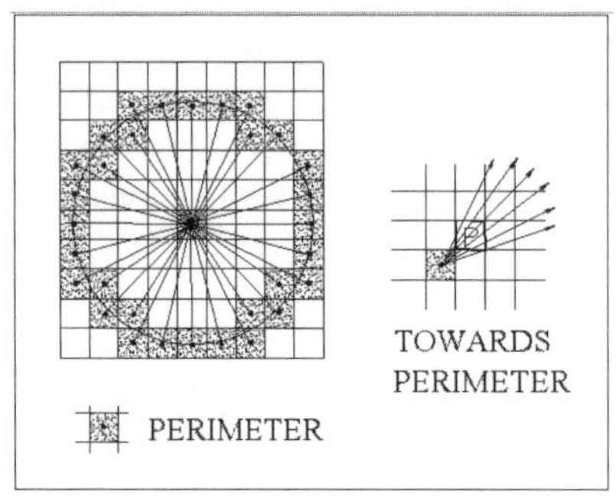

Figure 3: Perimetric examination algorithm.

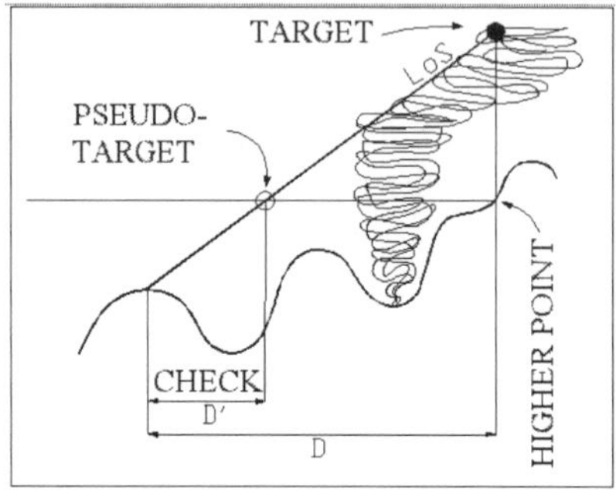

Figure 4: "Pseudo-Target" algorithm (the case of a target well above the relief).

3 Obtaining data for viewshed calculation

Algorithms are often compared and their results are also compared, without paying much attention to the quality of the elements used and to the way these are measured. With regard to viewsheds, three aspects are considered to be important and must be presented:

- How hypsometric information is obtained from a DTM.
- How the orientation of a viewpoint and a target point is carried out at a horizontal tracing position (and following this hypsometrically).
- How the decision if a target point is "Visible" or "Not Visible" is made.

3.1 Obtaining hypsometric information

A DTM is a hypsometric information registration in various points in space. A line of sight, when crosses a DTM, is nearly impossible to pass over only the points that have hypsometric information and are located either on the center of the pattern or on one of its edges. What is needed is to define the hypsometric location of the track of the line of sight on the DTM. There are many related techniques suggested and used by various scientists during the last years.

The gridded model technique was proposed by Yoeli in 1985 [3]. In this technique, the elevation of the DTM characterizes the centre of each pattern. The

Management Information Systems, C. A. Brebbia (Editor)

centers are interconnected horizontally and vertically (not diagonally) to a grid of sloped lines. The elevations of the track of the line of sight on the DTM are calculated at the intersections of the track line with the grid of the sloped lines (Figure 5). An interpolation is applied at the points of intersection in the sloped segments.

The Digital Elevation Model Triangulation Technique follows the logic of the gridded model technique, by extending the use in diagonal connections [3]. Diagonal connections are used, which intersect the line of sight laterally (Figure 6). The elevation values along the track of the line of sight are calculated at the intersection of the track line with the grid of these sloped segments.

The densified gridded model technique was proposed by Tomlin in 1990 [3]. She suggested the overlaying of a secondary grid of horizontal and vertical lines on the initial main grid, at the limits of the patterns (Figure 5). The elevation values in the connections of the secondary grid are calculated with interpolation in the elements of the main grid. The elevation values for the track of the line of sight over the DEM are calculated at its intersections with the main and secondary grid of straight - line segments.

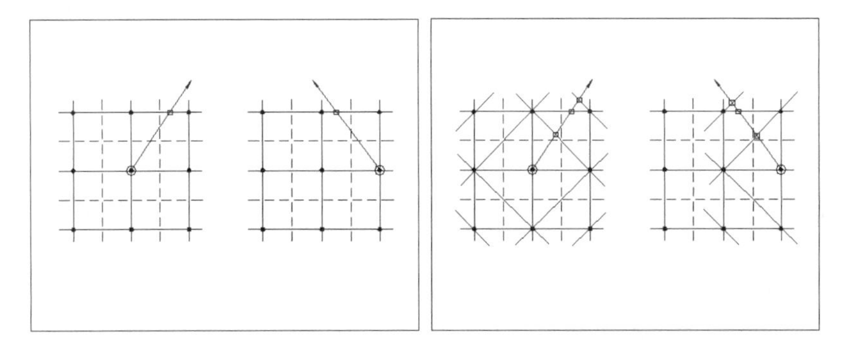

Figure 5: Gridded model technique.

Figure 6: DEM triangulation technique of normal structure.

This last technique addresses DEM as horizontal quadrangular surfaces, of the size of a pixel, which are located at the elevation that characterises them. This approach results in the DEM presenting “stairs-steps” at the limits of the pixels. The elevation values for the track of the line of sight are calculated at the points where the track meets and intersects these “stairs-steps”. Due to the discontinuity of the “stairs-steps”, the elevation of the “stairs” is recorded to be the higher of the two (Figure 8) [4]. This technique has been used by many scientists (Felleman, Travis / 1990, Teng, Davis / 1992, Burrough / 1986, see [3]) not only for visibility analysis purposes, but also for any kind of DEM management (discharge drainage, intersections, drainage networks, etc.).

In the first three techniques, progressing gradually from the first to the third, the extent of the viewshed is reduced. This is due to the fact that when progressing from the first to the third technique, the number of the points with an elevation, which should be compared to the elevations of the line of sight at the

Management Information Systems, C. A. Brebbia (Editor)
© 2004 WIT Press, www.witpress.com, ISBN 1-85312-728-0

corresponding locations, increases. The increase of these points increases also the possibility for the target point to be determined as "Not Visible", resulting to the fact that the total amount of "Not Visible" targets in a viewshed increases at the expense of the "Visible" targets.

The problem of the limited viewshed is even more intense in the fourth technique, that of the "stairs-steps". The projections created by the "stairs" increase the possibility of visibility obstruction, and therefore, of characterizing more target points as "Not Visible". A relevant examination by Fisher 1993 has shown that when using the fourth technique the extent of the viewshed is reduced by 75-95% compared to the first technique [3]. Therefore, the possible reduction is indeed significant and depends on the relief form, on which the viewshed calculation takes place.

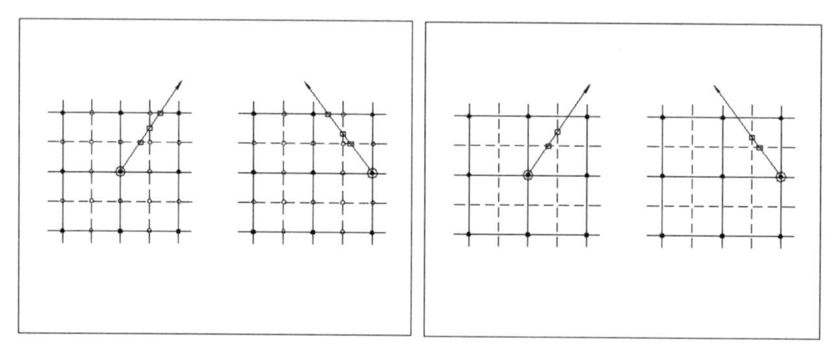

Figure 7: Densified gridded model technique.

Figure 8: "Stair-steps" technique.

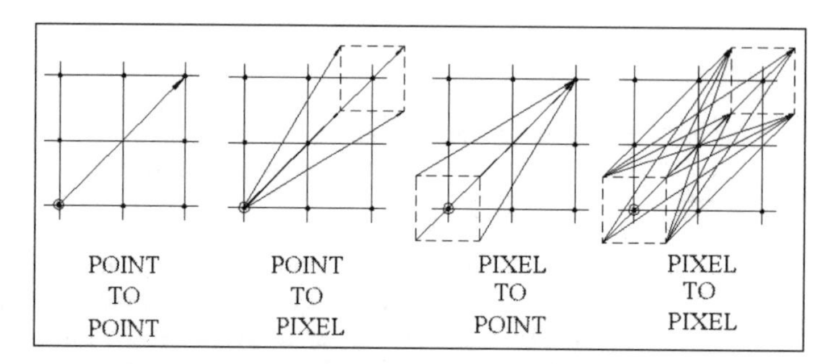

Figure 9: View and target point orientation.

3.2 View point and target point orientation

Since a normally structured DEM contains the notion of the pattern, we deal with the dilemma which location x, y, z on the pattern to define as the observation post. Therefore the following question arises: the observation post and the target are seen as points or as patterns occupying a certain surface area? And if they are seen as patterns, which of their elements shall be examined, given that patterns as surfaces have an indefinite number of points?

Management Information Systems, C. A. Brebbia (Editor)

Fisher assumes that the observation posts and the targets locations, if seen as patterns, are then represented by their four angles [3]. Therefore, there are four possible combinations to address the problem, which are presented in Figure 9.

Furthermore, the following rules / assumptions apply, in order to decide whether a target shall be included in the viewshed:

1. A Target point is defined as "Visible" from an Observation Post, if it fulfills the definition of visibility
2. A Target pattern is defined as "Visible" from an Observation Post, if at least one of its angles fulfils the definition of visibility
3. A Target point is defined as "Visible" from an Observation Pattern, if it fulfils the definition of visibility from at least one angle of the Observation Pattern
4. A Target Pattern is defined as "Visible" from the Observation Pattern, if at least one angle fulfils the definition of visibility from at least one angle of the Observation Pattern

It can be readily concluded that the first case will present much narrower viewshed than the fourth case, because in the fourth case the examination of visibility for the same pair of observation post – target post, is performed 4x4 times, i.e. 16 times, and one time is adequate for defining it as "Visible" so as to determine the target as "Visible".

3.3 Decision-making for "Visible" or "Not Visible" target

After having determined the techniques to obtain the previous data in the visibility examination, we must combine this data and compare the line of sight to its track over the DEM, so as to decide if the target is "Visible" or "Not Visible". This comparison can be carried out in many ways, which are summarized below, in Figure 10 [3].

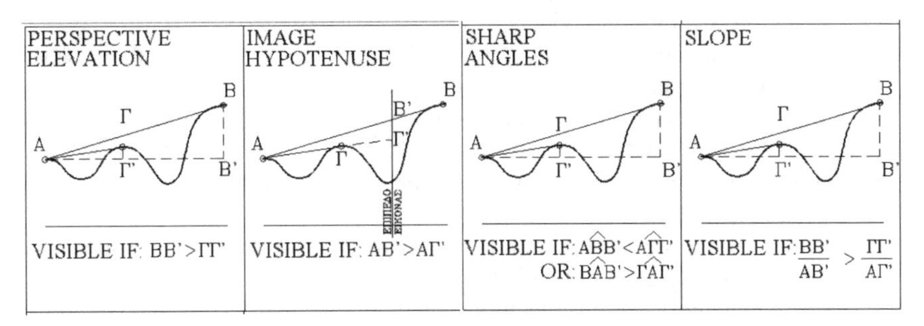

Figure 10: Decision making techniques for "Visible" or "Not Visible" target.

It should be stressed that these comparison methods constitute a purely mathematical process treatment of exactly the same data and they should give exactly the same result. This, however, is not really true, because the operations performed by the computer depend on the accuracy and the decimals that can be processed by the computer in question.

When comparing the viewsheds that derived from these four comparison techniques, the first one with a comparison of the perspective elevation and the

Management Information Systems, C. A. Brebbia (Editor)
© 2004 WIT Press, www.witpress.com, ISBN 1-85312-728-0

second one with a comparison of the inclination of the connection line to the observation post [3], it appeared that there is a difference between the two viewsheds as to their size. This is due to the fact that the inclination of the connection line involves more operations with “sensitive” numbers which affect the accuracy of the final results in the comparison.

4 Viewshed examination

The key factors that affect the accuracy of a viewshed are [5]:

- Hypsometric Information Accuracy
- Viewshed Calculation Algorithm
- Hypsometric Information Structure (DEM structure)
- Accuracy of the Horizontal tracing and Hypsometric Location of the View point and the Target point

Each of these factors affects, more or less, the accuracy of the viewshed and alters its form and shape. The most important factor is the influence of the accuracy of the information used [2, 17, 18, 19, 20], which is not usually known, and then the influence of the structure of the hypsometric information (in normally structured DEMs, the pattern’s size) [21]. These effects are not always linear and proportional. Randolph and Ray [16] found out in one of their studies that the larger the viewshed of a point, the smaller the influence of the DEM’s accuracy on its size.

One of the main problems regarding the viewshed calculation, in contrast to many other functions of GIS, is the ex-post examination of its quality, due to various factors [3]. These factors are summarized in two categories:

- Physical obstacles and physical deformities
- Error factors in the hypsometric information and in the simulation of the physical ground relief

The only way to control the viewshed calculated in relation to the actual viewshed is its perspective representation (3D) [21] and its comparison to what is seen in the viewshed in the specified observation post. This examination can be carried out either with an on-site visit in the physical relief, or with the use of photographs taken from the view point and the shooting data taken from the perspective representation.

It will be concluded that there are deviations between the two viewsheds, and this is due to the two categories of factors mentioned above.

The first category includes factors such as the vegetation covering the relief, the effect of the Earth’s curvature, the effect of the atmosphere on the visibility, the effect of the solar light reflection and the shading of the relief. The factors of the plant coverage and the Earth’s curvature may alter the viewshed upon its examination. The first factor may cause great alteration, while the second one a smaller one, as it affects only long distances from the view point. These effects are static and permanent. On the contrary, the factors of the atmosphere, of the solar light reflection and shading may affect the viewshed examination, to a smaller or larger extent. This effect is variable (dynamic), because it is a function

Management Information Systems, C. A. Brebbia (Editor)
© 2004 WIT Press, www.witpress.com, ISBN 1-85312-728-0

of time. These factors are not static and cause certain vagueness in the image the observer sees or shoots in the viewshed [1, 5, 10].

The second category of factors, which affect the calculated viewshed making it not comparable to the corresponding actual viewshed, includes the hypsometric data and physical relief description model (DEM, TIN, etc.).

The factor of the quality of the hypsometric information was presented in the previous paragraphs. Deviations of the hypsometric information from reality, of the order of 10m, 20m, 30m, [17] cause a variation in the viewshed and difficulties in its examination. The approach of the physical relief by a distinct model is another factor that makes the viewshed examination harder. However, we cannot avoid this factor.

The great influence of all the above-mentioned factors in the quality of the viewshed examination, in combination to the fact that these factors do exist at a large extent, leads us to the conclusion that viewsheds include a great deal of uncertainty [1, 8, 9]. This uncertainty guides research into new probabilistic approaches, such as the probabilistic viewsheds and the fuzzy viewsheds [3, 5, 8, 9, 10]. These approaches reject the use of a clear binary viewshed form (VISIBLE, NOT VISIBLE) and adopt milder forms (PROBABLY SEEN, SEEN WITH A p% PROBABILITY).

5 Comments – discussion

This paper attempts to present an aspect of the complexity and the significance of viewshed calculation and of the various approaches that have been developed and to some extent remain to be acknowledged. An attempt was made to present to the audience and the user of various viewshed calculation algorithms the fact that the viewshed is not just an examination of the hypsometric information of the type **"HIGHER" or "LOWER"** having direct result into the definition of "VISIBLE" or "NOT VISIBLE TARGET"; it is a more serious process requiring knowledge and skepticism. It is a process that does not allow the evaluation of its results, without prior knowledge of the parameters involved.

The presence of these algorithms in management information systems and, especially, in systems that manage environmental issues and problems constitute a significant and integral part of the approach. However, it involves risks in cases of ignorance, and given also the speed in obtaining the results using such systems, the results produced are either inadequate or false. Decision-making based on questionable results may complicate the problems and lead to their false management.

This paper attempted also to provide the reader with the dynamic of such processes. The word dynamic means that these processes cannot be regarded as static and therefore not susceptible of modification and improvement. On the contrary, they evolve and develop with the aim to optimize their result.

The concept of the accuracy of the result is included in all the forms of the results that derive from data processing. It is now a common place that each registration, processing and transformation of information is accompanied by an

Management Information Systems, C. A. Brebbia (Editor)
© 2004 WIT Press, www.witpress.com, ISBN 1-85312-728-0

error margin and a percentage of uncertainty. This error is transmitted to the information products, through the error transmission law.

Uncertainty leads to a probabilistic approach of the problem. In this way, one could imagine dealing with the perspective representation of a calculated viewshed; this representation is accompanied by other probable perspective representations of the same viewshed. Probabilistic and fuzzy approaches of the viewshed are still under research and will continue to be a matter of study for scientists in the forthcoming years.

What is of concern is the utilization of the results of these new approaches. How a probabilistic viewshed or a fuzzy viewshed will be utilized, given that the existing GIS software cannot manipulate probabilistic or fuzzy information. Furthermore, it is difficult to convince software development and marketing companies to integrate the results of the research in their systems, given the time-consuming and cost-burdening procedures.

Moreover, there is a great problem with all relative bodies that provide digital (or sometimes analogue) information to the public. These bodies are either unable to deliver digital information with its relative accuracy (at a spatial or not spatial level, as the case may be) or sometimes avoid doing so in order to prevent doubts as to the credibility and the quality of their research results.

Issues such as the one mentioned above should be specified and regulated by pertinent bodies and organizations, in order to allow users of the systems as well as of the information to have access to information about the accuracy of the final product, using sounded and widely acceptable processes.

References

[1] Fisher P. F., "Reconsideration of the Viewshed Function in Terrain Modelling", Geographical Systems, Vol. 3, 1996, pp. 33-58.

[2] Mikhail E., "Panel Discussion: The Future of DTM (presented at the ASP DTM Symposium, May 9-11, 1978, St. Louis, MO)", Photogrammetric Engineering & Remote Sensing, Vol. 44, No. 12, December 1978, pp 1487-1497.

[3] Fisher P. F., "Algorithm and Implementation Uncertainty in Viewshed Analysis", International Journal of Geographical Information Systems, Vol. 7, No. 4, 1993, pp. 331-347.

[4] Fisher P. F., "Stretching the Viewshed", 6th International Symposium on Spatial Data Handling, Edinburgh, UK, September 1994, pp. 725-738.

[5] Fisher P. F., "An Exploration of Probable Viewsheds in Landscape Planning", Environment and Planning B: Planning and Design, Vol. 22, 1995, pp. 527-546.

[6] De Floriani L., Magillo P., "Computing Visibility Maps on a Digital Terrain Model", Proceedings of European Conference, COSIT '93, Spatial Information Theory - A Theoretical Basis for GIS (editors: Frank A., Campari I.), Italy, September 1993, pp. 248-269.

[7] Magillo P., De Floriani L., Bruzzone E., "Updating Visibility Information on Multiresolution Terrain Models", Proceedings of European

Management Information Systems, C. A. Brebbia (Editor)

Conference, COSIT '93, Spatial Information Theory - A Theoretical Basis for GIS (editors: Frank A., Campari I.), Italy, September 1993, pp. 279-296.

[8] Fisher P. F., "First Experiments in Viewshed Uncertainty: The Accuracy of the Viewshed Area.", Photogrammetric Engineering & Remote Sensing, Vol. 57, No. 10, October 1991, pp. 1321-1327.

[9] Fisher P. F., "First Experiments in Viewshed Uncertainty: Simulating Fuzzy Viewsheds", Photogrammetric Engineering & Remote Sensing, Vol. 58, No. 3, March 1992, pp. 345-352.

[10] Fisher P. F., "Probable and Fuzzy Models of the Viewshed Operation", Innovations in GIS 1, (editor: Worboys M.), Taylor & Francis, London 1994, Chapter 12, pp. 161-175.

[11] Kim J. J., "High Target Visibility Analysis.", ASPRS / ACSM, Annual Convention & Exposition Technical Papers, Nevada, April 25-28, 1994, pp. 301-306.

[12] Eastman R. J., IDRISI V4.1 : Updated Manual, Clark University, IDRISI Project, September 1993, pp. 50-64.

[13] Erdas Inc., ERDAS: Field Guide, Second Edition, Atlanta, USA, 1991, Chapter 8, pp. 169-189.

[14] ESRI Inc., ARC/INFO User's Guide: Surface Modelling with TIN - Surface Analysis and Display, 2ond Edition, USA, March 1992, Chapter 6-11.

[15] Travis M. R., Elsner G. H., Iverson W. D., Johnson C. G., "VIEWIT: Computation of Seen Areas, Slope and Aspect for Land-Use Planning", General Technical Report PSW - 11 / 1975 (#Excerpt), Pacific Southwest Forest and Range Experiment Station, USDA, Forest Service, USA, 1975.

[16] Randolph Wm., Ray C. K., "Higher Isn't Necessarily Better: Visibility Algorithms and Experiments", pp. 751-770.

[17] Achilleos G., “DIGITAL TERRAIN MODELS: Hypsometric error: Error due to implementation of mutually parallel transformation in digital contours”, Technika Chronika. Scientific edition of the Technical Chamber of Greece, Vol. 17, Issue 3, Sept.-Dec. 1997, pp. 27-46 (in Greek).

[18] Nakos V., Digital Imaging of cartographic phenomena based on the theory of Fractal Geometry. Application to the topographic relief with the use of digital models, PhD. Dissertation, Dept. of Rural & Surv.Eng., NTUA, Athens 1990 (in Greek).

[19] Kumler M., "An Intensive Comparison of Triangulated Irregular Networks (TINs) and Digital Elevation Models (DEMs)", Cartographica, Vol. 31, No. 2, Summer 1994, Monograph.

[20] Robinson G.J., "The Accuracy of Digital Elevation Models Derived from Digitised Contour Data", Photogrammetric Record, 14(83), April 1994, pp. 805-814.

[21] Tsouchlaraki, A., Assessment methodology for Physical Relief Optical Value, PhD dissertation, Dept. of Rural & Surv.Eng., NTUA, Athens, 1997 (in Greek).

Management Information Systems, C. A. Brebbia (Editor)

Section 3
Remote sensing

Land cover classification of QuickBird multispectral data with an object-oriented approach

E. Tarantino
Polytechnic University of Bari, Italy

Abstract

Recent satellite technologies have produced new data as an alternative to aerial photographs. These Very High Resolution (VHR) data offer widely acceptable metric characteristics for cartographic updating and increase our ability to map land use in geometric detail and improve accuracy of local scale investigations.

Even if it is possible to distinguish spatial entities obscured in the past by low resolution sensors, high spatial resolution involves a general increase of internal variability within land covers. Moreover, accuracy of results may decrease on a per-pixel basis. If the full potentiality of the new image data for urban land use mapping is to be realised, new inferential remote-sensing analysis tools need to be investigated.

This study analyses a procedure of classification which aims to reduce negative impacts related to spectral ambiguity and spatial complexity of land cover classes of VHR imagery. In order to avoid such disadvantages, an object-oriented classification methodology on QuickBird multispectral data has been implemented on a peripheral area of Bari, in the region of Apulia (Italy). The analysis consists of two phases: multiresolution segmentation and nearest neighbour classification of the resulting image-objects. With this procedure the generated segments act as image objects whose physical and contextual characteristics can be described by means of fuzzy logic. Validity and limitations of such methodology for land cover classification is finally evaluated

1 Introduction

The applications of data obtained from satellite platforms at 300 to 600 miles above the earth have remained limited for urban areas until now for a variety of

Management Information Systems, C. A. Brebbia (Editor)
 ISBN 1-85312-728-0

reasons, including low resolution of images complexity of ground features, and technological differences with conventional photography. With satellite imagery technology the different bands are recorded synchronously so that their pixels may be precisely matched up and compared with their counterpart pixels in other bands. This means that spectral ("colour") differences can be used to identify urban features to the extent that colours are diagnostic - sort of a coarse spectroscopy from space. Spectral sensor technology, however, coupled with the complexity of ground features in urban areas, can make visual interpretations of satellite imagery both labour intensive and uncertain.

To implement image interpretation, commercial or free software can be used to "train" computers to recognise spectral signatures from sample locations. The software program will then read through an entire satellite dataset to tag other pixels that have similar characteristics [11]. This procedure works well with images of agricultural crops or forests in rural areas, but computers have had trouble distinguishing between different urban features. For example, roofs of building are small compared to satellite pixels and have many colours and shadings due to different materials used and to different orientation towards the sunlight. As a result, this approach has had limited success in dealing with data for urban environments [10, 13].

During the past few years improvement in the resolution of satellite images has broadened applications for satellite images to areas such as urban planning, data fusion with aerial photos and digital terrain models (DTMs), and the integration of cartographic features with GIS data. However, previous high-resolution satellites, such as 1-m resolution IKONOS, still could not replace the use of aerial photos, which have resolution as high as 0.2 to 0.3 m. The successful launch of QuickBird and its high resolution sensors has narrowed the gap between satellite images and aerial photos, permitting easy updating of urban information. The result is that the panchromatic resolution is increased from 1-metre to 70-centimetre and multi-spectral from 4-metre to 2.80-metre resolution. Panchromatic imagery is collected in 11-bit format (2048 grey levels) and delivered in 16 bit format for superior image interpretation (shadow detail, etc.), or 8-bit format (256 grey levels) supported on desktop GIS and mapping application packages. Four-band multi-spectral imagery consists of blue, green, red and near-infrared bands, delivered in 16-bit and 8-bit formats. Such data are likely to stimulate the development of urban remote sensing [9], but optimal results still usually need human visual interpretation of the computer enhanced images. For these reasons, much of the research in satellite data analysis has focused on agricultural and forested areas, where the spectral responses from large fields with homogeneous types of vegetation provide relatively uncomplicated landscapes.

If the full potential of the new image data for urban land use mapping is to be realised, new inferential remote-sensing analysis tools need to be investigated [5]. At the same time, it has been found that the land cover classification of VHR images results in a complex structural composition, which might inhibit recognition of distinct urban land use categories. High resolution of the data does not automatically lead to higher classification accuracy. This is due

Management Information Systems, C. A. Brebbia (Editor)

to the heterogeneity of objects within an urban area which leads to either misclassified pixels or unwanted details [8, 17].

In this paper a classification procedure aimed at reduction of negative impacts from land cover classification of VHR imagery was investigated. In order to avoid such disadvantages, an object-oriented classification methodology on QuickBird multispectral data has been implemented on the test area of Bari, in the region of Apulia (Italy). The analysis is conducted with eCognition software and consists of two phases: multiresolution segmentation and nearest neighbour classification of the resulting image-objects. With this procedure the generated segments act as image objects whose physical and contextual characteristics can be described by means of fuzzy logic. Validity and limitations of such methodology for land cover classification are finally drawn.

2 From pixel -based to object-based classification

In remote sensing techniques the informational classes of a thematic mapping are not directly registered, but must be derived indirectly by using evidence contained in the spectral data of an image. When standard procedures of per-pixel multispectral classification are applied to VHR data, the increase of spatial resolution leads to augmentation in ambiguity in the statistical definition of land cover classes and a decrease of accuracy in automatic identification. These sources of imagery are likely to generate other problems. Even if the radiometric resolution is enhanced (11 bits for QuickBird imagery), spectral capabilities are generally limited compared to those of the previous generation sensors (seven bands for the Landsat TM). Moreover, associated with an increase in spatial resolution there is, usually, an increase in variability within land parcels ('noise' in the image), generating a decrease in accuracy of land use classification on a per-pixel basis [16].

In order to solve such problems, some post-classification procedures were investigated on the basis of intrinsic contextual information of data [15]. Although a reduction of noise in the classified image was obtained, substantial improvements in overall accuracy of the results was not see. Moreover, a loss of meaningful information in classified data was shown because of geometric and dimensional non-correspondence of real elements with moving window implementation matrix (for example with *majority* logical filter). Such methods need extensive editing operations on classified images in order to be stored in GIS databases.

An alternative technique to per-pixel classification is the per-field classification (so called because fields, as opposed pixels, are classified as independent units), and this takes into account the spectral and spatial properties of the imagery, the size and shape of the fields and the land cover classes chosen. In fact, this approach requires a priori information on the boundaries of objects in the image. 'Field" or 'parcel' refers to homogenous patches of land (agricultural fields, gardens, urban structures or roads) which already exist and are superimposed on the image. Some studies [2, 7] indicate that this

Management Information Systems, C. A. Brebbia (Editor)
© 2004 WIT Press, www.witpress.com, ISBN 1-85312-728-0

methodology is contributing positively to the classification of remote sensing imagery of high to medium geometric resolution.

Problems arise in cases where no boundaries are readily available or when exactly those boundaries should be updated [6].

One solution is image segmentation. In many cases, image analysis leads to meaningful objects only when the image is segmented in 'homogenous' areas [3, 12]. Because an 'ideal' object scale does not exist, objects from different levels of segmentation (spatially) and of different meanings have to be combined for many applications. The human eye recognises large and small objects simultaneously but not across totally different dimensions. From a balloon for instance, the impression of a landscape is dominated by land use patterns such as the composition of fields, roads, ponds and built up areas. Closer to the ground, one starts to recognise small patterns such as single plants while simultaneously small scale patterns loose importance or cannot be perceived anymore. In remote sensing, a single sensor highly correlates with a specific range of scales. The delectability of an object can be treated relative to the sensor's resolution. A coarse rule of thumb is that the scale of image objects to be detected must be significantly bigger than the scale of image noise relative to texture. This ensures that subsequent object oriented image processing is based on meaningful image objects. Therefore, among the most important characteristics of a segmentation procedure is the homogeneity of the objects [6].

The resulting segmentation should be reproducible and universal, thus allowing application to a large variety of data [1, 3]. Moreover, multiresolution image processing based on texture and utilising fractal algorithms can alone fulfil all main requirements at once. Their 'fractal net evolution approach' uses a local mutual best-fit heuristics to find the least heterogeneous merge in a local vicinity following the gradient of the best-fit. Furthermore, their algorithm can be applied with pure spectral heterogeneity or with a mix of spectral and form heterogeneity [6].

3 Data and methods

The study area of this research is in Italy, i.e. part of the peripheral district of Bari, in the region of Apulia Italy). The flat geomorphology and settling features justify the choice of this area, as t permits testing of the validity of advanced classification methodologies on a over-regional scale for the new generation satellite sensor imagery.

The data sources acquired for the analysis consist of:

a) QuickBird multispectral images (Acquisition Date/Time: 2002-08-01 / 17:11:18). To reduce the size of image to be manipulated and, to maintain geographic consistency during analysis, equivalent sub-scenes were defined, fig. 1.
b) training set and the ground reference data obtained by land use survey (September 2002) and additional maps.

Management Information Systems, C. A. Brebbia (Editor)

Figure 1: QuickBird multispectral image processed in RGB mode- Bari (Italy).

3.1 Classification strategy

Object-oriented image analysis consists of two steps, segmentation and (fuzzy) classification of the resulting image-objects.

The multiresolution segmentation algorithm implemented in eCognition software extracts homogenous image-objects at different resolution hierarchy-levels respectively. For each segment (object) object-features are calculated and stored in a database. Compared to conventional pixel-based classification approaches, utilizing only the spectral response, image objects contain additional information, like, among others, object texture, shape, relationship to neighbours and sub-objects

Fuzzy classification is based on the different object-features available. In contrast to pixel-based statistic classifiers, fuzzy classification replaces the strict class definitions of "yes" and "no" by a continuous range where all numbers between 0 and 1 describe a certain state of class membership [4].

In this paper class description is performed using the fuzzy approach of nearest neighbour that supports a classification approach based on marking typical objects as representative samples. The nearest neighbour (NN) classifier is able to evaluate the correlation between object features and handles the presence of overlaps in the feature space much more easily. Moreover, it allows very fast and easy handling of the class hierarchy for classification (without class - related features).

With the aim of exploring the potentiality of multiresolution segmentation of QuickBird multispectral data, different scale parameters were implemented, by

producing three levels of spatial representation, as shown in figures 2, 3, 4. Such algorithms join neighbour regions that show a degree of fitting – computed with respect to their spectral variance and/or their shape proprieties – which is smaller than a pre-defined threshold (scale parameter). The choice of this parameter determines the number and size of resulting segments [14].

In order to assess the capability to discriminate urban features, level 1 of segmentation for the following classification was selected. Classes separated by object-features can represent parts of a thematic unit and can be combined into semantic groups. The analysed territory includes three main semantic groups of land use (urban settlements, agricultural fields and airport), which include low level objects. Following the classification methodology of eCognition software, the definition of classes is based on the possibility of distinguishing one class from another in certain features. This is realized by developing a knowledge base suitable for each classification strategy enabling description of each class. Seven land use classes for the study area were individuated (asphalt roads, country roads, buildings, meadows, uncultivated land, arable land, olive-groves). Generally, remotely sensed reflectance is related to land cover and not to land use, but in the present case each land use class was assumed to correspond to spectrally-separable land covers. Additional rule-set definitions were introduced, such as relations to neighbour objects and interpretation of deduced information (vegetation indices; band ratios; relevant land use, such as the airport complex). After this, the nearest neighbour (NN) classifier was trained on the representative sample of each of the land use classes, fig. 5.

Figure 2: Multiresolution segmentation of QuickBird multispectral data (level 1).

Management Information Systems, C. A. Brebbia (Editor)

Figure 3: Multiresolution segmentation of QuickBird multispectral data (level 2).

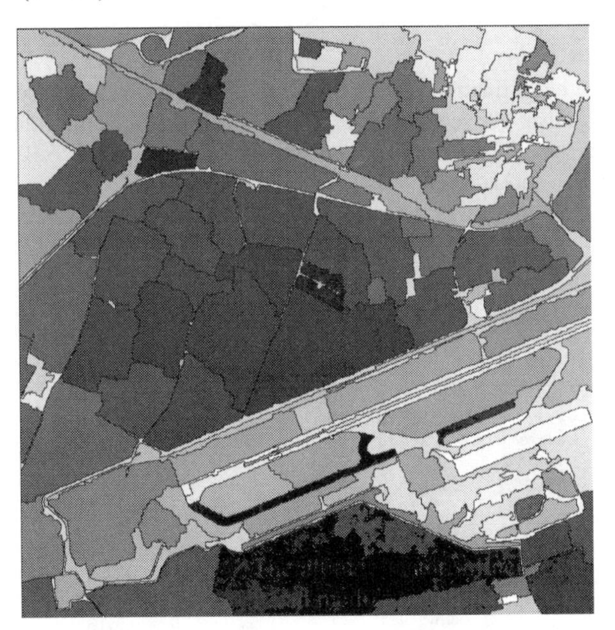

Figure 4: Multiresolution segmentation of QuickBird multispectral data (level 3).

Management Information Systems, C. A. Brebbia (Editor)

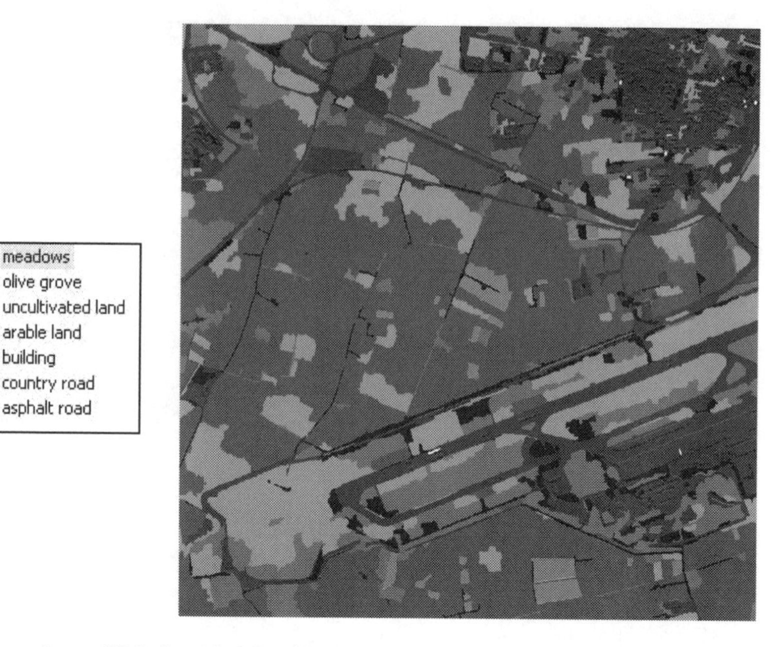

Figure 5: NN classification (level 1) of QuickBird multispectral data.

4 Results

The presented classification results show the ability of the so far developed rule-sets to distinguish between objects and to generate semantic classes. This is demonstrated by the high value of the overall accuracy (0,962) of the classification, as shown in table 1.

Table 1: Results of NN classification – level 1: error matrix and statistics.

User \ Referenc...	se...	arb...	inc...	terr...	edif...	stra...	stra...	Sum
Confusion Matrix								
seminativo	2837	0	0	0	0	0	0	2837
arborato	0	5620	0	0	0	0	0	5620
incolto	0	0	2932	0	0	0	0	2932
terreno	0	0	0	1145	0	134	0	1279
edificato	0	0	0	0	1427	0	0	1427
strada non asfaltata	0	0	0	0	0	399	0	399
strada asfaltata	0	0	0	0	504	0	1579	2083
unclassified	0	0	0	0	0	0	0	0
Sum	2837	5620	2932	1145	1931	533	1579	
Accuracy								
Producer	1	1	1	1	0.739	0.749	1	
User	1	1	1	0.895	1	1	0.758	
Hellden	1	1	1	0.945	0.85	0.856	0.862	
Short	1	1	1	0.895	0.739	0.749	0.758	
KIA Per Class	1	1	1	1	0.714	0.742	1	
Totals								
Overall Accuracy	**0.962**							
KIA	**0.952**							

There are various causes for obvious "classification" errors, with relevant difficulty in distinguishing between different urban features. For example, roofs of building have many colours and shadings due to different materials used and their different orientation towards the sunlight. These proprieties caused intense overlaps with the asphalt road class.

For spectral separable classes, such as natural features, like meadows, olive grove, uncultivated land a good recognition rate was obtained.
The arable land class shows a classification accuracy with light overlaps to country roads.

Semantic classes that present errors often contain different class-types causing a partly incorrect representation. For example a class for urban settlements can include trees or grass. Object-oriented classification distinguishes between these objects and classifies trees or grass (as meadow) depending on the object-level.

5 Conclusions

The semantic classes derived from satellite data cannot exactly fit the classes for large scale investigation, because of the resulting aggregation of information.

Very high resolution satellite data contain significantly more detailed information than previous satellites (e.g. Landsat TM) but problems arise from their high spectral variance. The so called "salt & pepper" effect is a handicap for the recognition of semantic classes.

Object oriented classification systems enable us to analyse high resolution data and avoid some of the above mentioned problems. Based on a preceding segmentation, a fuzzy approach of nearest neighbour is used to classify the QuickBird data.

The primary task is the creation of object lists that take the abilities of different remote sensing data into account. Major problems are the generation of meaningful objects that fit semantic classes of the object lists.

Classification rules for all object types and their representation in VHR data from different sensors and on varying acquisition times have to be developed.

The next step will be the assignment of the deduced classification rules on more complex scenes. Additional knowledge from thematic layers has not been used in this study so far but will be included in our future work.

References

[1] Andresen, T., Mott, C., Zimmermann S., Schneider, T., Monitoring of reed population on Bavarian lakes with high-resolution satellite data, *Proc. of ISPRS Workshop "High Resolution Mapping from Space 2001"*, Hannover, September 19-21, (CD-ROM), 2001.

[2] Aplin, P., Atkinson, P., Curran, P., Per-field classification of land use using the forthcoming very fine resolution satellite sensors: problems and potential solutions, Advances *in remote sensing and GIS analysis,* eds. Atkinson, P. & Tate, N., Wiley & Son, Chichester, pp. 219-239, 1999.

Management Information Systems, C. A. Brebbia (Editor)

[3] Baatz, M., Schäpe, A., Multiresolution Segmentation – an optimisation approach for high quality multi-scale image segmentation, *Angewandte Geographische Informationsverarbeitung*, eds. Strobl, J., Blaschke, Griesebner, XII, Wichmann-Verlag, Heidelberg, pp. 12-23, 2000.
[4] Baatz, M., Heynen, M., Hofmann, P., Lingenfelder, I., Mimler, M., Schäpe, A., Weber, M., Willhauck, G., *eCognition User Guide*. München, Definiens AG, 2000.
[5] Bauer, T., Steinnocher K., Per-parcel land use classification in urban areas applying a rule-based technique, *GeoBIT/GIS,* 6, pp. 12-17, 2001.
[6] Blaschke, T., Lang, S., Lorup, E., Strobl, J., Zeil, P., Object-oriented image processing in an integrated GIS/remote sensing environment and perspectives for environmental applications, *Environmental Information for Planning, Politics and the Public*, eds. Cremers, A., Greve, K., Metropolis Verlag, Marburg, 2, pp. 555-570, 2000.
[7] Caprioli M., Tarantino E., Accuracy assessment of per-field classification integrating very fine spatial resolution satellite sensors imagery with topographic data, *Journal of Geospatial Engineering*, 3 (2), pp. 127-134, 2001.
[8] Donnay, J.-P., Use of remote sensing in-formation in planning, *Geographical In-formation and Planning*, eds. Stillwell, J., Geertman, S. & Openshaw, S., Berlin, Springer-Ver-lag, pp. 242–260, 1999.
[9] Fritz, L. W., High resolution commercial remote sensing satellites and spatial information, 1999. www.isprs.org/publications/highlights/highlights0402/fritz.html.
[10] Gao, J., Skillcorn, D., Capability of SPOT XL data in producing detailed land cover maps at the urban-rural periphery, *International Journal of Remote Sensing*, 19 (15), pp. 2877-2891, 1998.
[11] Gong, P., & Howarth, P. J., An assessment of some factors influencing multispectral land-cover classification, *Photogrammetric Engineering and Remote Sensing*, 56 (5), pp. 597-603, 1990.
[12] Gorte, B., Probabilistic Segmentation of Remotely Sensed Images, ITC *Publication Series,* 63, 1998.
[13] Harris, P. M., Ventura, S. J., 1998. The integration of geographic data with remotely sensed imagery to improve classification on an urban area, *Photogrammetric Engineering and Remote Sensing*, 61(8), pp. 993-998, 1998.
[14] Schiewe, J., Tufte, L., Ehlers, M., Potential and problems of multi-scale segmentation methods in remote sensing, *GeoBIT/GIS*, pp. 34-39, 2001.
[15] Townshend, J. R. G., The enhancement of computer classification by logical smoothing. *Photogrammetric Engineering and Remote Sensing*, 52, pp. 213-221, 1986.
[16] Townshend, J. R. G., Land cover, International *Journal of Remote Sensing*, 13, pp. 1319-1328, 1992.
[17] Woodcock, C., Strahler, A., The factor of scale in remote sensing, *Remote Sensing of Environment*, Vol. 21, pp. 311–322, 1987.

Management Information Systems, C. A. Brebbia (Editor)
© 2004 WIT Press, www.witpress.com, ISBN 1-85312-728-0

Agrology in olive grove soils in the P.D.O. Baena (Spain)

I. L. Castillejo González[1], M. Sánchez de la Orden[1],
A. García-Ferrer Porras[1], L. García Torres[2] & F. López Granados[2]
[1]*Department of Cartography, Photogrametry, GIS and Remote Sensing, Córdoba University*
[2]*Instituto Agricultura Sostenible (CSIC)*

Abstract

Nowadays it is possible to observe that many zones present great imbalances in the use of soils with respect to the potential possibilities. These imbalances are very important because they produce a decrease of the crop production and a degradation of soil properties that lead to great environmental problems.

In this project, a methodology of work based on the Geographic Information System (GIS) has been developed for the study of the problems these situations are generating. The development of this GIS methodology has generated a useful tool for the evaluation of the present soil use to get its optimal productive capacity. Then, and with the data collected in this evaluation, it is possible to develop a planning and exploitation model that formulates the way to adapt the current agrarian use with their potential productive capacity.
Keywords: GIS, thematic cartography, territory planning

1 Introduction

The application of GIS technologies in the agriculture has supposed a revolution in this field because they allow to optimize the production from the knowledge of the physical, chemical and biological characteristics the soil presents. This fact helps to prevent the degradation of the environment caused by inadequate growing techniques. Therefore, the knowledge of certain factors like kind of soil, current use, slope,... and their combination allow to evaluate the agrarian possibilities of a certain area, showing imbalances when they exist.

Management Information Systems, C. A. Brebbia (Editor)

The aim of this project is to develop a territorial analysis to obtaining a planning and exploitation model of the different resources to formulate the more suitable system of growing.

The practical application of this methodology has been carried out in the Protected Designation of Origin (P.D.O.) Baena. This area extends for 84.983 hectares distributed in 6 municipalities located in the province of Córdoba, (Spain). The main cultivation of this area is the olive grove which gives a quality olive oil very recognized. Therefore, the constitution of the P.D.O. Baena has allowed the knowledge of its oil in national and international markets.

2 Methodology

The methodology developed in this study has five consecutive phases.

1. Compilation of information. For this study, the information has been mainly focused in the physical and agrarian characteristics of the studied area.

- Uses Map: Province of Córdoba to scale 1:50.000 based in the topographical of the MTA at 1:10.000. Dirección de la Producción Agraria of the Consejería de Agricultura y Pesca of the J.A. (Version 2000).

- Agrarian Classes Classification of the Province of Córdoba. Dirección de la Producción Agraria of the Consejería de Agricultura y Pesca of the J.A. (1999).

- Digital Terrain Model (DTM) of the Consejería de Medio Ambiente of the J.A.

2. Obtaining of the superficial distribution of different variables from the initial information.

- Distribution of the different uses from the Uses Map.

- Distribution of the Agrarian Classes from the Agrarian Classes Classification.

- Distribution of the slopes and altitudes obtained from the DTM.

3. Crossing of the different information coverings to obtain the interrelation among the studied variables.

4. Obtaining of the different imbalances grades starting from the study of the crossings carried out previously.

5. Indication of possible correctives measures that would diminish current imbalances.

3 Analysis of the cartographic synthesis

3.1 Olive grove distribution

The project has been centered in the study of the olive grove, because this is the main production of the area with almost 66% of the total studied surface. The rest of the territory has been grouped together in an only class that collect all the agrarian, forest and unproductive uses in the studied area (Figure 1).

Management Information Systems, C. A. Brebbia (Editor)
© 2004 WIT Press, www.witpress.com, ISBN 1-85312-728-0

Table 1: Olive grove distribution in P.D.O Baena.

Surface (ha)	Baena	Castro del Río	Doña Mencía	Luque	Nueva Carteya	Zuheros	Total	% Total
Total olive grove	23.853	14.380	1.050	9.141	6.546	865	55.836	100
% olive grove/ total olive grove	42,72	25,75	1,88	16,37	11,73	1,55	--	100
Total surface	36.246	21.990	1.518	14.081	6.917	4.228	84.984	100
% olive grove / total surface	65,80	65,39	69,17	64,92	94,64	20,46	--	--

The municipalities of more extension are also what more olive grove surface present. This is the case of Baena and Castro del Río olive grove that represents 43% and 26% of the whole studied crop. The rest of municipalities present less surface although it is necessary to emphasize that in all of them the olive grove take up more than 60% of the territory except in Zuheros where the uneven orography impedes the development of this grove.

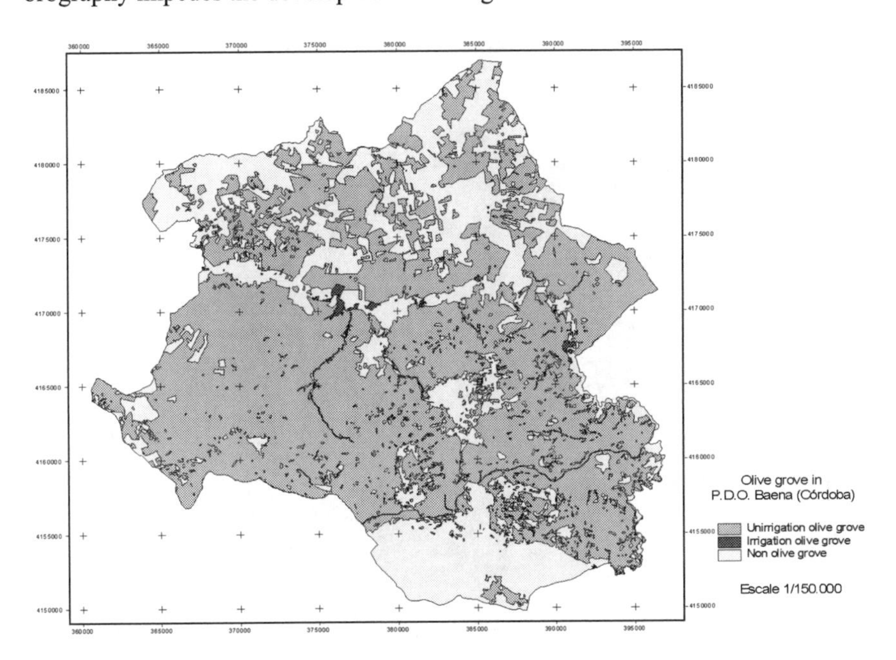

Figure 1: Olive grove distribution in P.D.O. Baena.

3.2 Agrarian Classes distribution

More than half of the surface, 54%, consists of class I, II and III soils with good aptitudes for the agricultural activity (Menéndez and Rodríguez [3]). Other 19% are IV class soils that only allow the growing of some species with lower yields.

The rest of the surface belongs to V, VI, VII and VIII classes which present edaphic and/or climatic limitations that restrict the agrarian uses (Figure 2).

Table 2: Agrarian classes distribution in P.D.O. Baena.

Surface (ha) / Classes	Baena	Castro del Río	Doña Mencía	Luque	Nueva Carteya	Zuheros	Total	% Total
I	315	0	0	79	0	0	394	0,46
II	6.670	3.463	254	2.228	1.365	454	14.434	16,99
III	11.336	8.692	809	5.698	2.922	2.479	31.936	37,59
IV	9.416	6.123	0	437	1	0	15.977	18,80
V	25	9	0	141	0	123	298	0,35
VI	7.745	3.351	190	2.912	2.338	324	16.860	19,85
VII	406	84	169	2.214	259	773	3.905	4,60
VIII	337	267	96	347	32	76	1.155	1,36
Total	36.248	21.989	1.518	14.056	6.917	4.229	84.959	100

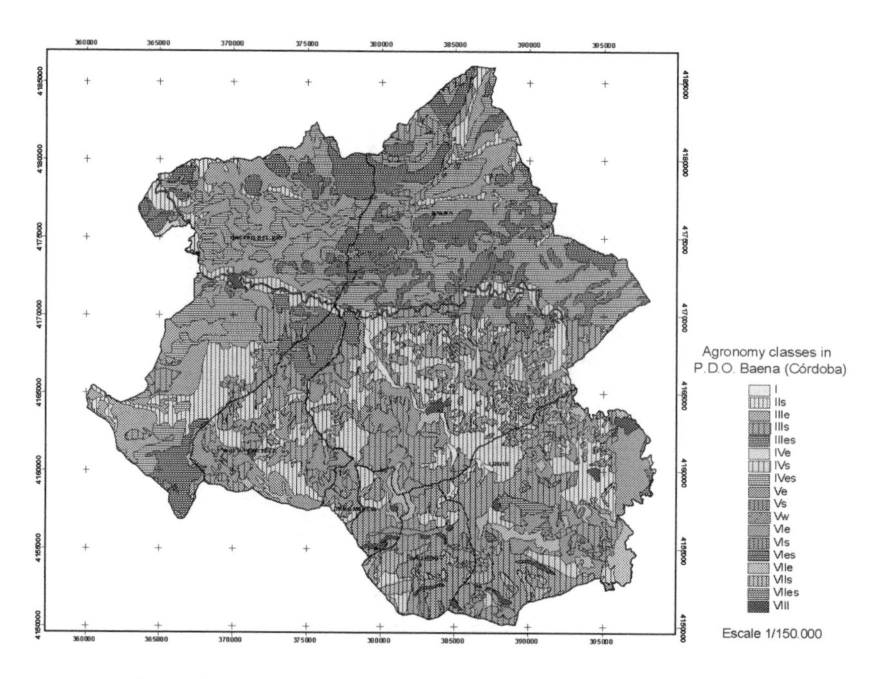

Figure 2: Agrarian classes distribution in P.D.O. Baena.

3.3 Slopes distribution

Most of the soils present slopes lower than 20%, therefore, they don't present great limitations for the agrarian activities. 16% of these soils don't need any conservation measure while 53% of them need advanced cogoverning measures

Management Information Systems, C. A. Brebbia (Editor)
© 2004 WIT Press, www.witpress.com, ISBN 1-85312-728-0

and 27% need severe and systematic measures. The remaining territory doesn't allow an agrarian use (Figure 3).

Table 3: Slopes distribution in P.D.O. Baena.

Surface (ha) / Slope	Baena	Castro del Río	Doña Mencía	Luque	Nueva Carteya	Zuheros	Total	% Total
<3%	5.891	5.272	22	1.759	569	287	13.800	16,24
3-10%	20.784	12.920	787	5.567	3.607	1.556	45.221	53,21
10-20%	9.274	3.737	478	5.185	2.623	1.638	22.935	26,99
20-30%	290	61	195	1.285	114	629	2.574	3,03
30-50%	10	0	35	285	4	118	452	0,53
>50%	0	0	0	0	0	0	0	0
Total	36.249	21.990	1.517	14.081	6.917	4.228	84.982	100

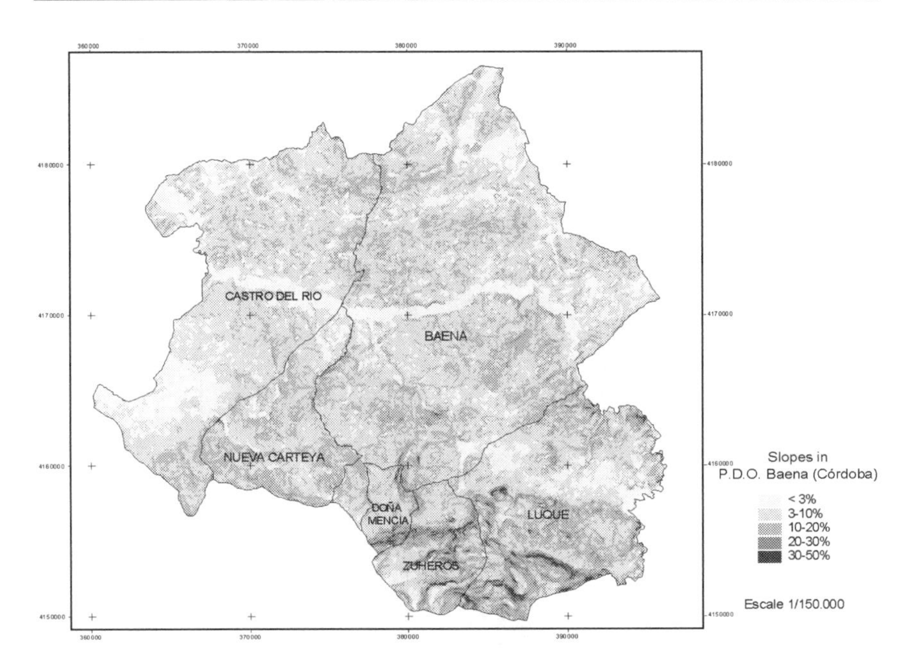

Figure 3: Slopes distribution in P.D.O. Baena.

3.4 Altitudes distribution

The altitude factor influences fundamentally in the viability of the crop because it can modify the climate (León [2]). 75% of the studied surface present altitudes lower than 500 meters. In this group, the most important stratums are the 300-400 m interval with almost 30% of the total surface and the 200-300 m one with 20%. The territories higher than 500 m don't present much surface (Figure 4).

Management Information Systems, C. A. Brebbia (Editor)
© 2004 WIT Press, www.witpress.com, ISBN 1-85312-728-0

Table 4: Altitudes distribution in P.D.O Baena.

Surface (ha) Altitude (m)	Baena	Castro del Río	Doña Mencía	Luque	Nueva Carteya	Zuheros	Total	% Total
<100	0	0	0	0	0	0	0	0
100-200	0	1.066	0	0	0	0	1.066	1,25
200-300	7.115	9.374	0	0	378	0	16.867	19,85
300-400	15.863	7.723	4	922	1.307	0	25.819	30,38
400-500	9.426	2.615	269	5.216	2.455	109	20.090	23,64
500-600	3.070	781	562	2.392	2.071	546	9.422	11,09
600-700	672	396	341	1.714	498	385	4.006	4,71
700-800	103	36	160	1.361	208	164	2.032	2,39
800-900	0	0	58	772	0	307	1.137	1,34
900-1.000	0	0	62	705	0	961	1.728	2,04
1.000-1.200	0	0	62	930	0	1.439	2.431	2,86
1.200-1.400	0	0	0	69	0	316	385	0,45
>1.4000	0	0	0	0	0	0	0	0
Total	36.249	21.991	1.518	14.081	6.917	4.227	84.983	100

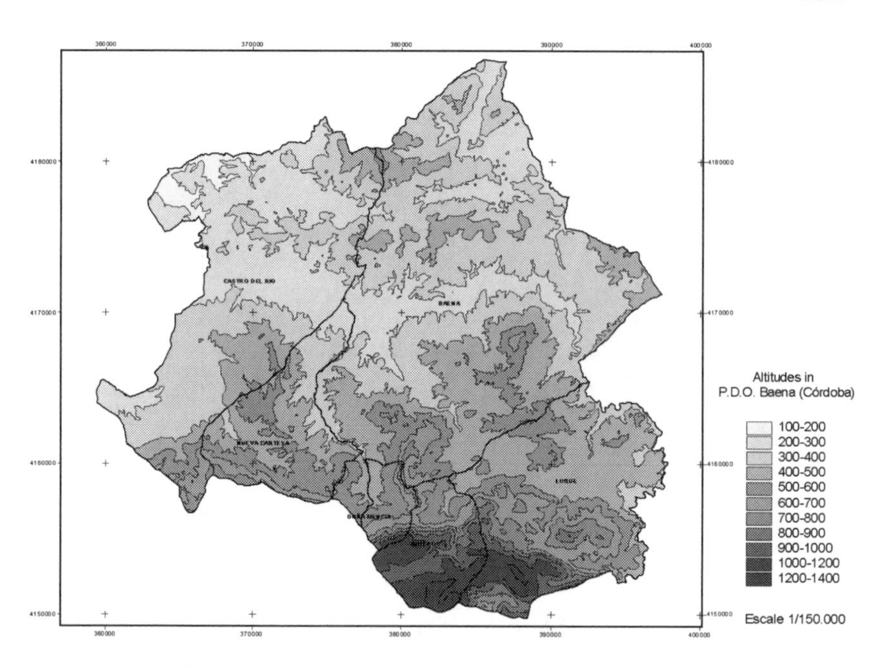

Figure 4: Altitudes distribution in P.D.O. Baena.

4 Comparative analysis between the current use and potential one

4.1 Evaluation of the olive grove

The agrarian evaluation of P.D.O. Baena has been made with the previously mentioned soil's variables. Although the definition of agrarian classes contains the slope, this factor has been studied independently because it influences directly in the erosion risk, soil wastage and mechanization (Figure 5).

The result of this analysis shows five imbalances grades:

A: Strong imbalances areas for higher growing intensity of use than potential.

B: Light imbalances areas for higher growing intensity of use than potential.

C: Areas without imbalances.

D: Light imbalances areas for lower growing intensity of use than potential.

E: Strong imbalances areas for lower growing intensity of use than potential.

4.2 Strong imbalances areas for higher growing intensity of use than potential

The olive groves of this group grow in soil of the agrarian classes number V, VI, VII and VIII. In them there are great edaphic limitations like the erosion. These zones present not much surface with slopes higher than 20%, slope limit for agrarian use, so the problems caused by the erosion and the soil wastage is not very stringent.

Table 5: A grade surface (ha) in P.D.O. Baena.

	A Grade		
	Surface (ha)	**% Olive grove**	**% Total**
Olive grove	14.283	25,58	100

4.3 Light imbalances areas for higher growing intensity of use than potential

This group contains the whole unirrigated olive grove located in the IV class and with slopes lower to 20% and the irrigated olive groves located in the III and IV classes with slopes lower than 10%. The first ones present limitations for the systematic mechanization although the occasional ones can take place.

Table 6: B grade surface (ha) in P.D.O. Baena.

	B Grade		
	Surface (ha)	**% Olive grove**	**% Total**
Olive grove	11.061	19,81	100

Management Information Systems, C. A. Brebbia (Editor)

4.4 Areas without imbalances

The grade C surfaces contain all the unirrigated olive groves growing in II and III classes and the irrigable ones located in I and II.

Table 7: C grade surface (ha) in P.D.O. Baena.

	C Grade		
	Surface (ha)	**% Olive grove**	**% Total**
Olive grove	30.347	54,35	100

4.5 Light imbalances areas for lower growing intensity of use than potential

The olive groves classified with grade D are those witch grow in the class number I because these soils present the best characteristics. In these soils, the crop can improve with irrigation.

Table 8: D grade surface (ha) in P.D.O. Baena.

	D Grade		
	Surface (ha)	**% Olive grove**	**% Total**
Olive grove	145	0,26	100

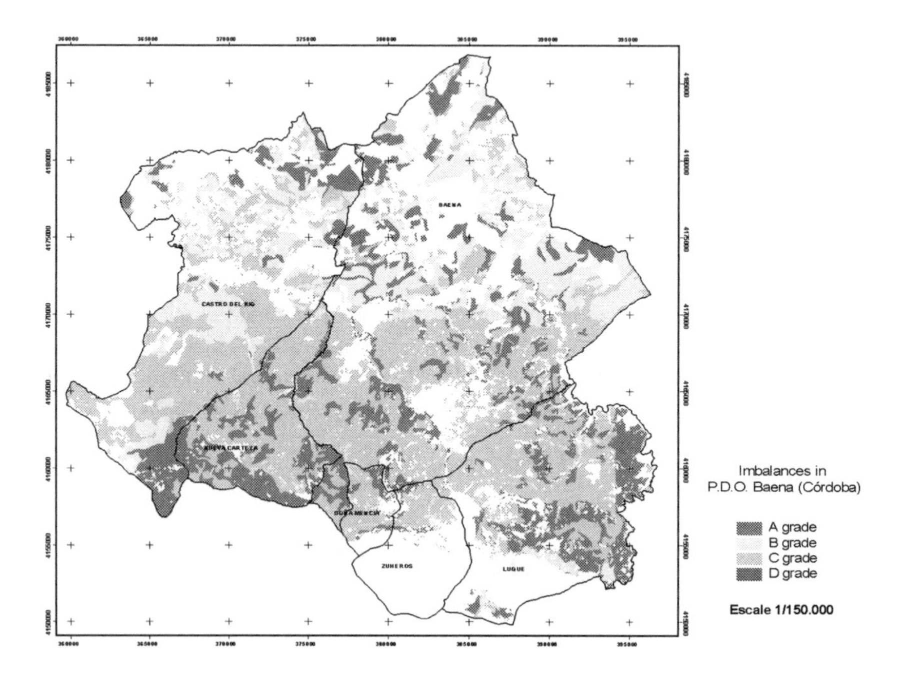

Figure 5: Imbalances distribution in P.D.O. Baena.

Management Information Systems, C. A. Brebbia (Editor)
© 2004 WIT Press, www.witpress.com, ISBN 1-85312-728-0

4.6 Strong imbalances areas for lower growing intensity of use than potential

This group contains all the non cultivated soils that grow in soils witch support a higher level of exploitation. As the study only consider the olive groves, it's impossible recognize these areas.

5 Corrective measures

5.1 Strong imbalances areas for higher growing intensity of use than potential

When the slopes are not very high, it is recommended the total abandonment of the mechanization and the establishment of a permanent vegetable cover or a seasonal one during the winter months if there are problems with the water (Pastor et al. [4]). For higher slopes it is recommended the reforestation. With irrigable olive groves is essential to apply minimum irrigation with trickle irrigation method (Figure 6).

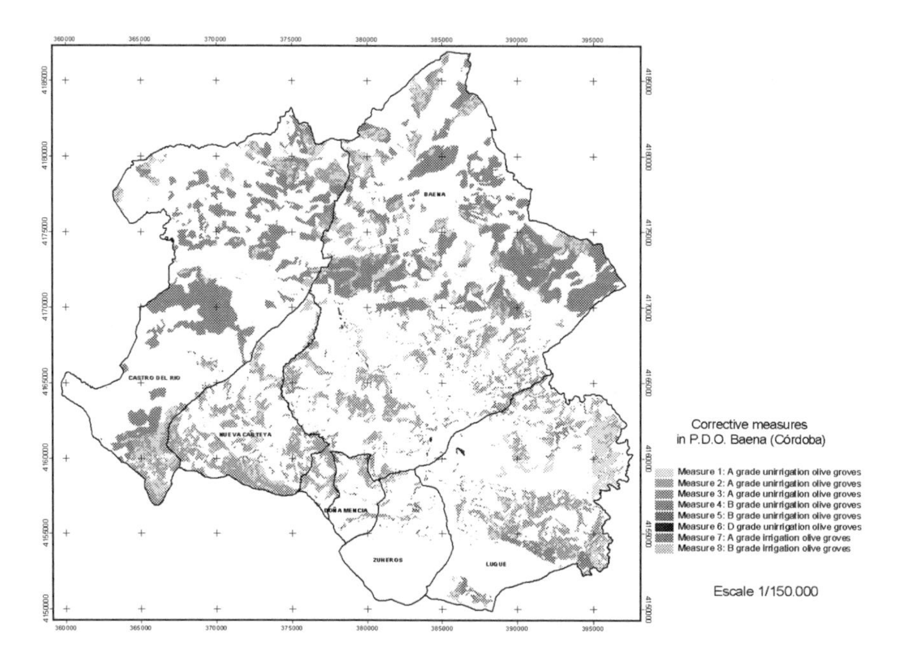

Figure 6: Corrective measures distribution in P.D.O. Baena.

5.2 Light imbalances areas for higher growing intensity of use than potential

In this case, the most advisable measure is to set a minimum mechanization with some mechanical measure of control. If the erosion problems are accused, it is

recommended the total abandonment of the mechanization and the establishment of a natural or artificial permanent vegetable cover, with leaves and rest of pruning of the own olive grove. Another solution is the application of herbicides in bare soils, although it is only advisable in soils with low slopes where the vegetable cover is not viable (Barranco et al. [1]). With the irrigation groves it's necessary to implant a trickle irrigation to diminish problems with the soil's fertility (Figure 6).

5.3 Light imbalances areas for lower growing intensity of use than potential

In this case is recommended an increase of the crop density with the installation of trickle irrigation. As cultivation system it's recommended a vegetable cover during the winter months if there is enough water or herbicides in bare soils (Figure 6).

6 Conclusions

The developed of this methodology obtains good results in the study of the imbalances among the current use of the soil and the potential one. The proof of this it's observed in the exposed example of the P.D.O. Baena where the differences imbalances grades can be shown. With that information it's possible to create a planning and exploitation model to correct the general problems and other more specific for each concrete situation.

Also, this methodology is open to the possibility of including any other type of agrarian information without modifying its basic structure.

References

[1] Barranco Navero, D, Fernández Escobar & R, Rallo, L., *El Cultivo del Olivo,* Consejería de Agricultura y Pesca de la Junta de Andalucía, Mundiprensa. Madrid, 1999.

[2] León Llamazares, A., *Caracterización Agroclimática de la Provincia de Córdoba,* Ministerio de Agricultura, Pesca y Alimentación, Madrid, 1989.

[3] Menéndez, M & Roth Rodríguez, J. C., *Caracterización del Uso Potencial del Suelo de la Provincia de Córdoba: Clases Agrológicas*, Consejería de Agricultura y Pesca de la Junta de Andalucía, 1999.

[4] Pastor, M., Castro, J., Humares, Mª D. & Saavedra, M., *La Erosión y el Olivar: Cultivo con Cubierta Vegetal*, Dirección General de Investigación y Formación Agraria, Junta de Andalucía, Sevilla, 1997.

Management Information Systems, C. A. Brebbia (Editor)
© 2004 WIT Press, www.witpress.com, ISBN 1-85312-728-0

Accuracy assessment of orthoimagery generation from QuickBird satellite data

M. Caprioli & E. Tarantino
Polytechnic University of Bari, Italy

Abstract

The use of satellite data, aimed at retrieving environmental and territorial information, has driven researchers to solve the problem of the geometric rectification in a determined reference system, by seeking suitable planimetric or altimetrical correction techniques. At the same time, recent diffusion of metric and sub-metric data has extended scientific interest towards cartographic updating and orthoimagery production. In the past, such application fields were limited by low spatial and radiometric resolutions for large area investigations. With second generation sensors, panchromatic data are the most suitable for cartographic production, because they permit greater detail and definition of geometric propriety than multispectral data. The most widespread mathematic methods of geometric processing are the parametric and non parametric models. Their use is connected to the availability of external sensor orientation parameters and the possibility to correct optic and terrestrial curvature distortions.

In this study QuickBird panchromatic satellite data with sub-metric spatial resolution were processed, in order to assess the accuracy in orthoimagery generation, with the aid of Ground Control Points (GCPs) and of a DTM with a 5 m spatial resolution. Moreover, the results of RMS obtained with consolidated geometric transformation models was evaluated (*5th order Polynomial Functions*, *Rational Function Model*, *Thin Plate Spline*), acquiring GCPs by means of GPS-RTK methodology (with GSM correction).
Keywords: QuickBird, GPS-RTK methodology with GSM correction, orthorectification accuracy.

1 Introduction

The recent technological improvement of the new generation satellite sensors has increased the use of VHR (*Very High Resolution*) satellite data in studies which require continuous updating and direct (metric and qualitative) information content, with consequent easy integration in GIS environments. At the same time, previous high resolution satellite data (e.g. 1 m spatial resolution of Ikonos data) have not permitted replacement of aerial photographs, nor have they provided the orbital parameters and stereo images that were promised with product distribution. The better spatial resolution of QuickBird data has narrowed the gap between satellite imagery and aerial photographs, allowing easier urban information updating [7].

Consequently, spatial resolution of panchromatic data has increased from 1 m to 70 cm and for multispectral data from 4 m to 2,8 cm. From the radiometric point of view, QuickBird panchromatic data are acquired in the 11 bit format (2048 grey levels) and distributed in the 16 bit format for detailed radiometric analysis, or in the 8 bit format (256 grey levels), easily managed in most GIS platforms and in cartographic applications. However, this imagery cannot be directly used because of its geometric deformations, and metrically corrected elements cannot be measured on it.

2 The orthorectification approaches

In order to execute digital rectification it is necessary to reconstruct sensor collection geometry for each line of the imagery. The classic photogrammetric approach is based on the extraction of the DTM from a stereo pair processing, followed by the orthorectification of one of the two images; it is also possible to use a pre-existing DTM, with a quality complying with the scale of the final product, and to introduce some GCPs [10], in order to ensure that the same reference system is used. This latter processing chain is the one normally followed for the processing of satellite data, since not all the satellite sensors have stereo pairs collection capabilities, and most cases require single frames processing [11].

The methods mostly used for rectification are based on sensor-related information, by using a sensor model that can be either physical or generic. The main difference is that physical models are rigorous, and require the knowledge of the specific sensor for which they have been designed; each parameter involved has a physical meaning. Generic sensor models, on the other hand, are sensor-independent, that is, knowledge of the sensor and of the physical meaning of the image collection process is not required.

A rigorous model produces an accurate three-dimensional description and orthorectification of the imagery. A generic sensor model supplies the relationship existing between the three-dimensional co-ordinates of an object, and the corresponding image co-ordinates in a generic mathematical form. *Polynomial Functions*, *Thin Plate Spline* (TPS) and *Rational Function Model* (RFM) can be considered the most generalised models which have been used for more than ten years instead of rigorous models [1, 6, 9, 15].

With *Polynomial Functions* the transformation model creates a new geocoded image space, where interpolated pixel values will be placed during resampling. The procedure requires that polynomial equations be fitted to the GCPs, using

Management Information Systems, C. A. Brebbia (Editor)
© 2004 WIT Press, www.witpress.com, ISBN 1-85312-728-0

least squares criteria to model the correction in the image domain, without identifying the source of the distortion. One of several polynomial orders may be chosen, based on the desired accuracy and the available number of GCPs. It is important to note that while a higher order polynomial will result in a more accurate fit in the immediate vicinity of the GCPs, it may introduce new significant errors in those parts of the image away from the GCPs. Therefore, errors may be introduced into the imagery that are greater than the initial errors.
Thin Plate Spline (TPS) provide an attractive alternative to the traditional polynomials. Thin Plate Splines are also global (i.e., all GCPs are used simultaneously to derive the transformation), but the derived functions have minimum curvature between control points, and become almost linear at great distances from the GCPs. The influence of individual GCPs is localized, and diminishes rapidly moving away from the points. The main disadvantage of TPSs, is that, to represent a warping transformation accurately, they should be constrained at all extreme points of the warping function. Moreover, when using TPSs to geo-register the image in rough terrain, it may be necessary to acquire hundreds of GCPs, since there should be a point at every extreme of terrain (peak or valley bottom), and along breaklines.

Rational Function Model (RFM) uses a ratio of two polynomial functions to compute image row, and a similar ratio to compute image column. All four polynomials are functions of three ground coordinates, namely latitude, longitude, and height (or elevation). A separate image geometry model is computed for each previously defined segment of a large image. Each polynomial has 20 terms, although the coefficients of some polynomial terms are often zero. Because rational polynomial considers elevation also, it is better than polynomial and thin plate spline methods. It is the best choice when the user has no information about the image where a true geometric modelling (or rigorous modelling) cannot be used. For aerial photos, it can be used when the user has no information about the camera. For satellite image, it can be used when satellite modelling cannot be applied because of no information about the satellite or some geometric correction already applied to the data [12, 13].

3 Data and methods

The experimentation was carried out on a test area, which included the north coastal zone of the municipal district of Bari (Palese – S. Spirito), with different data sources:

- raster data (0,70 m of spatial resolution) of the QuickBird panchromatic sensor (Acquisition Date/Time: 01-08-2002; 17:11:18);
- digital topographic map in a scale of 1: 5000 (Date 1998), for DTM processing (5 m of spatial resolution);
- ground survey (June- July 2003), by means of GPS-RTK methodology (with GSM correction), for the acquisition of 40 points, 32 of which were used as Ground Control Points (GCP) and 8 as Check Points (CP).

3.1 The survey of GCPs with GPS-RTK methodology and GSM correction

Past experiments carried out by means of acquisition of GCPs with GPS methodology, conducted on low and medium spatial resolution satellite data [5, 14], have demonstrated the effectiveness of such procedure, especially in irregular areas and with lack of base data.

In our case, the survey of GCPs was executed utilizing GPS-RTK methodology, in order to verify the effectiveness in acquisition and processing of the necessary points for the correct satellite data orthorectification.

As regards the Italian situation, from the end of 2002 there has been a TIM (Telecom Italia Mobile) Service called "GeoData". It permits correction of the topographical measurements using the GeoTIM network, constituted by 34 (23 certified) permanent GPS-stations, uniformly distributed over the whole national territory and inserted in the Italian GPS Reference Network (IGM95).

There are two possible ways of accessing the service:

1. Post Processing: data corrections are available on the Web. After survey with a GPS mobile receiver, memorizing data on magnetic support, it is possible to access to differential corrections on the Web. It is important to specify the permanent GPS-station, the date and the hours concerned, and the cycle-rate desired (1 second, 5 seconds, 15 seconds, 30 seconds). The data will be furnished in the RINEX standard data format. At this point corrections of the surveying data can be carried out using the common elaboration software on PC. It is possible to enter into the service through internet, using a password available through specific economic agreements (subscription, payment in case of need, etc.). Connection is easy and downloading, following simple and well organized instructions, is quick.
2. Real Time: correction is directly effected during the survey. A telephone number of radiomobile network (335 8820 YYY) is associated to each of the 23 permanent GPS-stations. In this way, while effecting the survey with the mobile GPS, the call must be started in data transmission, through GSM modem. VAS (TIM Service Centre) establishes the virtual connection with the selected permanent GPS-station, enabling it to transmit a flow of D-GPS corrections. Standard protocol RTCM for data transmission is used. This signal is elaborated in real time from the receiver‘s software, involving precision of the measurements

In this study, the GPS receiver adopted as Rover was "LEICA system 500-SR530" (L1 C/A code and L2 P-code), dual frequency antenna AT502, terminal TR500, to set parameters in the receiver and to steer the GPS, PC Card of 16 MB as the data storage medium (which was preferred to the Internal Memory), Wavecom GSM module with business SIM card for data transmission.

The survey consists in 40 points near IGM 95 vertex of Bari S. Spirito acquired with a GPS mobile receiver connected through analogic modem to the GSM network. In this way it is possible to take data corrections in real time from GeoTIM station of Bari Modugno (distance between 7 and 10 km), fig. 1.

After positioning the antenna, the GSM connection, through the call in data transmission to Bari TIM station, was started at each point. After a few seconds

Management Information Systems, C. A. Brebbia (Editor)
© 2004 WIT Press, www.witpress.com, ISBN 1-85312-728-0

the system elaborated the resolution of the ambiguity, so that accuracy was less than 1-2 cm. At this point it was possible to "occupy" the point (about 10 seconds) and to "store" it. Few minutes were enough for each point.
Satellite configuration was generally successful in our case study: there were only some problems in the centre of the towns (S. Spirito and Palese), because of unsuccessful ambiguity resolution.

The second phase in the survey consisted in transferring raw data from the sensor to PC, using the Leica SKI-PRO 2.5 software.

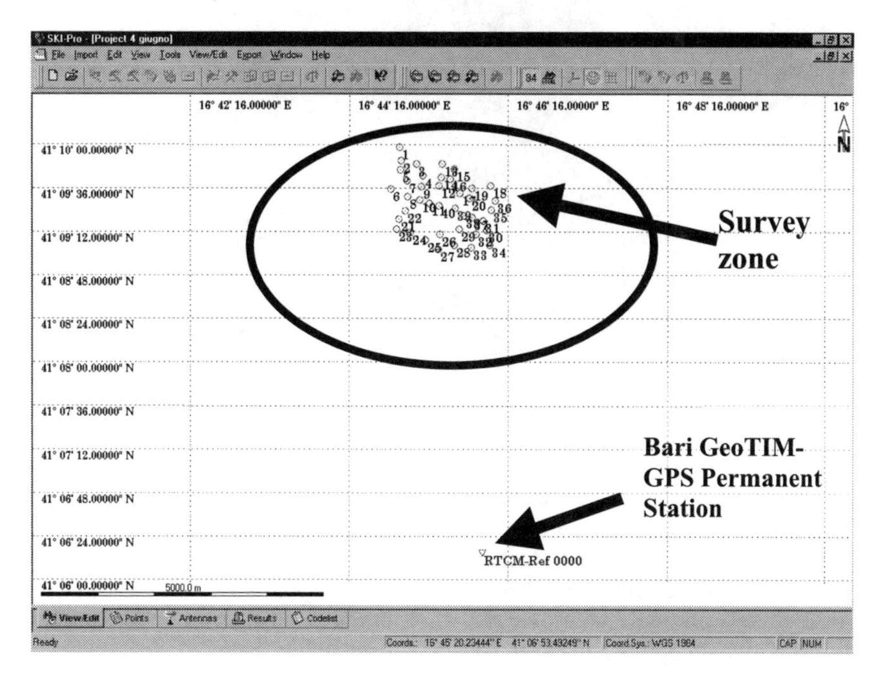

Figure 1: The survey zone and the Bari GeoTIM - GPS Permanent station.

3.2 Geometrical processing test

Orthorectification removes distortions in the imagery due to topography. The result is a map accurate product that can be used in GIS and other mapping applications. Depending on the collection geometry and availability of GCPs, a full 3D orthorectification of QuickBird images is possible, if it is combined with the type and accuracy of the used DEM.

In this case, a digital topographic map in a scale of 1: 5000 (Date 1998) for the DTM generation (5 m of spatial resolution) was processed.

The orthoprojection procedure was executed by utilizing different methods (*5th order Polynomial Functions*, *Thin Plate Spline, Rational Function Model*), provided in the PCI Geomatica V8.2 software, fig. 2.

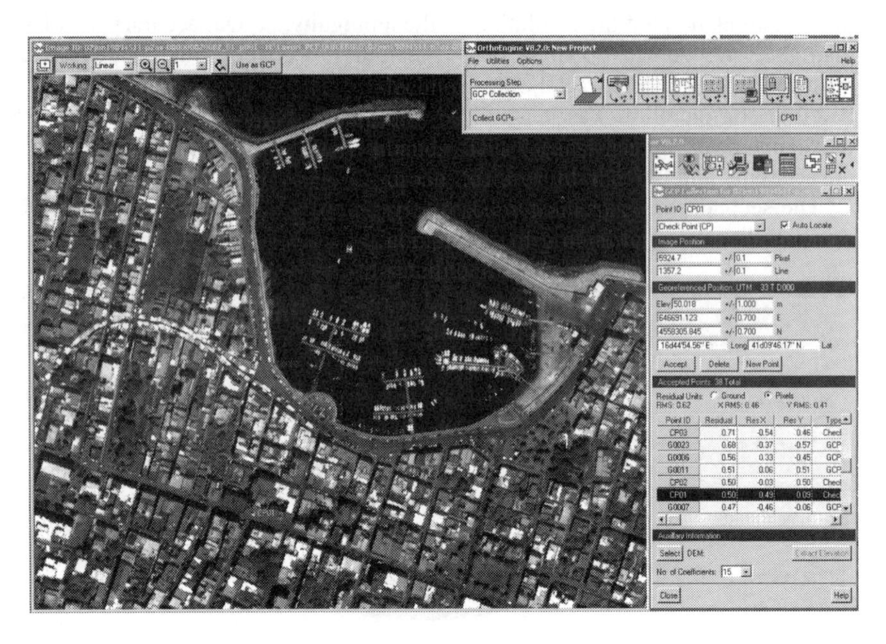

Figure 2: Orthoprojection procedure with GCPs identification.

The methodology followed for the test has been the same for all the methods. First of all, a dataset of points has been generated. For each point, 3D ground coordinates measurements (collected by GPS campaigns) and imagery sample/line coordinates, have been computed. The points have been selected in order to be well distributed over the whole frame, and to cover different height layers. Up to 40 points have been considered per single frame.
After the geometric transformation procedure, with which RMS values of 32 Ground Control Point (GCP) and 8 Check Points (CP) were calculated, the next step included the pre-processing of DEM (5 m of spatial resolution), in order to combine it with GCPs.

After completion of transformation procedures, the resampling process on the image was carried out, in order to determine radiometric values (DN - *Digital Number*) in the corrected output matrix. The choice of methods (*Nearest Neighbour, Bilinear Interpolation, Cubic Convolution*) depends on the final use of the corrected image. In this case the *Cubic Convolution* methods was the most suitable [8].

4 Results

The adopted geometric transformation procedures have confirmed the potential use of QuickBird data in contests in which direct metric information content of images is required, because the values of residuals result as being lower in comparison with the tolerance limits usually accepted for cartographic products at the scale of 1:10000 and 1:5000, as shown in table 1.

The results achieved demonstrate dependence of the resulting geometrical accuracy on the quality of the ancillary data used for orthorectification, and on the number of GCPs involved in the analysis. Moreover, in manual individuation of GCPs acquired through GPS network, most problems arose from the strong influence of this procedure on the final results, causing uncertainty connected to the user's choice in the corresponding pixel identification. As other studies have shown [3, 4], the sub-metric spatial resolution of QuickBird panchromatic imagery enables a remarkable precision in the phase of ground control points identification, when compared with previous generation satellite data. At the same time, the uncertainty of 1 pixel persists with the manual individuation of GCPs, representing a limit for the whole orthoprojection procedure.

Table 1: Planimetric errors (in m) of generated orthoimages.

Geometric transformation methods		**X RMS**	**Y RMS**	**RMS**
5th order Polynomial Functions	GCPs	0.60	0.60	0.84
	Check Points	0.64	1.52	
Rational Function Model (RFM)	GCPs	0.46	0.41	0.62
	Check Points	0.77	1.28	
Thin Plate Spline (TPS)	GCPs	0.66	0.70	0.93
	Check Points	0.90	1.73	

This analysis gave a strong indication of the metric integrity of QuickBird imagery, which augured well for the use of linear models in 3D geopositioning. The tests also demonstrated that in the presence of good quality control and DTM data (with moderately flat terrain), such imagery can readily yield an XY positioning accuracy of sub-pixels.

Moreover, the speed in the GCPs acquisition and processing with a sub-metric precision had verified the effectiveness of using GPS-RTK technology for geometric correction of VHR satellite data.

References

[1] Baltsavias, E., Pateraki, M., Zhang Li, Radiometric and geometric evaluation of Ikonos Geo images and their use for 3D building modelling, *Proc. of ISPRS Workshop "High Resolution Mapping from Space 2001*, Hannover, September 19-21, (CD-ROM), 2001.

[2] Barbarella M., Mancini, F., Zanni, M., Elaborazione di immagini Ikonos su aree in frana. Il caso di Sarno, *Bollettino della SIFET*, 1, pp. 23-36, 2002.

Management Information Systems, C. A. Brebbia (Editor)

[3] Caprioli, M., Gagliardi, S., Tarantino, E., Verifica delle caratteristiche metriche dei dati del satellite IKONOS, *Atti della 4ª Conferenza Nazionale ASITA*, Genova, October 3-6, pp. 347-352, 2000.
[4] Caprioli, M., Scognamiglio, A., Tarantino, E., Utilizzo della Tecnologia GPS per l'ortorettifica dei dati IKONOS, *Atti della 6ª Conferenza Nazionale ASITA*, Perugia, November 5-8, pp. 637-642, 2002.
[5] Gao, J., Non-differential GPS as an alternative source of planimetric control for rectifying satellite imagery, *Photogrammetric Engineering and Remote Sensing*, 67(1), p.p. 49-55,2001.
[6] Fraser, C.S., Baltsavias, E., Gruen, A., Processing of Ikonos imagery for submetre 3D position and building extraction, *ISPRS Journal of Photogrammetry & Remote Sensing*, 58, pp. 177-194, 2002.
[7] Fritz, L. W., High resolution commercial remote sensing satellites and spatial information, 1999. www.isprs.org/publications/highlights/highlights0402/fritz.html.
[8] Grodecki, J., Dial, G., Ikonos geometric accuracy, *Proc. of ISPRS Workshop "High Resolution Mapping from Space 2001"*, Hannover, September 19-21, (CD-ROM), 2001.
[9] Jacobsen K., Mapping with Ikonos images, *Proc. of 22nd EARSeL Symposium*, Prague, June 4-6, 2002.
[10] Manzer, G., Avoiding Digital Orthophoto Problems, *Digital Photogrammetry, An addendum to the Manual of Photogrammetry*, ed. Greve, C., American Society for Photogrammetry and Remote Sensing, Bethesda Maryland, pp. 151-155, 1996.
[11] Marsella, M., Volpe, F., Analisi delle potenzialità dei dati satellitari ad alta risoluzione per la generazione di ortofoto, *Atti della 6ª Conferenza Nazionale ASITA*, Perugia, November 5-8, pp. 1485-1490, 2002.
[12] OrthoEngine Satellite Edition, *User Guide, PCI Geomatica V8.2*, Richmond Hill, Ontario, Canada, 2000.
[13] Radicioni, F., Grassi, S., Mancini, F., Utilizzazione di immagini satellitari ad alta risoluzione a fini cartografici, *Atti della 6ª Conferenza Nazionale ASITA*, Perugia, November 5-8, pp. CXV-CXXVIII, 2002.
[14] Smith, D. P., Atkinson, S. F., Accuracy of rectification using topographic maps versus GPS ground control points, *Photogrammetric Engineering and Remote Sensing*, 67 (5), pp. 585-587, 2001.
[15] Tao, C. V., Hu, Y., A comprehensive study of the rational function model for Photogrammetric processing, *Photogrammetric Engineering and Remote Sensing*, 67 (12), pp. 1347-1357, 2001.

Management Information Systems, C. A. Brebbia (Editor)
© 2004 WIT Press, www.witpress.com, ISBN 1-85312-728-0

Remote quantification of chlorophyll in productive turbid waters

R. Khanbilvardi[1], Y. Yacobi[2], A. Gitelson[3] & B. Shteinman[1]
[1]*International Center for Environmental Research and Development, Department of Civil Engineering, City College of CUNY, USA*
[2]*Israel Oceanographic and Limnological Research, Yigal Allon Kinneret Limnological Laboratory, Israel*
[3]*School of Natural Resources, University of Nebraska-Lincoln, Lincoln, Nebraska, USA*

Abstract

The objectives of this paper were to study the spectral features of reflectance of inland waters in order to find spectral features, which are closely related to phytoplankton chlorophyll concentration, and to devise algorithms for chlorophyll estimation using remotely sensed data. The information gained from several spectral bands in the red and near-infra-red ranges of the spectrum were found to be sufficient for accurate retrieval of chlorophyll concentration in productive turbid waters. These algorithms were validated in Lake Kinneret, Israel as well as in other environments, with slight modification of the coefficients: the polluted water of Haifa Bay (Mediterranean Sea), fish ponds and wastewater reservoirs in Israel, and lakes with diverse trophic status in Israel, and lakes with diverse trophic status in northwestern Iowa and eastern Nebraska (USA).

1 Introduction

The Chl concentrations in Lake Kinneret, Israel, range from less than 5 mg/m^3 to hundreds of mg/m^3 [1]. The spatial distribution of Lake Kinneret phytoplankton is very heterogeneous, particularly when the dinoflagellate *Peridinium gatunense* in almost all years forms a dense bloom from February through May [2,3]. Funding considerations limit the sampling schedule to a few stations, and routinely to a single station, located above the deepest point of the lake. It is

Management Information Systems, C. A. Brebbia (Editor)

questionable whether this single station (or even a few stations) is truly representative the overall condition. To overcome this problem, we have initiated a program for remote sensing of chlorophyll in Lake Kinneret in October 1992. The primary aims were (1) to study the characteristics of the reflectance spectra during different seasons of the year, (2) to evaluate concepts developed for estimation of Chl concentration in inland waters [4, 5, 6] and (3) to devise algorithms for Chl estimation in Lake Kinneret from reflectance data. Following the establishment of algorithms for Chl detection in Lake Kinneret, an effort was undertaken to assess their validity in other productive and turbid water bodies in Israel and in the USA. In this paper, we present the summary of our work in those diverse water bodies, where remotely sensed data were used for the estimation of Chl concentrations.

2 Materials and methods

Data were collected in several locations in Israel and in the north central United States. A list of water bodies examined is given in Table 1. Descriptions of optical properties of these water bodies have been published previously: Lake Kinneret [7, 8, 9], Haifa Bay [10], Carter Lake [11], northwestern Iowa lakes [12], and wastewater treatment ponds [13, 14, 15]. In each experiment, upwelling radiance of water (L_w) and reference plate (L_{ref}) were measured using high spectral resolution spectroradiometers: LI-1800 (aquatic systems in Israel), Ocean Optics ST1000 (Carter Lake), and ASD (Iowa lakes). Reflectance spectra were then calculated as $R = L_w/L_{ref}$. Water samples were collected for analytical determination of Chl *a* concentration in the laboratory. Detailed descriptions of the analytical methods and data processing used for the treatment of the measured variables were presented in the mentioned above publications.

Table 1: Energy characteristics of the turbulent jet-flow in the Jordan River mouth.

Site	Period	Chlorophyll *a* (mg/m^3)	Dominant phytoplankton
Lake Kinneret	winter-spring	2.4 - 330	Peridinium gatunense
Lake Kinneret	summer-fall	3.8 - 26	Chlorophytes
Iowa Lakes, USA	summer	2.0 - 55	Diatoms
Carter Lake, USA	year round	20 - 280	Anabaena sp.
Fish ponds, northern Israel	winter-spring	2.1 - 674	Microcystis aeruginosa
Wastewater ponds, Israel	spring-summer	69 -2700	Chlorophytes
Haifa Bay, Israel	spring-summer	1 - 70	Dinoflagellates, diatoms

3 Results and discussion

Absorption by chlorophylls, carotenoids, dissolved organic matter in the blue range of the spectrum prevents to use this range for Chl retrieval in productive

Management Information Systems, C. A. Brebbia (Editor)

turbid waters. Thus, the main efforts were focus on finding spectral region where Chl has unique spectral feature allowing to use it for quantitative Chl estimation.

There are few special features in the red and NIR ranges. The trough near 670 nm is due to maximum absorbance by chlorophyll a in the red range of the spectrum. For a Chl concentration of more than 20 mg/m^3, the reflectance at 670 nm (R_{670}) almost does not depend upon Chl (Fig. 1) and primarily depends on the concentration of non-organic suspended matter [5, 6, 7, 8, 9, 16]. At this point, chlorophyll absorbance is offset by scattering of the cell walls and is a point of minimum sensitivity of reflectance to algal density and Chl concentration.

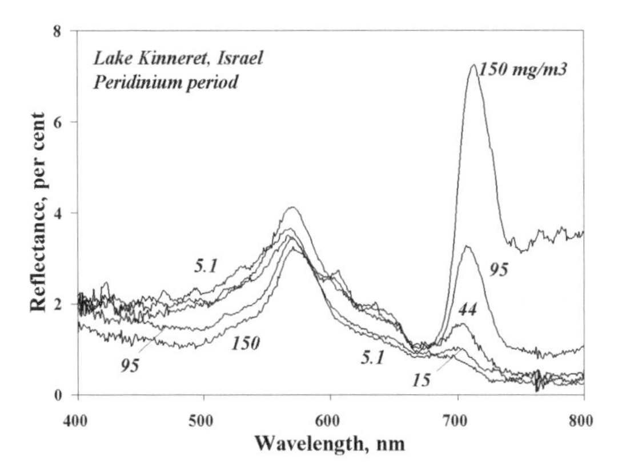

Figure 1: Reflectance spectra of Lake Kinneret, February 1994. In the range 400 to 550 nm, absorption by pigments was so strong that it was hard to distinguish between spectra with Chl ranging between 5 and 150 mg/m^3. Magnitude and position of the NIR peak depend strongly upon Chl concentration.

The magnitude of the peak near 700 nm, as well as its position, depends strongly on Chl concentration [7, 9, 11, 17, 18]. The peak magnitude depends on scattering by all suspended matter and, thus, increases with an increase in phytoplankton biomass. It correlated with Chl concentration *via* the link between Chl and algal biomass. As the algal biomass increases, or actually the active cell surface, scattering and the reflectance increase. In this range of the spectrum, combined absorption by Chl and water is minimal; therefore, the scattering surplus to the basic scattering by non-organic suspended matter may be attributed to phytoplankton cells surface [7, 9]. Chl has low but significant absorption between 690 and 715 nm. With increasing Chl concentration, the pigment absorption offsets cell scattering at progressively higher wavelengths, and the position of the peak shifts toward longer wavelengths [4, 5, 6, 9, 11, 12, 19, 20].

The algorithms developed and tested in our work are all based on the use of reflectance in the red and NIR range of the spectrum, since other portions of the spectrum are irrelevant in productive, coastal and freshwater ecosystems [4, 5, 6,

Management Information Systems, C. A. Brebbia (Editor)
© 2004 WIT Press, www.witpress.com, ISBN 1-85312-728-0

7, 9, 11, 16, 17, 19, 20, 21, 22, 23, 24, 25, 26, 27, 28, 29, 30]. The basic concept of those algorithms is the inclusion of the spectrum range which shows the maximal sensitivity to changes in Chl concentration and the range with the minimum sensitivity to variation of Chl concentration [4, 5, 6, 7, 9]. The latter accounts for non-pigmented suspended matter that causes variation in the reflectance. The magnitude and position of the reflectance peak near 700 nm were found to be the most sensitive variable for algorithms, and the reflectance at 670 nm was the least sensitive to changes in algal density, especially for Chl > 15-20 mg/m^3.

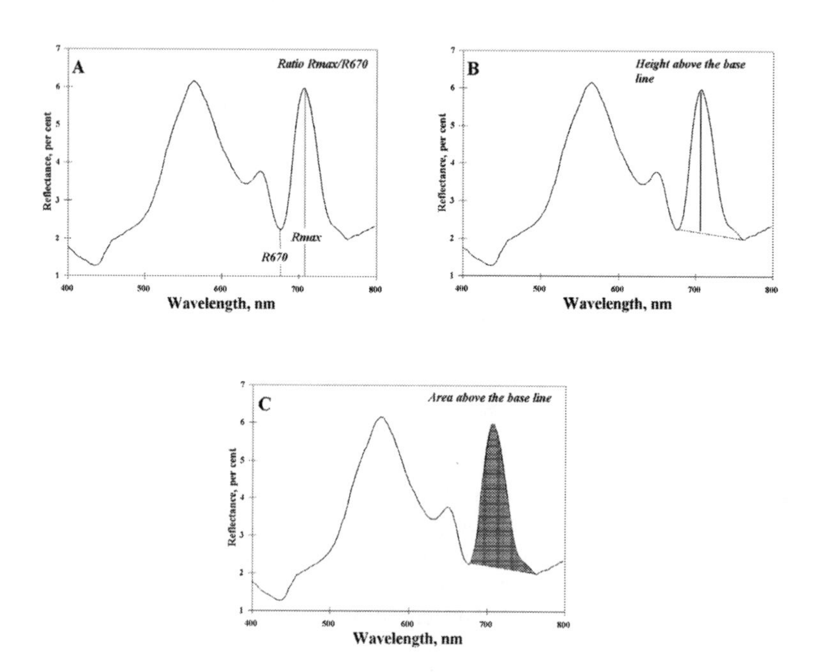

Figure 2: Algorithms for estimation of Chl concentration. (a) Ratio of the peak magnitude to reflectance at 670 nm; (b) Reflectance height above the base line between 670 and 750 nm; (c) Area above the base line between 670 and 750 nm.

A simple reflectance ratio R_{700}/R_{670} was first used as a predictor of Chl concentration [4, 5, 6, 21, 22]. The R_{700}/R_{670} ratio (Fig. 4a) was applied to our data obtained in highly diverse aquatic ecosystems dominated by different algal assemblages: Anabaena sp., Microcystis aeruginosa, and Peridinium gatunense. The relationship between the R_{700}/R_{670} ratio and Chl concentration was linear up to concentrations of approximately 180-200 mg/m^3, but turned exponential at higher concentrations. To estimate accuracy of Chl prediction in Lake Kinneret during Peridinium bloom, the combined dataset (reflectance spectra and Chl concentrations) was separated into model-development and model-testing subsets. For the model-development subset, data collected in March 1993 were

Management Information Systems, C. A. Brebbia (Editor)

used. Validation was done by data collected in April 1993. Predicted Chl concentrations were calculated using reflectance from model-testing subset with regression coefficients for the model-development dataset equation. The accuracy of Chl prediction was made against Chl concentrations actually measured did not exceed 18 mg/m^3.

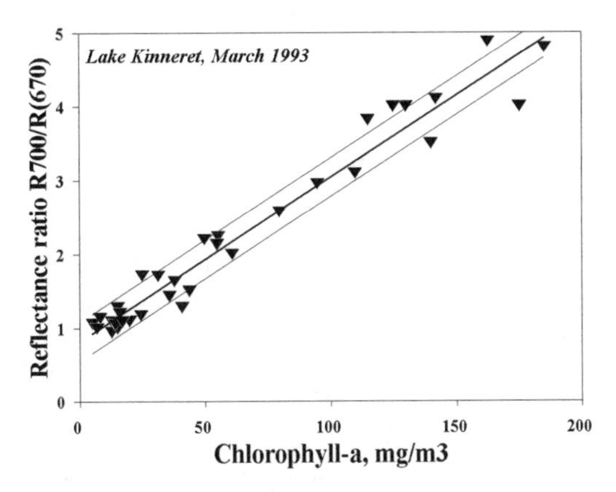

Figure 3: Performance of the reflectance ratio algorithm, R_{700}/R_{670}, for estimation of Chl concentration in lake Kinneret. Chl concentrations were predicted by ratio R_{700}/R_{670}, measured in April 1993. Relationship R_{700}/R_{670} vs. Chl (solid line), obtained in March 1993, was used to calculate Chl predicted. Then, Chl predicted was compared with actually measured Chl concentrations. Error of Chl prediction, determined as root-mean square deviation of predicted Chl values from actually measured, did not exceed 18 mg/m^3. The slope of the relationship R_{700}/R_{670} vs. Chl changed in various experiments. In the cases presented herein, slope of the relationship for Lake Kinneret was at least 50% higher than that of the fishponds.

Several other algorithms were developed for estimation of Chl concentration. They were based on the following variables: (a) the magnitude of the peak near 700 nm; (b) the height of the peak near 700 nm above the baseline drawn from 670 to 750 nm (Fig. 2b); (c) the area delimited by the reflectance curve and the mentioned baseline (Fig. 2c); (d) the position of the peak in NIR range. These algorithms were based on following concept. The trough at 670 nm is the wavelength of maximum absorbance by Chl in the red range of the spectrum. At this wavelength, Chl absorbance and scattering by cell walls are almost in equilibrium, and R_{670} shows minimum sensitivity to Chl concentration. For a Chl > 20 mg/m^3, R_{670} primarily depends on concentration of non-organic suspended matter [5, 7, 9, 11]. Reflectance beyond 750 nm depends on both organic and non-organic suspended matter concentrations and is insensitive to algal pigments [27]; the variation of R_{750} is comparatively small because of strong water

Management Information Systems, C. A. Brebbia (Editor)
© 2004 WIT Press, www.witpress.com, ISBN 1-85312-728-0

absorption in the NIR range. Thus, the slope of the baseline between 670 and 750 nm depends primarily on scattering by water constituents, but not chlorophyll absorption. With variation in the amount of non-organic and non-pigmented organic suspended matter concentration, the slope of the baseline changes but it has minimal influence on the height and area of the 700 nm peak above the base line. Therefore the height of the peak and the area above the baseline between 670 and 750 nm depends mainly on chlorophyll concentration and was used as its quantitative measure.

The regression of reflectance height at 700 nm above the baseline from 670 to 750 nm, as well as area above this baseline, and the position of the NIR peak [17,18], against Chl *a* concentrations yielded high correlation coefficients, always $r^2 > 0.90$ [7, 9]. The developed algorithms were validated by recurrent experiments in Lake Kinneret (Fig. 3), as well as in other environments, including Iowa lakes dominated by diatom algae and Haifa Bay with dinoflagellates and diatoms, both with small to moderate Chl concentrations (Table 1). The algorithms were also used to estimate Chl concentrations in wastewater ponds, where chlorophyte algae dominated with extremely high (70-520 mg/m^3) Chl concentration. In all aquatic systems studied, the algorithms proved expedient as tool for Chl concentration estimation.

For all inland waters studied, excluding wastewater ponds, the coefficients b in the relationships Chl vs. reflectance line height were in agreement (they ranged from 31.8 to 47.2 mg/m^3/% - Table 2). Minimal b value (31.8 mg/m^3/%) was found for Carter Lake, Nebraska in a full annual cycle. Taking into account extremely diverse trophic status of water bodies, the existence of robust relationship between Chl concentration and suggested variables of reflectance must to be taken as proven. It is worth to noting that the suitability of the algorithms to estimate chlorophyll concentration in water bodies dominated by blue-greens is of great importance, considering the significance of that algal group as potential environmental hazard.

Table 2: Comparison of regression statistics for several studies in which total chlorophyll was regressed against the predictor - reference peak height near 700 nm above baseline from 670 nm to 750 nm ($RLH_{670-750}$): $Chl = a + b*RLH_{670-750}$; r^2 is determination coefficient.

Site	Period	A	b	r^2
Lake Kinneret	March and April 1993	1.77	40.8	0.96
Lake Kinneret	February 1994	2.27	43.4	0.94
Haifa Bay, Israel	May 1994 and June 1995	4.90	47.2	0.93
Carter Lake, USA	April 1995 to April 1996	6.20	31.8	0.83
Waste water ponds, Israel	May 1996	1.20	18.0	0.88
Iowa lakes, USA	September 1996	2.30	36.0	0.86

Management Information Systems, C. A. Brebbia (Editor)
© 2004 WIT Press, www.witpress.com, ISBN 1-85312-728-0

Acknowledgments

The work was supported partially by grants from the Israel National Academy of Sciences and Humanities, Water Commissioner's Office, Israel Ministry of Science, US-Israel Bi-national Science Foundation and of the NASA.

References

[1] Berman, T., Yacobi, Y.Z. & Pollingher, U., Lake Kinneret Phytoplankton: Stability and Variability During Twenty Years (1970-1989). Aquat. Sci., 54, pp. 104-127, 1992.

[2] Pollingher, U., Phytoplankton Periodicity in a Subtropical Lake (Lake Kinneret, Israel). *Hydrobiol.*, **138**, pp. 127-138, 1986.

[3] Berman, T., Stone, L., Yacobi, Y.Z., Schlichter, M., Nishri, A. & Pollingher, U., Primary Production in Lake Kinneret: A long term record (1972-1993). *Limnol. Oceanogr.*, **40**, pp. 1064-1076, 1995.

[4] [4] Gitelson, A., Nikanorov, A.M., Sabo, G. & Szilagyi, F., Etude de la Qualite des Eaux de Surface par Teledetection. *IAHS Publications*, **157**, pp. 111-121. 1986.

[5] [5] Gitelson, A., Garbuzov, G., Szilagyi, F., Mittenzwey, K-H., Karnieli A. & Kaiser, A., Quantitative Remote Sensing Methods for Real-time Monitoring Inland Water Quality. *Int. J. Remote Sensing*, **14**, pp. 1269-1295, 1993a.

[6] Gitelson, A., Szilagyi, F. & Mittenzwey, K., Improving Quantitative Remote Sensing for Monitoring of Inland Water Quality. *Water Res.*, **7**, pp. 1185-1194, 1993b.

[7] Gitelson, A., Mayo, M. & Yacobi, Y.Z., Signature Analysis of Reflectance Spectra and its Application for Remote Observations of the Phytoplankton Distribution in Lake Kinneret. *Mesures Physiques et Signatures en Teledetection. ISPRS 6th International Symposium*, Val d'Isere, France, pp. 277-283, 1994a.

[8] Gitelson, A., Mayo, M., Yacobi, Y.Z., Parparov, A. & Berman, T., The Use of High Spectral Radiometer Data for Detection of Low Chlorophyll Concentrations in Lake Kinneret. *J. Plankton Res.*, **16**, pp. 993-1002, 1994b.

[9] Yacobi, Y. Z., Gitelson, A. & Mayo, M., Remote Sensing of Chlorophyll in Lake Kinneret Using High Spectral Resolution Radiometer and Landsat TM: Spectral features of Reflectance and Algorithm Development. *J. Plankton Res.*, **17**, pp. 2155-2173, 1995.

[10] Gitelson, A., Yacobi, Y.Z., Karnieli, A. & Kress, N., Reflectance Spectra of Polluted Marine Waters in Haifa Bay, Southeastern Mediterranean: Features and Application for Remote Estimation of Chlorophyll Concentration. *Israel J. Earth Sci.*, **45**, pp.127-136, 1996.

[11] Schalles, J.F., Gitelson, A., Yacobi, Y.Z. & Kroenke, A.E., Chlorophyll Estimation Using Whole Seasonal, Remotely Sensed High Spectral-Resolution Data for an Eutrophic Lake. *J. Phycol.*, **34**, pp. 383-390, 1998.

Management Information Systems, C. A. Brebbia (Editor)
© 2004 WIT Press, www.witpress.com, ISBN 1-85312-728-0

[12] Jones, J.R. & Bachmann, R.W., Limnological Features of Some Northwestern Iowa Lakes. *Proc. Iowa Acad. Sci.*, **81**, pp. 158-163, 1974.

[13] Oron, G. & Gitelson, A., Real-Time Quality Monitoring by Remote Sensing of Contaminated Water Bodies: Waste Stabilization Pond Effluent. *Water Res.*, **30**, pp. 3106-3114. 1996.

[14] Stark, R., *Real-Time Remote Sensing Monitoring Waste Water Effluent in Reservoirs*. M.Sc Thesis, Department of Geology, Ben Gurion University of Negev, 1997.

[15] Gitelson, A., Stark, R. & Dor, I., Quantitative Near-Surface Remote Sensing of Wastewater Quality in Oxidation Ponds and Reservoirs: A case study of the Naan system. *Water Environ. Res.*, **69**, pp.1263-1271, 1997.

[16] Dekker. A., *Detection of the Optical Water Quality Parameters for Eutrophic Waters by High Resolution Remote Sensing*. Ph.D. Thesis, Free University, Amsterdam, Netherlands, 1993.

[17] Gitelson, A., The Peak Near 700 nm on Reflectance Spectra of Algae and Water: Relationships of Its Magnitude and Position with Chlorophyll Concentration. *Int. J. Remote Sensing*, **13**, pp. 3367-3373, 1992.

[18] Gitelson, A., The Nature of the Peak Near 700 nm on the Radiance Spectra and its Application for Remote Estimation of Phytoplankton Pigments in Inland Waters. *Optical Engineering and Remote Sensing*, **SPIE 1971**, pp. 170-179.ov, 1993.

[19] Vos, W.L., Donze, M. & Bueteveld, H., On the Reflectance Spectrum of Algae in Water: The Nature of the Peak at 700 nm and Its Shift with Varying Concentration. *Communication on Sanitary Engineering and Water Management*, Delft, The Netherlands, Technical Report, pp.86-22, 1986.

[20] Rundquist, D.C., Schalles, J.F. & Peake, J.S., The Response of Volume Reflectance to Manipulated Algal Concentrations Above Bright and Dark Bottoms at Various Depths in an Experimental Pool. *Geocart Int.*, **10**, pp. 5-14, 1995

[21] Mittenzwey, K.-H. & Gitelson, A., In-situ Monitoring of Water Quality on the Basis of Spectral Reflectance. *Int. Revue Ges. Hydrobiol.*, **73**, pp. 61-72, 1988.

[22] Mittenzwey, K-H., Ullrich, S., Gitelson, A. & Kondrat'ev, K. Ya., Determination of Chlorophyll-a of Inland Waters on the Basis of Spectral Reflectance. *Limnol. Oceanogr.*, **37**, pp. 147-149, 1992.

[23] Millie, D.F., Baker, M.C., Tucker, C. S., Vinyard, B.T. & Dionigi, C. P., High-resolution Airborne Remote Sensing of Bloom-forming Phytoplankton. *J. Phycol.,* **28**, pp. 281-90, 1992.

[24] Quibell, G., Estimation Chlorophyll Concentrations Using Upwelling Radiance from Different Freshwater Algal Genera. *Int. J. Remote Sensing*, **13**, pp. 2611-2621,1992.

[25] Goodin, D.G., Han, L., Fraser, R.N., Rundquist, D.C., Stebbins, W.A. & Schalles, J.F., Analysis of Suspended Solids in Water Using Remotely Sensed High Resolution Derivative Spectra. *Photogram. Eng. Rem. Sens.*, **59**, pp.505-10, 1993.

Management Information Systems, C. A. Brebbia (Editor)

[26] Matthews, A.M. & Boxall, S.R., Novel Algorithms for the Determination of Phytoplankton Concentration and Maturity. *Proc. of the Second Thematic Conference on Remote Sensing for Marine and Coastal Environments*, 31 January-2 February 1994, New Orleans, Louisiana, USA. Environmental Research Institute of Michigan, **1**, pp. I-173-I-180, 1994.

[27] Han, L., Rundquist, D.C., Liu, L. L., Fraser, R.N. & Schalles, J.F., The Spectral Responses of Algal Chlorophyll in Water with Varying Levels of Suspended Sediment. *Int. J. Remote Sensing*, **15**, pp. 3707-3718, 1994.

[28] Richardson, L.L., Buisson, D. & Ambrosia, V., Use of Remote Sensing Coupled with Algal Accessory Pigment Data to Study Phytoplankton Bloom Dynamics in Florida Bay. *Proc. of the Third Thematic Conference on Remote Sensing for Marine and Coastal Environments*, 18-20 September 1995, Seattle, Washington, USA. Environmental Research Institute of Michigan, **2**, pp.125-134, 1995.

[29] Rundquist, D.C., Han, L., Schalles, J.F. & Peake, J.S, Remote Measurement of Algal Chlorophyll in Surface Waters: The Case for the First Derivative of Reflectance Near 690 nm. *Photogram. Eng. Rem. Sens.*, **62**, pp.195-200, 1996.

[30] [30] Schalles, J.F., Schiebe, F.R., Starks, P.J. & Troeger, W.W., Estimation of Algal and Suspended Sediment Loads (Singly and Combined) Using Hyperspectral Sensors and Integrated Mesocosm Experiments. *Proc. of the Fourth International Conference on Remote Sensing of Marine and Coastal Environments*. 17-19 March 1997, Orlando, Florida, USA. Environmental Research Institute of Michigan, Ann Arbor, MI, **1**, pp. 247-258, 1997.

Management Information Systems, C. A. Brebbia (Editor)
© 2004 WIT Press, www.witpress.com, ISBN 1-85312-728-0

Section 4
Information system strategies and methodologies

Information systems design: a procedural approach

G. Haramis[1], G. Pavlidis[2], Th. Fotiadis[1], Ch. Vassiliadis[1] & Ch. Tsialtas
[1]*University of Macedonia*
[2]*University of Patras*

Abstract

The procedure of Information Systems (ISs) design aims to assist in the discovery of that solution which shall both cover in the best possible way the users' demands and also be feasible within the framework of the data processing environment of the corporation.

In general, a good ISs design should possess the following characteristics: to be acceptable by all users, to be auditable, to be functional, to face problems "head on", to be satisfactory, not to be costly, to be easy in its development, maintenance and operation, to be will documented, and to be easily evaluated.

It is believed that the above presuppositions are covered, to a great extent, by the proposed procedure.
Keywords: information systems, procedures, development, maintenance, operationability.

1 General comments, purpose

It is important for the ISs design to start with an itemization of all the necessary steps which will be taken for its implementation (initial outline design), because this will ensure a large probability for the design to be both correct and satisfactory.

The itemization of these steps—i.e. the outline, or scenaric design—must be based upon:

- the grouping of the new system's requirements,
- the detection of the necessary presuppositions,
- the classification of cases into categories:

Management Information Systems, C. A. Brebbia (Editor)
 ISBN 1-85312-728-0

- required (must...)
- desirable (should...)
- welcome (could...)

- the consideration of the effect upon the input/processing/output as well as upon the method and the working procedures of users.

The sooner the initial outline design commences, the greater the probability of achieving:

- the (final) selection of the main issues and goals of the system before the accumulation of the main volume of information, the "study" of which shall cover all "available" time,
- the examination, of the already defined requirements of the new system, prior to the design job, which is time-consuming.
- the users' instruction on the effects of the system upon their work procedures and methods.

The consideration of all the above shall be of help in "ensuring" that the user requirements have been correctly defined before proceeding with the "detailed" design of the system.

The purpose of the Design phase is the recording of the specifications, which shall enable the full and exact system implementation without the need for further user help in providing supplementary information during its development.

This does not mean that during the implementation phase contact with users is cut off, since "implementation" does signify more than the realization of the system's programs.

Finally, the "basic" design of the new information system should have a" top-down approach, i.e. start with the "larger issues" and head to the "smaller" ones (top-down approach).

2 Systems design

The Design phase of ISs Development is usually split in two distinct stages; the stage of External or Logical Design and the stage of Internal of Physical Design.

2.1 External design

The goal of the External Design is the detailed specification of the necessary actions and procedures, as well as the data needed for the implementation of the business process.

Out of these procedures, it is necessary to determine which ones will be finally performed by the computer, and, therefore, which data should be "kept" in Data Bases and/or Traditional Data Files.

Some of the activities of this phase involve a more detailed "re-examination" of the output of the systems requirements definition. As a result, possible errors or omissions should be corrected/amended and approved before the main systems design.

The principal output of the External Design consists of:

- The definition of new procedures, as previously stated.
- The definition of file records and messages. In particular, the definition of:

Management Information Systems, C. A. Brebbia (Editor)
© 2004 WIT Press, www.witpress.com, ISBN 1-85312-728-0

input documents and input screens,
the status of output lists and screens,
the contents of file records,
"internal" messages of the system, and
"dialogues".

- The outline definition of databases.
- Data descriptions, which, at this stage, should cover:

the field name,
its size (number of characters)
a brief description of its purpose, and its possible significance within the business process,
the values (decoding) and their meaning in case the field is a code,
the indication (cross-reference) of the files and messages in which the record appears, and
the method of the creation of a record, of it is the result of processing of other records.

The definition of the business factors, upon which the decision-making process is based. Such factors, may be quantities, discount limits etc. and should be "itemised" and defined independently of the programs which shall make use of them.

- The outline definition of the various systems procedures.
- The authoring of the Users' Manual. This should comprise:
  - a general report on the purpose of the system,
  - its description,
  - system operating instructions,
  - the design of input forms, the design of output reports,
  - the description of message to and from the computer, and
  - the design of screen formats.

2.2 Internal design

The goal of the Internal Design is the exact definition of the manner of implementation of the procedures "designed" in the previous phase of external design. The output of Internal Design should comprise:

- The System Flow Chart Design, which should be detailed for batch systems.
- The Specifications of System Transactions.
- The Program's Specifications.
- The Physical Data Base Design and Definition.
- The realization of the Systems Test Plan.
- The realization of Systems Test Data.
- The recording of the System Operations Manual.

3 Information systems design procedure

3.1 Introduction

Any Information Systems Design Procedure should describe in detail the necessary tasks for its implementation, locate the most important design criteria

Management Information Systems, C. A. Brebbia (Editor)
 ISBN 1-85312-728-0

of each task and suggest a method of breaking down the various requirements, so as to facilitate both the work and its evaluation.

In practice, a design procedure is "iterative", as one is forced (due to new data, resulting mostly from users' omissions) to return to already completed tasks, in order to enhance or correct them. This "iteration" is continued until the users' final requirements of the new system are fully met.

Subsequently, The Flow Chart of such a Procedure is presented, and immediately following, its sixteen steps are described. Adhered to it, one can be assured of a proper System Design.

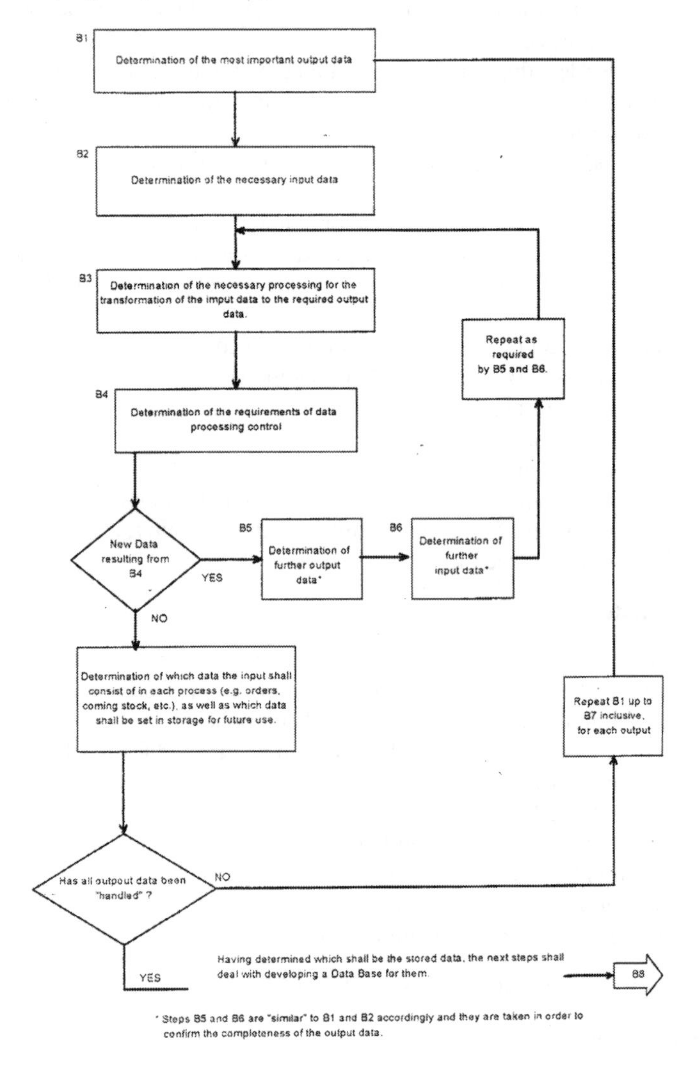

Figure 1: Flow of the steps (B1-B16) of an information system design.

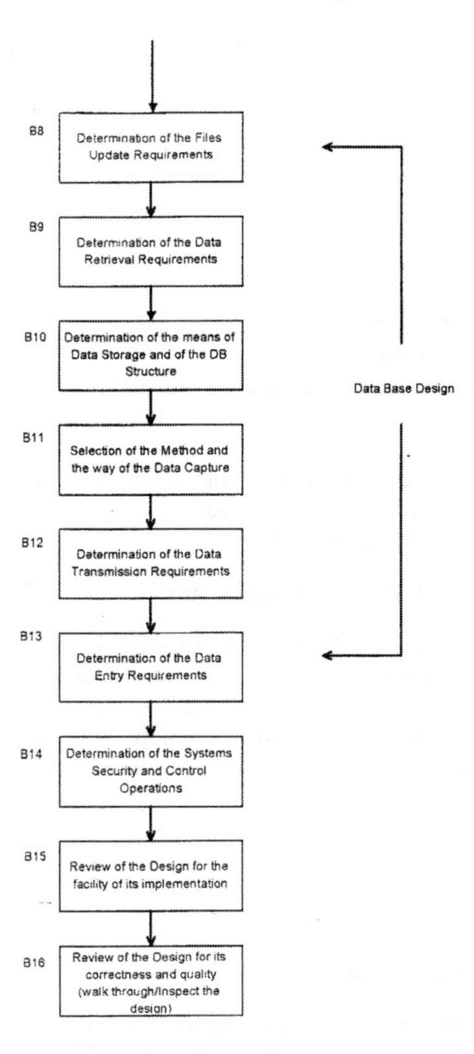

Figure 2.

4 Description of the steps (B1-B16) of information systems design

4.1 Determination of the most important output data

While planning the systems output, its goal should be determined, the user requirements should be defined and its content should be described.

4.2 Determination of the goal

The specific goal of the output should be exactly defined.

Management Information Systems, C. A. Brebbia (Editor)
© 2004 WIT Press, www.witpress.com, ISBN 1-85312-728-0

Using the "verb-noun approach" (Top-Down Analysis) one can be offered of significant help. For instance:
- Authorize (Action),
- Confirm (Accuracy),
- Notify (Situation),
- Report (Progress),
- Order (Action),
- Register (List),
etc.

All the above are indicative and the parentheses should include as many descriptive words as possible.

The following, whatever is necessary to support the basic goal of the output, should be studied.

That means:
- what kind of actions are expected (by the users) with the output delivery,
- what kind of commands are required, and
- what kind of information is needed.

4.3 Determination of the user requirements

At this stage, the perception about the output has to be "cleared out", taking into consideration two possible aspects, from the users' point of view:
- the probability that the output will cause the receiver a favorable situation, and
- the receiver's possibility to respond accordingly to the requirements.

This, of course, demands that the following possible levels of the receivers have been considered:
- intelligence,
- education,
- training, and
- experience.

4.4 Definition of the content

Following the above mentioned, the informative data can be defined, i.e. the data which are required for the "formation" of the output. This effort should terminate in the naming of the data groups and probably in the fields which will be included in these groups.

These three stages, (determination of the most important output data, determination of the user requirements, definition of the content) are required for the procedure of the whole systems design but they are inadequate for the whole design of the output data. That is why they should be "completed" with the work for:
- a unified format
- the creation of a distribution plan, and
- the evaluation in case of any restrictions.

Management Information Systems, C. A. Brebbia (Editor)

4.5 Determination of the necessary input data

The amount of the required data for the system's input might be bigger than the one for the output.

The effort for this determination can be significantly assisted, if the classification is as follows:

- identification data,
- descriptive data,
- measurement data,

The first ones are the "means" to find a specific item, that is, the key, such as, the number of the material, in case the item is a kind of material.

The second ones explain the item and in that way they "confirm" the identity.

The third ones quantify the items and the transactions' results.

This classification is a small part of the Data Analysis Procedure and helps in the decision of which data:

- should be stored
- will be simply used by the system
- will be used by a specific process of the system.

4.6 Determination of the necessary processing for the transformation of the input data to the required output data

During this step, the "processing" has to be defined, that is, the functions which will have to be performed for the conversion of the input data into the required information output.

Probable functions are:

- calculations,
- classifications,
- sorting,
- summarizing and
- reproducing.

Apart from these functions, the following should be defined as well:

- capturing,
- verifying,
- storing,
- retrieving,

but these will be explained during the next steps of the procedure.

4.7 Determination of the requirements of data processing control

It is will known that there exist declinations between the desired situations and those which occur mainly during the initial stages of an informative system.

Therefore, the probability of these declinations should be restricted and at the same time they should be located and restored whenever they occur.

Thus, at this step, those controls will be defined which are needed for the confirmation of accuracy and validity of the processing functions.

This is the reason why the usage of programmed (through computer programs) controls has to be studied, such as:

Management Information Systems, C. A. Brebbia (Editor)

- limit control,
- numerical evidence,
- identification,
- sequence,
- etc.

4.8 Determination of which data the input shall consist of in each process (e.g. orders, incoming stock, etc.) as well as which data shall be set in storage for future use

Practically the procedure of data analysis will have to be done before any design procedure starts.
At this step, the following should be confirmed:
- definition,
- classification,
- selection and
- coding

of the data which have to be "stored".

Therefore, the decision to be taken on which data should be 'stored', will be assisted by taking into consideration the following:
- quantity/time,
- quality and
- cost.

4.9 Quantity/time

Certainly, data cannot be "stored" infinitely in a database, therefore, a selection is required.

Through the quantity/time inspection the following ways can be studied:
- acquisition,
- retrieval, and
- data retention.

Acquisition
This will influence and will be influenced by the method of data capture (Bll).
At this stage the time of data collections, enlistment and coding has to be known.

Retrieval
This is referred to the way and the time of data "retrieval" from the database for the creation of the information output.

Retention
This is referred to the period during which the data must be kept in the database.

4.10 Quality

There are some very significant factors which will have to be studied during the selection of the data to be stored, such as:
-validity,
- accuracy

of the data, to such an extent that they ensure the confidence about the information output.

4.11 Validity

This is referred to the data's "ability" to create the information output, which will solve a problem or make a decision.

4.12 Accuracy

It is known that the data collection costs and that any decrease of their quantity contributes to the decrease of the cost which is required in order to collect and diminish the number of errors to be listed. The accuracy of the data depends on the way they are collected as well as the frequency they are updated.

4.13 Cost

The data storage, in the way that the computer operation demands for a quick and effective approach, costs too high.

One of the factors that affect the decision for such a storage is the necessity to use the existing storing possibilities in the most effective way.

4.14 Database design

Generally, for the data base design, there are two logically divided goals:
- The design of a data structure, which can approach its real structure, as practically as possible.

This design is defined mainly by the following actions: » entity analysis, « data modelling,

* determination of the files update requirements, (Step 8),
* determination of the data retrieval requirements (Step 9), and
* determination of the means of data storage and the DB Structure (Step 10).

- The design of effective ways of collecting data and their location in the database, as far as the cost is concerned. This design is done in 3 stages (Bll, B12, B13). Nevertheless it may occur that an action during one of these steps limits of cancels part or all the work of other steps.

4.15 Data capture

At this step, the method of "capturing" the data, the moment they are created, has to be decided. Therefore, the following two factors should be considered: the kind of input which is required for the work to be done and the kind of controls which are necessary to ensure the correct operation of the system.

4.16 Data transmission

The requirements for the data transmission can be studied during other working phases, such as, the decision phase for the processing through remote terminals. Of course, if this has not been done yet, it should be, during this step.

Management Information Systems, C. A. Brebbia (Editor)

4.17 Data entry

The input procedure of the data in the computer, must be examined, and mainly for the following cases:
- the direct entry procedure to the computer (mainly by the users), and
- the "off line storage", mainly by the data entry operators for a further processing.

4.18 Systems security and control

The system has already been designed, and although it is on a general base, it is "capable" of creating the required information output.

At this step it should be confirmed that the system:
- is capable of resisting to pressures which could hinder the production of its results to the desired levels, and that
- during its operation, it will not cause any problems to the operation of other systems.

Also, at this step, the systems integrity should be confirmed, by inquiring into:
- the cases of possible errors,
- the following actions for their corrections, and
- the cases of "securing" the system from any possible malevolent actions.

Consequently, during this step, it is necessary for all the cases which have not been faced yet to be added to the design up to the point of:
- the system control and the error detection/correction, as well as
- the system's security operations, which will isolate its delicate points and which will permit the access to authorized people only.

5 Review of the design for the facility of its Implementation

The last step on the procedure of the design is the confirmation that the system covers all the presuppositions for an easy realization and implementation, that is, Programs, Installation, Operation.

This will require the full understanding of:
- the system's goals and activities,
- the administration of the means to be. used,
- the system's quality controls,
- the time schedules which will have to be kept,
- the training which will be required for the operation of the system, and
- the requirements of the system's physical installation.

6 Walkthrough/inspect the design

At this step, one of the following two typical techniques can be used:
- walkthrough or
- inspection

Management Information Systems, C. A. Brebbia (Editor)

for the confirmation of the design's accuracy and quality, in the way that it has been realized up to that point of the system's development.

6.1 Walkthrough

During a walkthrough the analyst describes the operation of some parts of the designed system to 2-3 other colleagues and to one analyst and a programmer of the project team. This description is done by the usage of sample data and the simulation of the system's operations. That means that it is done b a dry run of the system.

The aim of the walkthrough is to be proved that the system really covers the requirements for which it is developed. Any differentiation from the development and installation standards have to be noticed, but the effort should be directed mainly to the error discovery. It is obvious that the error correction will be left to the analyst for the time after the Walkthrough, otherwise, the walkthrough will never end.

6.2 Inspection

This differs from the walkthrough mainly because of the presence of a Judge who operates as an independent coordinator of the inspection.

The design is described in the inspection as well, but by a member of the team, who has not contributed to it. In case of the opposite situation, the inspection does not differ significantly from the walkthrough.

6.3 Comparison between walkthrough and inspection

Generally an inspection is more formal than a walkthrough and has less chances to e degenerated to a comedy.

It is sure that the inspections are more effective in the error detection; they require though a better training and preparation and probably they do not fit to small Data Processing Centres.

7 Epilogue

It is better for the accuracy and the wholeness of the design to be checked continuously while it is being built, than to leave this control to the formal Walkthrough and Inspections.

A series of questions which will be referred to the requirements of every step of the design procedure, will be helpful for the confirmation that the design has been realized correctly.

For every part of the design, questions which begin as following could be used:

What? Why? Who? Where? When? How?

in order to confirm that the system's basic requirements have been satisfied.

The above procedure can be expanded in more details, through the evaluation of each part of the design, in relation to its:

Management Information Systems, C. A. Brebbia (Editor)
© 2004 WIT Press, www.witpress.com, ISBN 1-85312-728-0

operation, its goal, performance, reliability, maintainability, flexibility, facility of implementation, facility to confirm its validation, requirements as far as the staff's training is concerned, etc.

Since every part of the design, immediately after its realization has been judged satisfactory under the above criteria, it is almost sure that the new IS will have been designed correctly.

References

[1] Alter, S. (1992). Information Systems: A Management Perspective. Reading Massachusetts: The Benjamin / Cummings Publishing Co., Inc.

[2] Blackman, M. (1975). The Design of Real Time Applications. London: John Wiley & Sons, Ltd.

[3] Brathwaite, K. (1985). Data Administration/Selected Topics of Data Control. New York: John Wiley & Sons.

[4] Cutts, G. (1991). Structured Systems Analysis and Design Methodology. Oxford: Blackwell Scientific Publications.

[5] Edwards, P. (1993). Systems Analysis and Design. New York: Mitchell McGraw-Hill.

[6] Gillenson, M. & Goldberg, R., (1984). Strategic Planning-Systems Analysis & Data Base Design. New York: John Wiley & Sons.

[7] Hawryszkiewycz, I. (1988). Introduction to Systems Analysis and Design. New York: Prentice Hall.

[8] Hoffer, J., George, J., & Valacich, J. (1996). Modern Systems Analysis and Design. Reading / Massachusetts, New York: The Benjamin / Cummings Publishing Co., Inc.

[9] IBM, Direct Access Storage Devices and Organization Methods. M. C20-1649.

[10] IBM, Systems Analysis, M. BI04.

[11] Kroenke, D. (1992). Database Processing: Fundamentals-Design-Implementation. New York: Mackmillan Publishing Company.

[12] Martin, J. (1976). Principles of Data Base Management, New Jersey: Prentice Hall, Inc.

[13] Martin J. (1977). Computer Database Organization. New Jersey: Prentice Hall, Inc.

[14] Ross, S. (1994). Understanding Information Systems. Minneapolis / St. Paul: West Publishing Co.

[15] Steig, D., Improving Commercial Software Design, Data Processing Magazine, v. 13, nr 9, pp. 40-42.

[16] Sutherland, J. (1975). Systems Analysis, Administration and Architecture, New York: Van Nostrand Reinhold Co.

Management Information Systems, C. A. Brebbia (Editor)

A study of customer relationship management in financial services on the Web

J. Lawler, D. Anderson & E. Rosenberg
School of Computer Science and Information Systems, Pace University, USA

Abstract

In the current period of constrained economic conditions, this study initiates an analysis of customer relationship management on the Web sites of financial service businesses. Customer relationship management is a critical differential that effects competitive edge for businesses focused on the affluent customer market. Though investment in marketing, sales and service innovation is limited under the existing financial and global political conditions, the analysis of the study on large financial businesses in the United States indicates that the businesses enable higher commerce, content and context, but lower and generally inadequate communication, community, connection and customization design on their Web sites. The analysis contributes an important insight into the competitive dynamics of customer relationship management for Web empowered financial service businesses striving to service the demanding affluent market. This study furnishes an expanded framework to research customer relationship management of financial service businesses competing in the paradigm of the Web.
Keywords: customer relationship management (CRM), customer service, e-banking, financial services, wealth management.

1 Introduction

The World Wide Web is an important consideration in the growth of businesses. From its inception, the Web has initiated unprecedented change in effecting fast and innovative channels of business with customers. From customer-centric channels to customer-focused processes, businesses have employed innovation in customer relationship management (CRM) to enhance their business models

Management Information Systems, C. A. Brebbia (Editor)

(Sweat [1]). New methods for marketing, sales and service on Web sites have contributed benefits to 23 million customers and to the businesses in the United States Tubin [2]. The Web is a key factor in the customer relationship management strategy of businesses.

The importance of customer relationship management on the Web to businesses in the financial services industry is unclear in the academic research. Investment in not only customer relationship management, but also in the Web, is constrained because of the economy. Financial justification of complex innovation in customer relationship management is difficult, due to dissatisfaction from frequent failure in technological implementation. Though innovation on the Web is beneficial for businesses in financial services and their affluent customers, the Web sites in this industry generally lack the advanced functionality of those in other industries that integrate their online and offline commerce strategies (Peppers and Rogers [3]). The financial services industry is highly cautious about investment in customer relationship management on the Web (Schmerken [4]).

Studies from non-academic sources indicate that this caution is contributing to online functionality of customer relationship management in financial service businesses that is inconsistent from the offline (Tubin [2]). The experience these institutions effect for affluent customers is considered impersonal, intrusive and passive (Punishill [5]), in contrast to the personal office experience. Functions on the sites of the businesses are perceived to be time-consuming (Mearian [6]) and discouraging (Freed [7], Kahn [8], Scheier [9] and Verma [10]). Only 15% of financial service businesses focused on affluent customers contend that they are enabling innovative customer relationship functionality beyond the expectations of these customers (Punishill [5]). The high technology and high touch channel of customer relationship management on the Web is apparently elusive on financial service business sites (Cawthon [11] and Ulfelder [12]).

Half of the adult population in the United States actually favors an offline office. Though the offline channels of the office and the telephone are considered desirable in effecting experience for the affluent customer (Rupp and Smith [13]), the online channel of the Web continues as an important factor in interactivity experienced by the customer (Coviello et al.[14]). The experience on the Web in its negative or positive impact influences future customer relationship (Kirby et al. [15] and Haubl and Trifts [16]). Financial service institutions that affluent customers believe are highly competent in customer relationship management on the Web are considered ideal business choices for the customers. The competitive differential is effectively equivalent experience throughout the channels. Financial service businesses are clearly challenged in customer relationship management on the Web.

The focus of the study in short is to introduce and analyze the effectiveness of innovation in customer relationship management on Web sites in the financial services industry. The study will attempt to demonstrate empirically that the Web is critical for the future of financial service businesses that compete for affluent customers. This study is timely, in that important functions effecting

Management Information Systems, C. A. Brebbia (Editor)

customer relationship on the Web need to be clarified for decision makers in the industry, in order to invest intelligently in this virtual marketplace.

2 Financial services market

The financial services market is in its adolescence in Web channeling, comprising 20% of all sized financial service firms having commerce sites in the United States. This market contrasts with a high of 40% in Finland and Norway, 20% in United Kingdom, and a low of 10% in France, Germany, Italy and Sweden (Rupp and Smith [13]). Consumer demand for functionality in online service is growing however, as the number of banking consumers in the United States on the Web increased to 23 million in 2003 from 13 million in 2001 (Conrad [17]). Studies indicate the number of commerce or transactional financial service sites domestically and internationally increasing significantly by 2010 (US Banker Weekly Bulletin [18]).

The large business market for high affluent consumers in this current study consists of customers in the United States having invested assets of $1 million or higher (Schmerken [4]), of which 50% has $5 million to $50 million in assets (Punishill [5]). These customers consist of advisor-directed, who have their financial advisors manage their finances, advisor / self-directed customers, who have the advisors manage their finances but share in the management, and self-directed customers, who manage their finances themselves. 50% of the market consists of advisor / self-directed customers, and 30% consists of self-directed customers (Bruno [19]), most of which are older and if managing their finances frequently manage from the sites of the businesses.

50% of the customers consider the sites effective in customer relationship, 25% define effective sites in advanced customer management functionality that facilitates relationship, and most of the customers define advisor availability through the sites as important in their relationship with the firms (Punishill [5]). Expectations of customers in media-mediated interaction continue to be high (Morrell [20]).

Important in customer relationship is the indication that millionaires are 30% more likely than non-millionaires to manage their finances from the Web (Cawthon [11]), though self-management is less likely self-directed if the financial plan increases in complexity and more likely advisor / self-directed and helped in Web interaction. 20% of households that have advisors are already affluent and older, but are helped on the Web (Clemmer [21]).

Of final interest to financial service firms is the impact of mass affluent and emerging entrepreneurial, younger, affluent customers having invested assets of less than $1 million (Schmerken [4] and Carroll [22]). These customers will be increasingly high affluent, interested in an advisor relationship, and more frequently managing their finances from the Web than older customers. Financial service businesses are focused on the older affluent customers. Indications of $20 billion in revenue from younger affluent customers are not to be ignored in the financial services market (Bank Technology News [23]).

Management Information Systems, C. A. Brebbia (Editor)

3 Characteristics of financial service Web sites

The characteristics of customer management on the sites of financial service businesses analyzed in the study are elements of e-Commerce customer interface that influence the relationship effectiveness of a Web site. Rayport and Jaworski introduce these elements as context, content, communication, connection, community, customization and commerce (Rayport and Jaworski [24]), enhanced in this study to include the following framework of constructs:

- *Context*, defined as depth of comfort and ease of navigation in foreign language, help, hyperlink and search, real time and single sign-on on the site;
- *Content*, depth of information in account aggregation detail, account analysis, account product holding, at a glance dashboard, document imaging, frequently asked product questions and glossary, household summary, performance reporting, planning tutorials, portfolio tools, publications, research, transaction history and watch list;
- *Communication*, degree of dialogue in alert, chat, co-browsing, collaboration, conferencing, form and screen pushing, e-mail link, e-mail notification, instant messaging, live Web help, telephone, telephony and video, white-boarding and wireless;
- *Connection*, degree of connecting to affiliated financial industry market news and quotes, business news, featured domestic and international news, life style news syndicates and links to preferred merchant sites;
- *Community*, degree of investor interaction in customer bulletin boards, continuing education forums, discussion groups, non-charitable and charitable event sponsorships and expert investment seminars;
- *Customization*, degree of explicit and implicit personalization of site in account preferred dashboard, alert, e-mail notification, my profile and watch list, forums and seminars, industry, life style and merchant links, recommendations in products and services and planning tutorials and portfolio tools; and
- *Commerce*, extent and speed of commercial interaction in account and fund transfer, automatic investment, bill payment and presentment, brokerage services, confirmations, direct deposits, pre-filled application forms, reorders, privacy and security services and tools and transaction on wireless.

These constructs include highly advanced and fundamentally basic functionality on the Web, in order to facilitate customer relationship management, which is defined in this study to be the following:

Customer Relationship Management - *the strategy of optimizing the processes of financial service businesses, which enable the marketing, sales and service of products, though innovation of integrated customer channels including*

Management Information Systems, C. A. Brebbia (Editor)

that of the functionality of the Web, and which effect customer loyalty and financial service profitability.

4 Research methodology

The research methodology of the study employed a content analysis of a sample of large-sized financial service businesses consisting of banks, brokerages, investment banks and mutual fund firms, in three iterative stages of analysis. In stage 1, thirteen businesses in the United States were selected based on their reputation in the practitioner literature, such as the Gartner Group, as leading business-to-consumer (B2C) investor sites. The sites selected were American Express, Bank of America, Citibank, E*Trade, Fidelity, Merrill Lynch, J.P. Morgan, Northern Trust, Paine Webber, Salomon Smith Barney, US Trust, Vanguard and Wells Fargo. These sites were analyzed by 51 adult graduate students, aged 26-38 years, who as information systems majors but as realistic Web consumer subjects, received logon to the customer sites and potential services in a *Customer Relationship Management: Processes and Technologies* course at Pace University, in New York City. The sites were analyzed anonymously and independently by the students in a nine-month period of the spring, summer and winter 2003 semesters of the course.

The sites were analyzed for their relationship effectiveness in the constructs of *context*, *content*, *communication*, *connection*, *community*, *customization* and *commerce*, applying a three-point scale of 1=low, 2=intermediate or 3=high functionality in technology. The constructs were defined from a checklist multi-page instrument of the 78 content-specific functions indicated in the last section, which spanned depth of comfort to transaction on wireless, with low functionality including no functionality. The students were motivated to help as analysts, as the analysis was an extra credit assignment in the course. The data was then collected by the authors at the conclusion of the semesters. In stage 2, the data from stage 1 was analyzed in Proc Mixed SAS to determine construct mean scores and statistical significance. This analysis included unadjusted averaging, that was deemed better than an advanced analysis due to ease of interpretation and robustness, and though averaging limited the degrees of freedom available for analysis, adequate strength was available to distinguish between the constructs. In stage 3 of an advanced study, the sites will be analyzed in future semester surveys of the customer relationship management strategies of the financial businesses.

5 Analysis of financial service sites

The analysis from stages 1 and 2 indicated that most of the financial service business sites in the sample have higher *context*, *content* and *commerce*, but lower *communication customization*, *connection* and *community*, as summarized in Table 1. That is, context, content and commerce constructs as a group were high in statistical significance (2.6, 2.4 and 2.2 scores), communication, customization and connection as a group were low (1.7, 1.7 and 1.6), and

Management Information Systems, C. A. Brebbia (Editor)

community was the lowest group (1.2). Pair comparisons of constructs were unadjusted but were significant at P>=.01 level. Communication, customization and connection were not significantly different statistically for each other, but the other constructs were significantly different at the P>=.01. The constructs were assumed to share the same overall variance, given student analysis of all constructs, rigorous function criteria and the diminutive functionality scale.

Table 1: Summary analysis of financial service sites.

Financial Service Firms				
	Average Score	*High* Functionality Frequency * Score 3	*Intermediate* Functionality Frequency * Score 2	*Low* Functionality Frequency * Score 1
Constructs				
Context	2.6	413	242	8
Content	2.4	309	316	38
Commerce	2.2	264	269	130
Communication	1.7	63	307	293
Customization	1.7	56	350	257
Connection	1.6	35	311	317
Community	1.2	7	102	554

* n =13 Firms x 51 Students (18+21+12)

Throughout this analysis, the interpretation was that the individual sites are highly functional, informative and navigable in fundamental services. For an affluent customer, these features are not fully enhanced on the sites to include implicit customization of recommended products and services. Only explicit but limited personalization of site tools or stock tracking is the frequent feature on the sites. Features of customer communication and connection to financial advisors and service representatives are also limiting to an affluent customer, in skeletal functions of the telephone that do not leverage the live help and interactivity of the Web. Functionality of interactivity with other customers, investment experts in discussion forums or event sponsors is essentially non-existent on the sites.

The dimension of customer relationship management is inevitably elusive on the sites of these businesses. In the best of the sites effecting the closest in relationship to their customers, the interaction is indicated to be faster in commerce and content inquiry, but slower in business communication and connection. In the worst of the sites, the interaction is slower in commerce and content searching, but still slower in communication and connectivity. None of the on-line sites in the analysis excel in customer relationship management. The study from stages 1 and 2 indicates lower relationship but higher usability on these sites.

The analysis confirms the practitioner literature and proposes the following:

Management Information Systems, C. A. Brebbia (Editor)

PROPOSITION 1. Large financial service businesses have higher context interactivity, informational content and commerce functionality on their Web sites, increasing the potential of relationship effectiveness of the sites.

PROPOSITION 2. Large financial service businesses have only intermediate or lower advisor communication, affiliated financial connection, customized feature and investor community functionality on their sites, decreasing the potential of relationship effectiveness of the sites.

These propositions, and the importance of the constructs in customer relationship management, will be analyzed further in stage 3 of the advanced on-going study, based on an instrument that will survey decision-makers in the financial service businesses on their strategies on the Web. Though the results of the expanded study will not be final until completion of stage 3, the results of the current study are helpful in initially analyzing the financial service business market.

6 Implications of study

The results of the sample size of the study confirmed non-academic sources that indicated the constructs of interaction on financial service sites are higher in *commerce*, *content* and *context* features, but lower in the *communication*, *connection*, *community* and *customization* functions that facilitate customer relationship on the Web. The focus of the sites analyzed is in the marketing and sales functionality of customer relationship management, not in the service functionality (Mc Kendrick [25]). This focus is not fully effective in facilitating service, a fundamental ingredient in customer relationship management (Newell [26] and Sterne [27]). Service in commerce functionality that is further collaborative, contextual in information and customized in advisor and colleague investor interaction (Cawthon [11]) is limiting on these sites. Though the bulk of these financial service businesses are focused on advisor interaction in offices, complementing advisor desirability, the benefit of effecting increased relationship in customer interaction from the sites is elusive to the businesses.

The criticality of financial advisors in the currently constrained economy is apparent from the impetus in consumers considering alternatives to their dwindling investments (Carroll [22]). The human help of advisors in brick offices is an important channel of interaction. Sensitivity to affluent customers interfacing concurrently or alternately on the Web channel is additionally important to self-directed and also advisor / self-directed and advisor-directed customers. Self-directed and advisor / self-directed customers frequently insist upon advanced *content* functionality, such as personalized stock option tools on the Morgan Online site and tax services on the Fidelity and Schwab sites (Ulfelder [12]). In contrast, advisor-directed customers insist upon basic *communication* functionality in advisor e-mail and live Web help and fundamental *content* functionality (Canfield [28]) in simple *customization* tools and tutorials. Results of the current study indicate that the functionality on the

Management Information Systems, C. A. Brebbia (Editor)
© 2004 WIT Press, www.witpress.com, ISBN 1-85312-728-0

analyzed sites is inadequate not only in advanced but in basic interaction, despite evolution in Web innovation.

This lack of full functionality on financial service sites is further important in future strategy. Forecasters indicate that emerging and mass affluent consumers, who include 2.5 million investors of $6 trillion in assets in this decade Clemmer [21], will depend upon advisor help in Web-initiated interaction or self-service on the Web. Their expectation in interaction will be in immediate and advanced advisor *communication*, *community, connection* and *customization* functionality enabling customer relationship management that is not often apparent today. Though businesses, such as Merrill Lynch, are implementing strategy in which less highly affluent customers will be directed not to financial advisors, but to sites or customer service representatives (Saccocia [29]), their self-service on the Web is limiting. The functionality of financial service sites is homogeneous in innovation.

The implied interpretation of limited interaction on the sites is that online financial service businesses continue to be indistinguishable in functionality (Koller [30]). The interest is not in customer benefit, but in cost containment (Bruno [31]). Studies indicate that few of these businesses frequently follow up in evaluating the satisfaction of their customers in the functionality of the sites (Kirby et al. [15]). Other studies confirmed in this analysis indicate slowed response in service on the sites (Bass [32]). Relationship is not cultivated on the Web sites.

To effect relationship management on the Web, financial service businesses that initiate innovation will focus on advisor friendly wealth management technologies (Schmerken [4] and Punishill [5]) and customer relationship management tools (Logue [33], Schmerken [4] and Mc Kendrick [25]), such as those of the Merrill Lynch Global Private Client Group (Ulfelder [12]). The inclusion of these tools will be important in enabling enhanced functionality in *communication*, *community*, *connection* and *customization* for future emerging, mass and high affluent consumers. Advisor interaction in the office or on the Web will similarly satisfy in servicing the customer. The businesses will be challenged in deciding on wealth management technologies, in that the field is limited in success stories (Schmerken [4]). Still, those financial service businesses that initiate technological transformation of their sites on the Web will have an edge in serving the affluent market, an imperative implied in this study.

7 Conclusion

Impacted by the constrained economy and defined business initiatives, large financial service businesses in the affluent market are currently focused less on enhanced *communication*, *community*, *connection* and *customization* and more on the fundamental *commerce*, *content* and *context* functionality enabling customer relationship management on their Web sites. To compete effectively for an expanding marketplace of affluent consumers, and improve perceptions of their B2C sites, in the United States, the businesses need to innovate further in

Management Information Systems, C. A. Brebbia (Editor)

transforming the site touch point to provide superior service to their customers. Further research is needed to explore the current and future strategies of the financial service businesses in their functionality of customer relationship management on the Web, and this study is introducing a new framework.

References

[1] Sweat, Jeff, Keep them happy. *Information Week*, 28 January, pp. 55-60, 2002.

[2] Tubin, George, Tower Group review of top US online banking sites exposes gaps in functionality and usability. *US Banker Weekly Bulletin*, 7 October, p. 2, 2003.

[3] Peppers, Don & Rogers, Martha, The state of one to one online: sites for sore eyes. *Peppers and Rogers Group Report*, May, 1999.

[4] Schmerken, Ivy, One size does not fit all. *Wall Street and Technology Online*, 17 January, pp. 1-3, 2003.

[5] Punishill, Jaime, Virtual family offices blossom. *The Forrester Report – Action, Analysis and Interviews*, February, pp. 1,2,3-4,9, 2001.

[6] Mearian, Lucas, Banking sites show big gaps in performance. *Computerworld*, 11 August, p. 15, 2003.

[7] Freed, Larry, Banks lag online. *Windows on Financial Services*, October, pp. 6-7, 2003.

[8] Kahn, Munger, Online advice finds few takers. *The New York Times*, 2 March, p. 6, 2003.

[9] Scheier, Robert L., Know thy customer. *Computerworld*, 24 March, pp. 25,28, 2003.

[10] Verma, Gaurav, CRM market strides in self-service. *Information Week*, 18 February, 18 February, pp. 67-70, 2002.

[11] Cawthon, Raad, Taking financial web sites to the next level. *Bank Technology News*, June, pp. 45-48, 2001.

[12] Ulfelder, Steve, Catering to the wealthy. *Computerworld*, 10 November, pp. 41-42, 2003.

[13] Rupp, Bill and Smith, Alan, Emerging trends in e-commerce: e-lending in an information intensive society. *Proceedings of Southeast Decision Sciences Institute*, ed. Emmanuel Omojokun: Williamsburg, Virginia, pp. 248-250, 2003.

[14] Coviello, N., Milley, R. & Marcolin, B., Understanding IT-enabled interactivity in contemporary marketing. *Journal of Interactive Marketing*, 15(4), pp. 18-33, 2001.

[15] Kirby, Jennifer, Wecksell, Joel, Janowski, Walter & Berg, Tom, Value of customer experience. *Gartner Report*, 10 March, pp. 1,17, 2003.

[16] Haubl, G. & Trifts, V., Consumer decision-making in online shopper environments: the effects of interactive decision aids. *Marketing Science*, 19(1), pp. 4-21, 2000.

[17] Conrad, Lee, E-bank client counts soar, but questions persist. *Bank Technology News*, October, p. 12, 2003.

Management Information Systems, C. A. Brebbia (Editor)

[18] ----------, Europeans hot for Internet banking. *US Banker Weekly Bulletin*, 1 April, p. 2, 2003.
[19] Bruno, Mark, Growing ranks of the young and the wealthy. *US Banker*, February, p. 56, 2003.
[20] Morrell, S., Benchmarking customer service technology: how companies are embracing their customers. *Reuters Business Insight*, 2000.
[21] Clemmer, Kenneth, Why investors seek advisors. *The Forrester Technographics Report – Consumer Landscape*, February, pp. 1-2, 2001.
[22] Carroll, John, High net worth customers are turning to advisors in droves. *Bank Technology News*, November, p. 38, 2002.
[23] ----------, By the numbers. *Bank Technology News*, July, p. 14, 2002.
[24] Rayport, Jeffrey F. & Jaworski, Bernard J., *Introduction to e-Commerce*, McGraw-Hill / Irwin: New York, pp. 184-187, 2002.
[25] Mc Kendrick, Joseph, Should they stay or should they go. *Bank Technology News*, November, p. 33, 2002.
[26] Newell, Amy, Tools of the online trade. *Bank Technology News*, May, p. 30, 2002.
[27] Sterne, Jim, Customers spoke loudly. *Business 2.0*, 12 December, p. 150, 2000.
[28] Canfield, Todd, Know thy customer. *Business 2.0*, 24 October, p. 248, 2000.
[29] Saccocia, Cynthia, By the numbers. *Bank Technology News*, March, p. 22, 2003.
[30] Koller, Lynn, Going after the rich. *Bank Technology News*, January, p. 22, 2002.
[31] Bruno, Maria, Chasing. *Bank Technology News*, June, p. 14, 2002.
[32] Bass, Alison, Online big names say do not disturb. *CIO*, 15 January, pp. 28-30, 2003.
[33] Logue, Elizabeth, Advisors finally waking up to the Internet's useful influence. *Bank Technology News*, September, p. 9, 2002.

A proposed theoretical strategic framework for the successful entrance of new products of high technology

Th. Fotiadis[1], G. Haramis[1], G. Siomkos[2], Ch. Vassiliadis[1] & Ch. Tsialtas
[1] *University of Macedonia*
[2] *Athens University of Economics and Business*

Abstract

An innovative decision-making procedure is expounded in the present paper. This procedure concerns decision-making about New Hi-Tech Products and their introduction into the uncertain, turbulent, volatile, and tumultuous environment of the market through the systemic approach/operational coexistence of Marketing with Research and Development (R and D).

Upon the completion of two parts, the input, coordination, evaluation and circulation of information is optimized, while quality and quantity aspects are expanded. Also, the aspect concerning the market-orientated use of information is developed.

This procedure makes the effective distribution of resources and time possible, and optimizes the learning experience curve of the company guaranteeing that pointless and insubstantial actions are reduced and that errors are incorporated into the decision-making algorithm. In addition, it brings down the imminent features of uncertainty and cuts the introduction time of New Hi-Tech Products into the market.

The strategic orientation of the suggested theoretical framework, the emphasis laid on the market-orientated approach as well as the software—which was developed for the purposes of the present theoretical framework—support and encourage the reasoning of the decision-making procedure concerning the introduction of New Hi-Tech Products into the market. At the same time, it includes—within the decision-making mechanism—the parameter of the achieved completion degree.

Management Information Systems, C. A. Brebbia (Editor)
 ISBN 1-85312-728-0

1 Objective

The objective herein is to create a procedure for the development of New Products of High Technology (NPHT) for a more thorough and well-founded entrance of NPHT to the market, which shall encompass in its mechanisms the effective cooperation of the functional departments (mainly) of Marketing (MKT) and Research and Development (R&D), and which:
-will contribute to the integration (mainly) of Marketing and R&D, but will also sustain the comprehensive interdepartmental cooperation with other functional departments of the enterprise
-will be market-oriented (i.e. oriented to consumers and competitors)
-will facilitate the rational management of the NPHT portfolio of High Tech enterprises that adopt it
-will contribute to the improvement of the enterprise's learning experience curve
-will contribute to the reduction of uncertainty
-will contribute to the enhancement of the functional specialisation of MKT and R&D departments to the direction defined by the common organisational objective, through their integration
-through the aforementioned, will contribute to the shortness of the time required for NPHT development.

2 Approach-structure of the model in stages

2.1 Establishment of teams

I. Core Team (CT)
The MKT Manager and the R&D Manager determine the executives from their functional departments respectively who will compose the Integration Core. (By integration core is implied that mixed team of MKT and R&D executives that has the responsibility for coordinating functional competencies, for approaching the organisational objective and for interfacing specialisations). In this way, a basis for communication is created, a communication channel that allows accessing the tools of work, the methodology of thought, the procedures-routines and the mentality of relevant departments. The aim is to achieve an (at least gradual) increase in the information of the non-integrated but specialised executives (who are the cores of the intradepartmental forces that improve-increase the degree of functional specialisation qualitatively).

On one hand, CT has the responsibility for providing the direction towards the common organisational objective, and on the other hand it defines and determines the strategic coordination of the functional departments' inflow and outflow information. The latter is achieved as a result of the dual role of CT regarding information. It simultaneously functions both as information source of potential performance and viability but also as a recipient of partial information from other departments that mostly shapes the estimate of performance and viability.

Management Information Systems, C. A. Brebbia (Editor)

II. Basic Team (BT)
Subsequently, CT contacts other executives in order to identify capabilities-functions-characteristics-sources-procedures-knowledge and their accordance with the desired objective or potential areas of contribution.

Following the aforementioned identification of important departments-executives, a wider team is formed; the Basic Team, enfolding CT. BT has the responsibility to supply CT with relevant information-knowledge-competencies and to identify teams-bodies inside or/and outside the enterprise that shall function as information channels. BT also includes the functional cores of specialisation.

III. Expanded Team (ET)
Lastly, the ET is formed (enfolding the BT and consequently the CT), which includes the Top Management and the stakeholders, i.e. the individuals or teams that have an influence upon the enterprise or its decisions, and/or are directly or indirectly the final recipients.

2.2 Collection and organisation of Information

A SWOT analysis -with the contribution of all 3 teams- is performed. Herein, the aim is to identify, in both external and internal environments of the enterprise, those variables composing the Strengths, Weaknesses, Opportunities and Threats. At this stage, is identified the set of significant factors that are crucial. Their identification and knowledge but also their common analysis and labeling as S.W.O.T., define the activation field and the terms of actions.

Next, weighted importance is recorded and depicted diagrammatically; team members rate the performance and also the Importance Weight—given by each team separately—for the Strengths, Weaknesses, Opportunities and Threats, the resultants of which were identified in SWOT analysis.
It can be produced as follows:

A matrix is constructed, upon which the identified variables are organised and their weighted importance is given and the performance is derived, in the manner that those factors are conceived by the team members of CT, of BT (OT – BT) and of ET (ET – BT), and additional weighting of the product of the aforementioned sets, depending on the team which brought them in. Variables that were identified and recorded as crucial are organised in horizontal rows. They are recorded and organised as crucial variables of the Opportunities–Strengths–Threats–Weaknesses. Moreover, opportunities and strengths are grouped as variables, the size of which will determine the axes coordinates of the matrix of diagrammatical illustration, which will be described and developed straight after. Opportunities and Strengths will be depicted on the dimension of the horizontal axis, while Weaknesses and Threats will be depicted on the vertical.

CT, for each Opportunity–Strength–Threat–Weakness (i.e. for each of the variables that have been grouped under the label of the abovementioned descriptions of member variables), does the following activities:

It assigns relevant importance from 1 to 10 to all variables capable of being described as factors of Opportunities (and the Strengths–Threats–Weaknesses),

Management Information Systems, C. A. Brebbia (Editor)

based on the perceived/identified magnitude of each variable. The range of the assigned importance 1-10 is indicative, but the total of the Importance rates to be given to the factors composing the Opportunities (and the Strengths–Threats–Weaknesses) as sum should not exceed the upper limit of the range of rating options (e.g. for the range 1-10, the sum of rates of Opportunities' variables and then the sum of rates of Strengths' Variables etc. should not exceed 10) for the weighting to be valid.

Then, it rates from 1 to 5 (or in any range of quantitative measure is considered helpful) the perceived performance upon the enterprise of each variable. At this point, the cancellation of weighting (i.e. for the range 1-5 the sum of the performance rates to be 5) is not helpful since the performance assignment of each factor upon the enterprise may be equivalently good (or equivalently low respectively) for more than one variable or even for all. Of course, the perceived performance is assigned to all the variables of Opportunities, Strengths, Threats and Weaknesses.

Then, it calculates the product of the two previous activities, which in essence reflects the importance of each variable through the perceived size of influence upon the enterprise, in proportion with the commonly expressed variables that were produced in the previous stage.

Then, it produces the product of the weighted total of the previous activity with an importance factor (weighted), which indicates the importance of the expressed opinion of the members that make up the Core Team.

Similarly, the Basic Team conducts the same activities (of course without the participation on the evaluation procedure of the Core Team members, who have already submitted their evaluations). Here, the factor of weighted importance may not necessarily be the same.

The Expanded Team performs the same activities, without the participation of BT members, for similar reasons to the ones described in the previous paragraph. The three weighted totals (with the relevant importance assigned to the opinions of each team individually) will sum up to the upper limit of the range of the quantitative measure selected.

Lastly, the weighted totals of the (weighted) totals of all three teams are added horizontally and are recorded in the column "Total of totals" of the matrix. The vertical sum of the "Total of totals" of Opportunities and Strengths gives the position upon the horizontal axis, while the vertical sum (of the subset of the "Total of totals") of Weaknesses and Threats, gives the position upon the vertical axis. These coordinates are recorded upon the matrix as the sum of the total of weighted totals (weighted) of opportunities and strengths, and as a sum of the (weighted) total of totals of treats and weaknesses. Thus, the vertical column of the total of totals depicts:

I. The total of weighted totals for each factor that participates in the Opportunities, Strengths, Threats and Weaknesses respectively, from all the teams.

II. The total of weighted totals as the sum of the evaluation overall of the factors that compose respectively the SWOT, and which was done by the teams.

Management Information Systems, C. A. Brebbia (Editor)

III. The sum of the total of weighted totals of opportunities and strengths, i.e. the abscissa.
IV. The sum of the total of weighted totals of threats and weaknesses, i.e. the ordinate.
V. Moreover, a second column is created (under the general title total of totals), which gives the total of weighted importance (of the already weighted variables) that is attributed to each team (factors α, β, γ for CT, BT, ET respectively).

2.3 Illustration of outputs upon matrix and bar diagrams

A matrix is created having four quadrants (squares) that derive from the interface of all possible High and Low levels of the total of the variables depicted on the Horizontal Axis (of course after the weighting process described before) and of Low - High levels of the total of variables depicted on the Vertical Axis (following the weighting process).

A point of the matrix is given by the combination of the abscissa and the ordinate, the calculation of which was described earlier.

A circle can be produced having centre the point defined on the matrix; the area of the circle is reversely proportional to the difference that emerges from the product of weighted importance and the performance rate assigned by CT to crucial variables (that have been grouped on the matrix as SWOT) and the product of weighted importance and performance rate assigned by BT to the corresponding crucial variables. (That is alternatively said, the area of the circle is proportional to the agreement in the opinions of the two teams). It should be stressed that the abovementioned difference of products is always a positive number, (is in absolute value), and that whenever BT is mentioned, it is implied BT-CT.

The aforementioned depiction of the products' difference in a circle is prepared as follows: Consider circle of radius P, where P is the known number that corresponds to the minimum difference of products, i.e. it corresponds to the perfect agreement between CT and BT. However, to maintain the proportionality of the area, every corresponding magnitude of the absolute value of the difference CT-BT is converted into the corresponding radius length.
This becomes feasible with the following deduction: Let us consider the area of the circle as known (say α square units, where α is defined arbitrarily), which defines the absolute value of the product difference CT-BT, which gives the minimum difference in the opinions (or the perfect agreement) of the two teams. Of course, this area may correspond to a radius of the same circle say ρ1 (since it is commonly known that $E\kappa=\pi\rho^2$ and $E\kappa= \alpha$ s.u., and $\pi=3.14$). Then, radius ρ2 that will define (or more correctly will be defined from) the area of the circle that the opinion difference will correspond to (or respectively the remainder of the agreement), e.g. β will have a length equal to:

$$p_2 = \sqrt{\frac{\beta}{\alpha}}$$

Management Information Systems, C. A. Brebbia (Editor)

The difference of the circles' areas will represent in that sense the difference of the products. If the absolute value of the two products is the maximum (perfect disagreement between the opinions of the two teams), the second circle will become a point, i.e. a circle with radius 0.

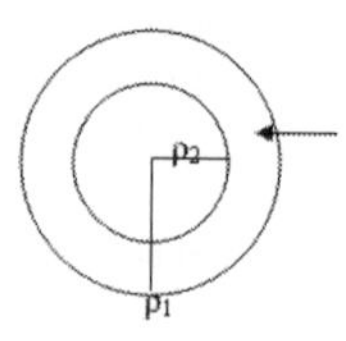

It should be stressed that the proportionality of the area of the circle that represents maximum agreement between the two teams and the circle that represents the achieved agreement, determines the radius ρ2 that corresponds to the area of the circle of the achieved agreement.

An angle is formed upon the aforementioned circle, (epicentre), which is reversely proportional to the difference that occurs from the product of weighted importance and performance rate given by BT (where BT = BT – CT), and the product of weighted importance and performance rate given by ET (alternatively said, the higher the degree of agreement—or the lower the difference between BT and CT (and indeed in absolute value)—the larger the epicentre angle)
In this case, we compare BT with ET. We seek to depict the difference in their opinions, a difference (of the corresponding products) that is in absolute value.

The abovementioned depiction of the difference with an epicentre angle is performed as follows:
360° correspond to perfect agreement of the opinions (100% of the circle) between BT and ET. For an opinion difference of magnitude let's say γ, the formula that gives the degrees is:

$$x^0 = 360\frac{\gamma}{100}$$

As the products' difference gets larger (difference in the opinions between ET and BT, in absolute value), the area of the epicentre angle apparently gets smaller.

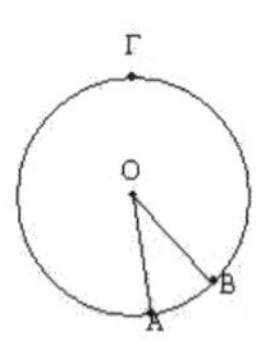

Therefore now upon the matrix, the relevant position is illustrated as a point defined by the perceived/identified assigned significance of influence (of all variables labeled as Opportunities–Strengths and Threats–Weaknesses) and the level of agreement in the opinions of CT and BT and of BT and ET, upon the aforementioned significance of the variables' influence.

It should be stressed that the advantage of the aforementioned depiction upon the matrix is considered to be the easy perception of the produced factors within described. Hence, the relevant position as a function of opportunities–strengths and weaknesses–threats, is complemented also by the description of agreement of opinions for the crucial variables between teams, and all these are demonstrated easily, briefly, and minimally. Therefore the main advantage is the comprehensiveness of all those three attractiveness criteria for NPHT development, from all participators independently their knowledge background, simultaneously.

Furthermore, the whole procedure of rating and assigning relevant importance, promotes the integration of different departments in the same way that SWOT analysis also contributes positively to that direction.

Anyway, also the use of the opinion difference as a criterion of attractiveness for the NPHT under development, i.e. as a means for measuring the integration of departments (at least upon their expressed opinions, and the way – of thinking – of argumentation used for assigning weighted importance and performance rate on the crucial—and commonly identified as Opportunities–Threats–Strengths–Weaknesses—variables) signifies the key role of integration regarding NPHT development, which also appears as a result of activities stemming from inside the enterprise.

Another advantage is that if the aforementioned procedures are repeated for some other NPHT being discussed, then the one's attractiveness (with relation to the criteria in use) will be directly and easily comparable to the other.

The table used in this stage, as well as the matrix with the progressive steps for completion are found in graphs 1,2,3,4 that follow. The numbering of the vertical and the horizontal axis of the matrix depends on the limits used for weighted rating and on the scale of performance assignment.

A measurement of the difference is performed between the weighted (with α, β) products of weighted importance upon the performance rating of CT with BT (where BT = BT – CT). Thus, the comparison of opinions between CT and BT is repeated, with the difference that now CT is weighted with α and BT with β. The factors α, β, γ attribute the relevant importance of opinion of each team. The same procedure is also repeated for finding the weighted product difference of the weighted importance times the assignment of BT and the corresponding magnitude of ET. The differences here are also in absolute values. We seek to find the weighted (with the special weight of each department) distance of opinions between the CT and BT and CT + BT and ET. The weight factors α, β, γ (that can only belong to the range 0-1 and can sum up to 1) represent the relevant significance of the opinion of each team, depending on the way Top Management (which ultimately has the responsibility of the final result) judges the manner teams contribute to the development process. In this way, the potential approach of Moenaert and Souder, who argue that it is the most effective way for the organisation of technological innovations (the potential approach takes into consideration the level of particularity of development teams) but also the "manipulation" of the organisational climate (according to Agarwala and Rogers it is the most effective mechanism for achieving harmony, trust, extroversion and finally integration) is left to the Top Management since this is considered appropriate by Rogers and Agarwala .

Management Information Systems, C. A. Brebbia (Editor)
© 2004 WIT Press, www.witpress.com, ISBN 1-85312-728-0

Graph 1

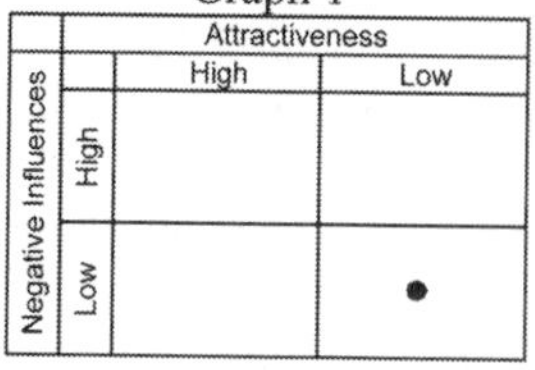

		Attractiveness	
		High	Low
Negative Influences	High		
	Low		•

Graph 2

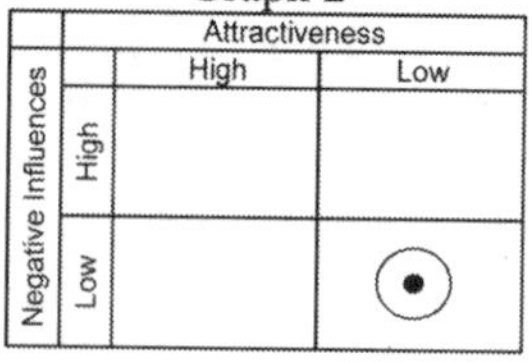

		Attractiveness	
		High	Low
Negative Influences	High		
	Low		⊙

Graph 3

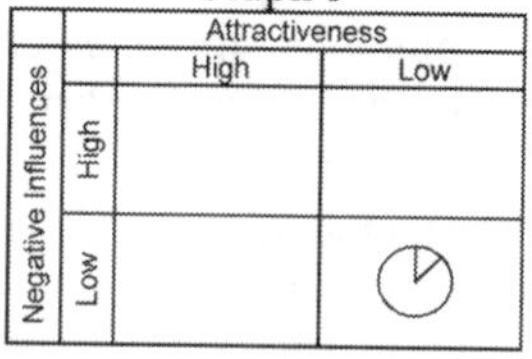

		Attractiveness	
		High	Low
Negative Influences	High		
	Low		

Graph 4

	Variables	C.T. Weight A_1	C.T. Relevant Signif B_1	C.T. $A_1 X B_1$ Γ_1	Weighting(α)	B.T. Weight A_2	B.T. Relevant Signif B_2	B.T. $A_2 X B_2$ Γ_2	Weighting (β)	E.T. Weight A_3	E.T. Relevant Signif B_3	E.T. $A_3 X B_3$ Γ_3	Weighting (γ)	Total sum (weights): Sum of vars	Total sum (weights): Total sum of vars
Horizontal axis	Opportunity														
	Streangth														
														Sum	
Vertical axis	Threat														
	Weakness														
														Sum	

The differences discussed above will be illustrated in Bar Diagrams, in order to have the condition (and the ratio) of linearity in the picture.

In this way, we have the knowledge of opinion difference (integration) also from the dimension of relevant importance of the opinion expressed by each team.

Also, it should be pointed out the fact that the comparison of the total of weighted with α, β, γ respectively of CT and BT with ET is advisable, in order to reveal the discordance of opinions between the functional departments and top management mainly, hence to also indicate indirectly the degree of its expected support.

3 Evaluation

At the end of the 3-stages procedure we have the following data/facts:
The total of the variables perceived composing the SWOT.

The degree of importance for each variable from each team, and the corresponding attributed performance indicator.

The total of the perceived influence (product of the weighted importance of each variable times the size of its performance), for each group of variables from all the teams (CT, BT, ET).

The perceived (by the members of each team) influence of each variable, either it has been grouped as an Opportunity, or Strength, or Weakness, or Threat, by the members of CT, BT and ET.

The total of the weighted influence of each variable from all the teams.

The total of the weighted influence, for each group of variables from CT, BT and ET.

The degree of attractiveness, i.e. the total of the weighted sum (with the α, β, γ weight factors, which give the importance of the opinion of CT, BT and ET, respectively) of the weighted (with weight factors the relevant importance of each factor, as perceived by the group members) influences of each variable that belongs to the Opportunities – Strengths by the total of the members that compose the teams.

The degree of negative influences, i.e. the total of the weighted sum (with weight factor the relevant importance of each factor, as perceived by the team members) of influences of each variable that belongs to Threats – Weaknesses.

The combination of the attractiveness described with the negative influences described, as a point upon a matrix (graphically).

The opinion difference about the influence of variables between the CT and BT, which can be graphically depicted as a circle area having as its point the combination of negativity and attractiveness (BT = BT – CT).

The opinion difference about the influence of variables between the BT and ET, which can be graphically depicted as an epicentre angle of the circle defined in 10.

The weighted (with α, β) opinion difference about the influences of variables between the CT and BT (where BT = BT – CT), which can be linearly depicted in a Bar – Diagram.

The weighted (with α, β, γ) opinion difference about the influence of the variables between the CT + BT and ET, which can be depicted in a Bar – Diagram.

It is important to stress that the aforementioned data derived from a simple procedure, which totally included the opinions of all executives who are (directly or less directly) involved in the decision-making process for the new product, and which substantially through its mechanisms directly contributed to integration, but also indirectly through its well-documented way of participation as a factor for the attractiveness of NPHT.

The evaluation of the deduced factors for the NPHT, takes place in two stages.

The first stage is the evaluation of the deduced conclusions for the NPHT, as these are depicted either proportionally in the matrix that includes the circle and the sector of the epicentre angle, or linearly and proportionally in Bar-Diagrams.

The second stage is that of the comparison of the NPHT (regarding its potential for success from the viewpoint of the proposed framework) with an alternative (or more) NPHT (of course after it undergoes the same procedure).

The first stage gives the measure that is advisable for the NPHT to pass at the first stage of its entrance, having as common denominator the significant role of integration (in the theoretical model) and the systemic benefits of the symbiotic interrelation between the departments it describes. From this viewpoint, both the framework in which integration of departments is deployed and the method for its measurement now become clear.

Management Information Systems, C. A. Brebbia (Editor)
© 2004 WIT Press, www.witpress.com, ISBN 1-85312-728-0

Through their comparison, the fields where the one is better than the other will become clear. In that way, the selection process is simplified, based on the adopted criteria of the framework, and the dominant—in many aspects—option is demonstrated.

4 Contribution of the model

After developing special software, through which the proposed processes and calculations are automated, the model was tested in Greek enterprises of High Technology, (see Fotiadis Thomas, doctoral dissertation, University of Macedonia, 2004 [1]).

The proposed theoretical framework contributes to the integration of the functional departments of Marketing and Research & Development. The integration, in the concept adopted in the scope of the work herein, is the systemic approach of the competencies of the functional departments of high tech enterprises. This functional interface and the common utilisation of the specialised functional competencies, offers to high tech enterprises benefits that exceed the sum of benefits of the individual efforts of the departments of high tech enterprises.

The theoretical framework, through the mechanism of establishment of teams, adds to the interface of the separate specialised functional competences which is the most referred acknowledgment-command for the smooth functioning – effective entrance of NPHT in the literature of journals, e.g. and the required condition for the success of the innovation in order to create the appropriate conditions for the production of additional benefits. It gives particular importance to the combination of competences of the Marketing and Research & Development departments, however without leaving with no responsibilities—and mainly contribution—the other functional departments (the interdependence relationship of MKT, among which also the importance of the net of their interrelations, although it has attracted little attention, it has enormous importance) and other top management executives (whose lack of support comprised 42% of the answers in a research by Ashok, Gupta, Willemon [7] for the reasons of entrance delay of new high tech products) of the enterprise. It does not abolish the functional cores of specialisation. It considers development of specialisation necessary (as it is repeatedly stressed also in the literature, departments should have the ability to maintain their individual identity, while simultaneously they should be encouraged to work, abolishing the conceivable limits of specialisation, in order to move effectively towards the achievement of the common aim). Non-alignment and endistancement from the direction of the organisational objective, act reductively in achieving it effectively, as stressed by the literature.

Therefore, apart from the coordination of the functional competences of operational departments, the required degree of integration is both supported and aligned with any requirements, under the direction provided by the integration of departments. In that way, the avoidance of consuming valuable resources (financial, executives' efforts) is ensured, and the probability to have cost from pointless activities (entrance of an undesirable product, wrong time of its

Management Information Systems, C. A. Brebbia (Editor)
© 2004 WIT Press, www.witpress.com, ISBN 1-85312-728-0

entrance, investment in technology strange to the requirements of consumers, financial failure, etc) is minimised.

The shortness of time and of resources comes as a result of coordination and of the more efficient flow of information, in a systematic way. It should be stressed that both shortness of time (that is equivalent to the fastest possible entrance of the product) and efficiency of the distribution of resources (that is equivalent to respective decrease in the uncertainty of resources), have been highlighted many times for being crucial characteristics for the successful outcome of the entrance of new products of high technology. Especially regarding the efficient transfer- flow- distribution of information in a systematic way, i.e. through the establishment of teams, through the way of coordination, through the recording of variables on the matrix and the completion of their evaluation, and through the evaluation/feedback, it is observed that both the SWOT analysis and the organisation of the quantitative parameters of relevant importance and performance, apart from providing a supporting framework for integration, are channels of efficient transfer of information. With SWOT analysis, both the qualitative and the quantitative supply of information increase. The volume of the received flow of information increases due to the fact that more information becomes known to more people (members of CT, ET and BT) simultaneously, and due to the fact that the dialectics and the exchange of opinions during the time of recording information brings out further information, revealing factors (or sources of information) that up until now have not been realised/ defined/ mapped out in detail and due to the fact that their systematic registration "in the egg" rules out overlooking them.

The qualitative dimension is improved; because their accuracy and compatibility is crosschecked by the total of the members, and there is a validation of the usability of information regarding the objective that is called on to satisfy. In that way is accomplished a deep and wide enhancement of the "tools" to be used for decision-making, but also the reduction of resultants that compose uncertainty; hence the innovative effort has higher probabilities of success .

Reduction of variability is achieved, since this term refers to the number of special cases the high tech enterprise is challenged with, since SWOT records the components of special cases, hence in the course of time, there is a continuous decrease in their potential number (of the unknown special cases).

In the same attitude, analysability also increases due to the fact that in the course of time there is an increase in the degree in which the procedures that determine the sequence of the steps to be done are known, but also the procedure itself improves due to the fact that the factors that compose the processes are identified and change in parallel with the environment of the enterprise.

And if we accept Perrow's opinion that from the reduction of uncertainty viewpoint the main challenge is the maximum identification of the relevant potential uncertainty (decrease in variability) and the identification of tools for uncertainty reduction (increase in analysability) then the role of SWOT analysis becomes clear.

Management Information Systems, C. A. Brebbia (Editor)

Souder and Chakrabarti observed that through the reduction of uncertainty we get to improve decision-making and a better implementation of innovative programmes.

In addition, the thought mechanism for identifying a factor as a crucial variable appears. The different knowledge background (in which a large part of the different conceivable structure of the way of thinking is attributed) becomes to some extent a common domain, or at least its determinants become clear. Through the direct comparison of ideas (through which anyway the composition derives and therefore the appropriate conditions for integration are created), the appropriate grounds are created for the more effective and more efficient evaluation and utilisation of data.

The transfer takes place on a systematic basis—not in periods of crises, and acts increasingly in integration and creates in a dynamic way continuously improving conditions for the establishment of the advantages of integration.

The participation of the teams assigned in the NPHT development process, leaves little margins for misunderstandings or/and one-sided misjudged ascription of responsibilities. Therefore, it strengthens the climate of cooperativeness also in an indirect way, since it eliminates or does not allow creating conditions which would strengthen factors that alienate departments, such as negative stereotypes, mitigation of the way of thinking-perceiving, attribute of high-low importance, etc. Besides, it contributes to achieving the principle of Millman [15], who considers that it is a common responsibility of the MKT and R&D departments to safeguard the maintenance of their good relations, and the efforts they put to ensure that any conflict is solved before it turns into a permanent characteristic of the way the enterprise works. Apart from that, it also contributes positively to the reduction of the recorded-by-the-bibliography resultants of disharmony between MKT and R&D. The composite procedure for decision-making is also related to the degree of participation in decision-making and –forming and the hierarchy of power. Apart from the prevention of the after-cast consequences of power concentration (decision-making done only by higher units), through the participation of members of the teams described before, are also prevented the reductions of credibility and functionalism of the results of MKT research, which was reported from the empirical research of Deshpande and John and Martin, but also the problems reported by Wind to occur in centrally-oriented structures.

The framework contributes to the improvement of the learning experience curve of the enterprise. Crucial factors – assumptions and variables as well as their originations are recorded systematically. In that way, in the case of a potential recurrence of the procedure upon the decision for the entrance of some other new high tech product, apart from the fact that the background of knowledge and assumptions already exists, this decision is both approachable and accessible. The decision-making mechanism improves, as it is a direct result of the enrichment of the learning experience curve which is re-supplied by the framework, mobilising in this manner a two-way relation of continuous improvement of the one factor (knowledge–experience) through the other (decision-making). Thus, it becomes clear that the cost of possible pointless

Management Information Systems, C. A. Brebbia (Editor)
© 2004 WIT Press, www.witpress.com, ISBN 1-85312-728-0

activities as a result of wrong assumptions—wrong identification of variables, misjudgement of their importance, etc.—is minimised, in view of the fact that even the error contributes positively to the enrichment of the procedure which, as already mentioned, is dynamically self-improving through the addition of further knowledge and experience to its mechanism (hence addition also in learning and minimisation of the possibility of the same or similar error occurring again).

The strategic orientation of the model is indicated from the use of the extended SWOT analysis, which also assists to: a) Setting new objectives—priorities for new products, b) giving birth to new ideas for new products, c) supplying information about competitors, d) supplying information about consumers.

It also contributes, apart from the rational decision making process, to the rational management of the product portfolio of high tech enterprises. The final stage of the procedure, if done for more than one product, is in position to provide answers regarding the different degree of attractiveness of each product. It is important to stress that the proposed theoretical framework encompasses also the criterion of the achieved degree of integration, and adds it to its mechanism as an innate variable of attractiveness.

The framework contributes to the reduction of factors of uncertainty. The uncertainty of market, technology, competition, and consumers, decrease as a result of systematic extraction and accumulation of information, knowledge and experiences through the methodological tools that were developed and which helped in obtaining a more efficient flow and administration of information—but also a qualitative and quantitative increase.

Through the reduction of the abovementioned factors of uncertainty, occurs the reduction of the deriving from the above uncertainties, uncertainty of resources. The effective reduction of time comes as a consequential reaction to the efficiency and effectiveness of the mechanism of processes and their way of functioning.

Finally, it should be stressed that the theoretical framework satisfies the condition of simplicity, indeed with the facilitation of its application through special software that simplifies the process of entering variables and their relevant weightings and the brief and graphical output of results.

References

[1] Th. Fotiadis, Doctoral Dissertation, University of Macedonia, Macedonia Greece 2004.

[2] Edward B. Roberts, "Managing Invention and Innovation", Research Technology Management, 31/1 (January/February): 27.

[3] Takeuchi and Nonaka, The New Product Development Game", Harvard Business Review (January/February).

[4] "Speeding New Ideas to Market", Fortune, March 2, 1987.

[5] Bela Gold, "Approaches to Accelerating Product and Process Development", Journal of Product Innovation Management, 412 (June 1987): 81-88.

[6] "How managers can Succeed Through Speed", Fortune, Feb 13, 1989.

[7] Ashok, Gupta, and David Willemon (1990), "Accelerating the Development of Technology Based New Products, California Management Review (Winter) 1990.
[8] "Manufacturers Strive to Slice Time Needed to Develop Products", Wall Street Journal, February 22, 1988, p. 1.
[9] Robert G. Cooper, and Elko J. Kleinschmidt, "An investigation into the New Product Process: Steps, Deficiencies, and Impact", Journal of Product Innovation Management, 312 (June 1986): 84.
[10] Robert T. Hise, Larry O'Neal, James U. McNeal and A. Parasuraman, "The Effect of Product Design Activities on Commercial Success Levels of New Industrial Products", Journal of Product Innovation Management, 6/1 (March 1989): 43 – 50.
[11] Joseph L. Bower and Thomas M. Hout, "Fast Cycle Capability for Competitive Power", Harvard Business Review (November / December 1988), pp 110-118.
[12] Robert G. Cooper, "A process Model for Industrial New Product Development", IEEE Transactions on Engineering Management, 30/1 February 1983),: 2-11.
[13] Souder, W. E. and Chakrabarti, A. K. (1978), ''The R+D /marketing interface: results from an empirical study of innovation projects'', IEEE Transaction on Engineering Management, E. M. – 25, 4, 88-93.
[14] Lawrence, P. R. and Lorsch, J. W. (1967), ''Differentiation and Integration in complex organizations'', Administrative Science Quarterly, 12, June, 1-47.
[15] A. F. Millman, "Understanding Barriers to Product innovation at the R+D/ Marketing Interface", European Journal of Marketing, 16, 5.
[16] Advisory Council for Applied Research and Development (ACARD), Publication, Industrial Innovation, HMSO, December 1978.
[17] Cooper, R. G. (1979), ''The dimensions of Industrial new product success and failure'', Journal of Marketing, 43: 93-103.
[18] Maidique, M. A. and Zirger, B. J. (1984), "A study of success and Failure in product innovation: the case of the U. S. electronics industry". IEEE Transactions on Engineering Management EM – 31 (4): 192-203.
[19] Rothwell, R. Freeman, C., Horlsey, A., Jervis, V. T. P., Robertson, A. B. and Townsend, (1974) J. SAPPHO updated – project SAPPHO phase II. Research Policy 3: 258-291.
[20] Souder, W. E. (1987), Managing new product innovations. Lexington, MA: Lexington Books.
[21] Souder, W. E. and Chahrabarti, A. K. (1979), Industrial Innovations: a demographical analysis. IEEE Transactions on Engineering Management – 26 (4): 101-109.
[22] Souder, W. E. and Chakrabarti, A. K. (1980), "Managing the coordination of Marketing and R+D in the innovation process", in Dean and Goldhar (Eds), TIMS Studies in the Management Sciences, Vol. 15, Amsterdam: North – Holland, 135- 150.

Management Information Systems, C. A. Brebbia (Editor)

UML for data warehouse dimensional modeling

Y. Mai[1], J. Li[1] & H. L. Viktor[2]
[1] Concordia University, Canada
[2] University of Ottawa, Canada

Abstract

Dimensional modeling is a common modeling technique in data warehousing. It reflects a simple logical view of a data warehouse system. It can be easily mapped to a physical design. Traditional dimensional modeling is data-oriented and semantically informal. From a software engineering perspective, the informal notations and *data-oriented* feature are insufficient to tackle the complexity of large data warehouse projects. UML, with its well-defined semantics, is now a standard modeling language that is used to model the entire life cycle of a software system. UML has rich and extensible semantics. The combination of the knowledge in standard object-oriented modeling and dimensional modeling add variable semantics to dimensional modeling without losing its understandability. This paper proposes a metamodel for data warehouse dimensional modeling using UML. Based on this metamodel, we illustrate how to model the business process and data marts of a large mobile telephone company.
Keywords: data warehouse, dimensional modeling, UML, metamodel.

1 Introduction and related work

In recent years, data warehouses have gained increasing popularity and are becoming a business growth strategy. A data warehouse is essentially a data container, which contains complete and historical business data from numerous operational sources. The data, as contained in a data warehouse, are used to analyze business, help predicting the organizational growth and improve customer relationships. In essence, a data warehouse is a queryable data source that exists to answer questions people have about the organization. These queries thus reflect the way that managers think about their organization and assist them to make sense of the data, form policies and to make informed decisions.

Management Information Systems, C. A. Brebbia (Editor)

From the technical perspective, data warehouse modeling is critically important in a data warehouse system design. One key factor in ensuring the success of a data warehouse is to create the right model that reflects the business needs and covers all business requirements. Dimensional modeling [1, 2], popularly known as the Star Schema approach, is a widely used technique for modeling the logical aspects of a data warehouse system according to business views. The main idea is to logically design a data warehouse as a set of incrementally designed data marts, each representing a view of a business process. A single dimensional model consists of fact tables and dimension tables. The data warehouse bus architecture [1, 2] then glues all data marts into a logical data warehouse. This is accomplished through the use of conformed facts and conformed dimensions, which ensure that the grain of the various data marts are compatible.

Recently, there are some attempts to model a data warehouse with Unified Modeling Language (UML) [3, 4, 5, 6, 7, 8]. UML is an OMG standard modeling language that has been widely used in object-oriented system modeling. It defines a common vocabulary for communications among designers and users. There also have been many discussions on modeling database systems using UML [9, 10, 11], focusing more specific on databases (especially for relational databases) rather than data warehouses. OMG defines a Common Warehouse Metamodel [12] that can be used to guide common warehouse modeling. It contains limited metamodel definitions for data warehouse design. But it does not contain a metamodel for multidimensional modeling. UML notations are rich in semantics but they may be complex for database modelers. Traditional data warehouse modeling techniques are straightforward for database teams and reflect the ways that people think about a database system. But since the traditional notations have little semantics and are only data-oriented, they are insufficient to tackle a complex data warehouse modeling. Thus, in this paper, we provide a combination of both, that is, using certain UML notations to represent dimensional modeling. The paper differs from previous work in 3 major areas:

- We propose a UML metamodel for data warehouse dimensional modeling. Although there are many object-oriented dimensional data models, we haven't seen any UML metamodels for data warehouse dimensional modeling so far.
- We believe UML is useful in modeling the entire software lift cycle of the data warehouse. One section in the paper shows the business process modeling.
- We define UML notations to model data marts, data warehouse bus matrix, table grains, foreign keys, and table relationships.

The paper is organized as follows. Section 2 defines a metamodel for UML dimensional modeling. This metamodel defines semantics for the basic data warehouse components including fact tables, dimension tables and their attributes. Based on this metamodel, Section 3 illustrates how to model a data warehouse involving a large mobile telephone company using UML. Section 4 concludes the paper and highlights some future research directions.

Management Information Systems, C. A. Brebbia (Editor)
© 2004 WIT Press, www.witpress.com, ISBN 1-85312-728-0

2 A UML metamodel for dimensional modeling

A metamodel is a model that describes the syntax and semantics of a modeling [13]. It contains descriptions that define the structure and semantics of a model. Metamodels [12, 13, 14, 15] have been widely used to define models. The metamodel in this paper is defined under the UML definitions and is derived from the concepts of adaptive object model [14, 15]. We follow the structure and semantics of data warehouse dimensional modeling as described in [1, 2].

Figure 1 shows the data warehouse core metamodel. `Table` defines the basic construct in the data warehouse. The `Table` defined in the meta-level contains `Attribute` that describes the properties of the `Table`. The metamodel for `Table` and `Attribute` is defined in the meta-level layer. Any data warehouse table is an instance of `Table`. For example, tables such as Product, Customer, and Date are all instances of `Table`. A `Table` instance consists of a set of concrete properties such as name and location. Each property has a type, such as the `ConcreteAttribute` shown in the figure. The `ConcreteAttribute` is an instance of `Attribute`. The relationship between a concrete table and its attributes is shown in the base-level layer.

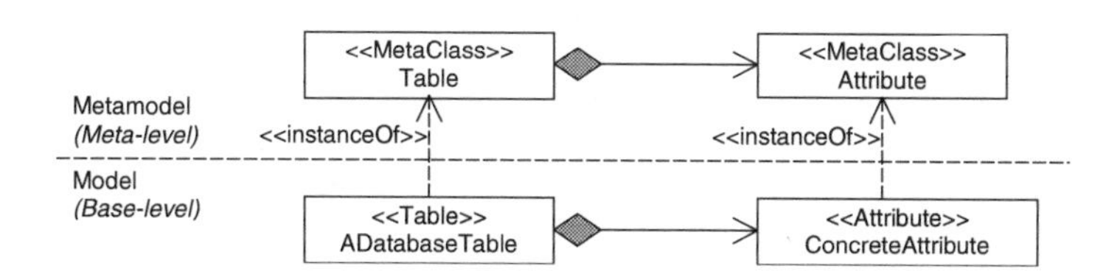

Figure 1: Data warehouse core metamodel.

`Fact` tables and `Dimension` tables are two basic kinds of tables in the data warehouse. `Aggregate Fact` tables or `Factless Fact` tables are two special cases of `Fact` table [2, 3, 16]. They differ from a normal fact table in that an aggregate table contains derived facts (or measures) for performance improvement purpose, while a factless fact table is a kind of fact table without facts. The relationships among these tables can be organized into a hierarchy tree as shown in Figure 2. Both `Fact` table and `Dimension` table inherit from the `Table` metaclass. `Aggregate Fact` table and `Factless Fact` table are subclasses of `Fact` table. UML stereotypes can be used to describe table types. We use ≪Fact≫, ≪Dimension≫, ≪Aggregate Fact≫, and ≪Factless≫ to represent tables Fact, Dimension, Aggregate Fact, and Factless Fact, respectively.

Each table consists of attributes. An attribute has three members: a string representation name, a data type, and a boolean indicating whether this attribute represents a grain or not. More details about the data type can be found in the UML metamodel [17]. A grain is the level of granularity (detail) of all properties in a table. It is crucial that every row in a fact table be recorded at exactly the same

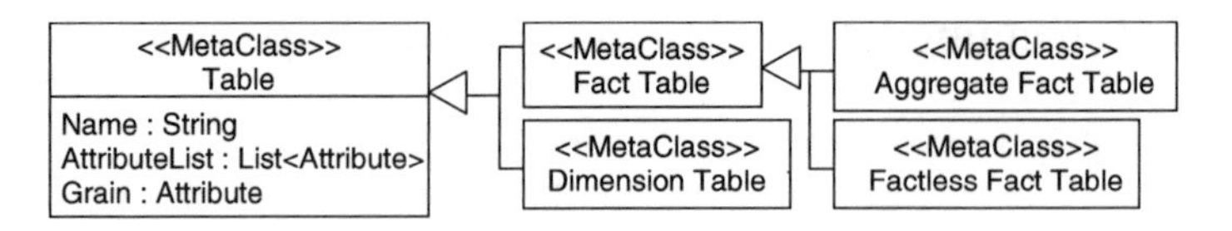

Figure 2: Metamodel hierarchy of the data warehouse tables.

level of detail at which measures will be recorded. Figure 3 shows the meta-level description of the attributes in the dimensional data warehouse.

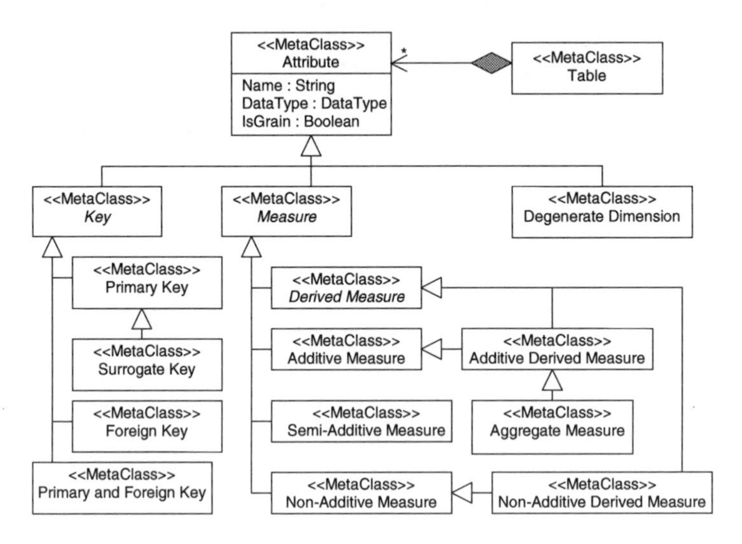

Figure 3: UML meta-level description of data warehouse attributes.

In addition to normal attributes in a database table, there are three special kinds of attributes: (1) An attribute can be a key: primary key, surrogate key, foreign key, or a combination of primary and foreign key. The left-hand side inheritance tree in Figure 3 shows the attribute key metaclasses. (2) An attribute in a fact table can be a measure: additive, semi-additive, non-additive, both derived and additive, aggregate, or both derived and non-additive. (3) In a fact table, an attribute can be a degenerated dimension. A degenerated dimension is a special attribute that is lack of properties and can be combined with a fact table. Figure 3 shows the entire attribute hierarchy tree. We use stereotypes (enclosed in "≪≫") to represent items (1) and (3); Constraints (enclosed in "{}") to represent item (2) except derived measure which is adorned with "/".

This section presented the UML metamodel for data warehouse dimensional modeling. The metamodel describes the object-oriented model that we propose for dimensional modeling. In the next section, we will illustrate how to model the data warehouse based on this metamodel.

Management Information Systems, C. A. Brebbia (Editor)
© 2004 WIT Press, www.witpress.com, ISBN 1-85312-728-0

3 Data warehouse modeling with UML

Data warehouse is designed to answer questions being asked throughout the business on an every day basis. These questions do not focus on individual transactions but on the overall process and are mainly used to determine trends and bottlenecks in the organization. To answer these questions, the design of the data warehouse should directly reflect the way that managers perceive their business [16]. That is, it should capture the measurements of importance to the business, and the parameters by which these measurements are viewed. We will first discuss the business process modeling, and then the data mart modeling.

Throughout this section, we will use a mobile telephone company data warehouse as a running example. The purpose of this data warehouse is to provide a repository of data regarding the customers and the telecommunications network, including diverse aspects such as call duration, peak call times, sales of contracts, sales of mobile phones and customer profiles, amongst others.

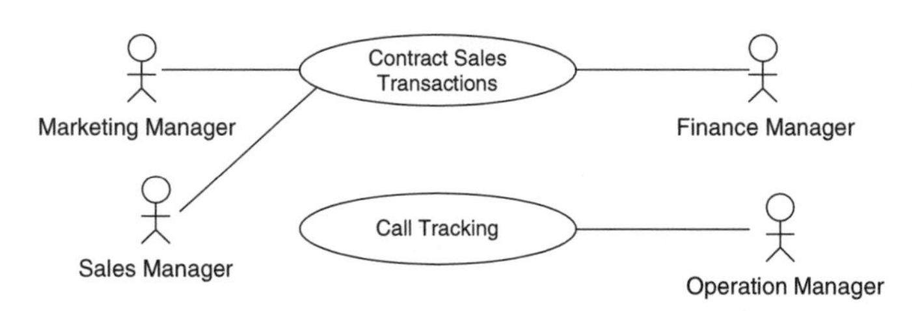

Figure 4: Modeling business process functionality.

3.1 Business process modeling

Business process modeling contains 3 steps:

- To capture the stakeholders. In UML, stakeholders are modeled as actors, i.e. roles that are outside the system but that closely interact with the system [18]. Actors are not part of the system. In a data warehouse system, they can be organization executives, different levels of managers, data warehouse administrators, etc. They can also be data sources including operational or transactional systems.
- To capture business requirements. A business process can be represented as a use case in UML. Each business process can have a number of actors interacting with it. The use case (business process) provides necessary functionality for the actors to fulfill some specific tasks. All use cases put together comprise the entire functionality that the data warehouse provides. We use UML Use Case diagrams to model the business processes (and requirements) in a data warehouse system. Figure 4 shows an example. On the other hand, a business process can be *realized* by a data mart, a collection of data

Management Information Systems, C. A. Brebbia (Editor)

used by this single key business process. we model data marts with UML Collaborations and apply Realization relationship to modeling the relationship between the data mart and the business process. Figure 5 shows that `Customer Call Activity` data mart, which is used to model the business process use case `Customer Call Activity`.

- To model data warehouse bus matrix. Table 1 shows a simplified data warehouse bus matrix for the mobile phone company, created for the use case diagram shown in Figure 4. The first column in the table represents the data marts, and the others represent dimension tables. A cross represents that a dimension participates in a data mart. A corresponding UML representation is showed in Figure 6.

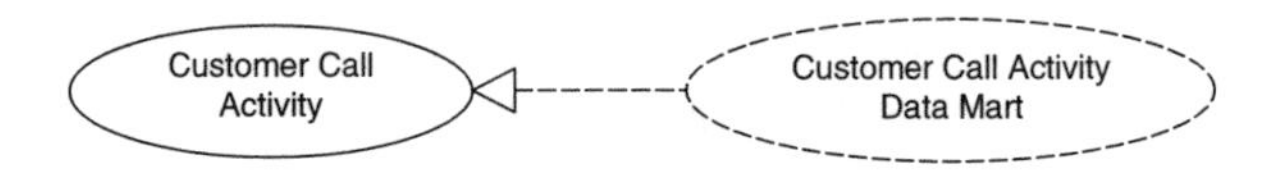

Figure 5: Modeling relationship between the use case and the data mart.

Table 1: Data warehouse bus matrix.

	Date	Customer	Product	Sales Rep	Store	Promotion	Transaction
Contract Sales	X	X	X	X	X	X	X
Call tracking	X	X	X				
... ...							

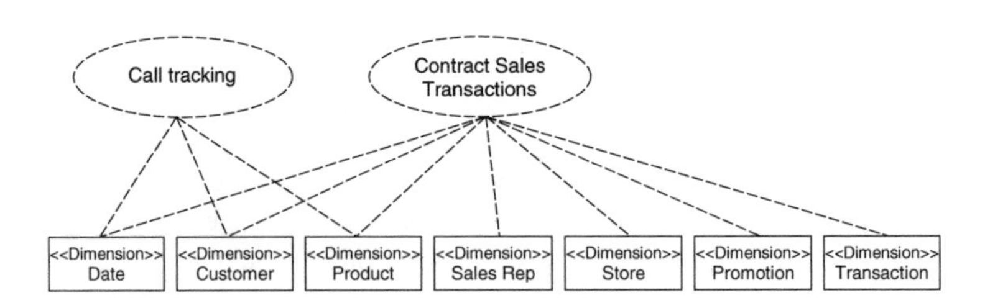

Figure 6: Modeling data warehouse bus matrix.

3.2 Modeling the data mart

A data mart reflects a process view of the data warehouse. It implements a data warehouse use case. A data mart contains a set of tables, for example, a fact table,

Management Information Systems, C. A. Brebbia (Editor)

aggregate tables (if any), and dimension tables, that are organized together into a star schema to represent a specific business process. Since the modeling of the fact and dimension tables in the data mart are straightforward based on the metamodel definition, we only show an example of basic modeling of the data mart and then discuss a few special cases in the data mart modeling.

Figure 7 shows an example of modeling of the `Contract Sales` data mart. The fact table is modeled as a composite class in the diagram. It has shared-aggregation relationships with the corresponding dimension classes. The role of a dimension class represents a reference from the fact class to a dimension class. The role has a name, which is the name of the surrogate key of the dimension class (or the foreign key name in the fact table for that dimension table). Fact class `Sales` links to `Time` dimension class which plays the role of "*time_key*" in the fact table. The "*time_key*" is also the surrogate key name in the `Time` dimension. The grains of the dimension and fact tables are shown in separate compartments of the class icons. The cardinality for the relationships between the fact class and the dimension classes are marked besides the table links. "1" represents exact one. "*" represents zero or more.

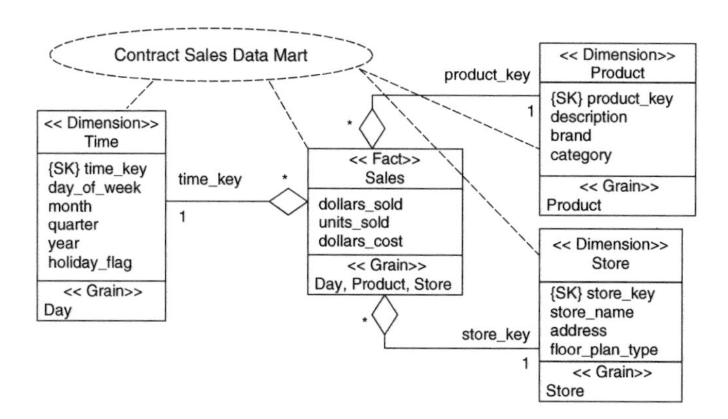

Figure 7: Modeling contract sales data mart.

We cover 3 special cases of data mart modeling:

- Model multiple foreign key relationships. We represent each foreign key as a shared-aggregation relationship. Figure 8 shows the shipment fact class has two foreign keys, *shipment_date_key* and *order_date_key*, each representing as a shared-aggregation relationship linking to the `Date` dimension.
- Model many-to-many relationship between tables. Figure 9 shows an example of using a bridge table `Account-Customer` (adapted from [2]). This table contains a composite primary key with one *account_key* refers to the account table and the other *customer_key* refers to the customer table. In such way, the `Monthly Account Balance` table refers to the `Account-Customer` bridge table through the *account_key* and the *customer_key* in

Management Information Systems, C. A. Brebbia (Editor)

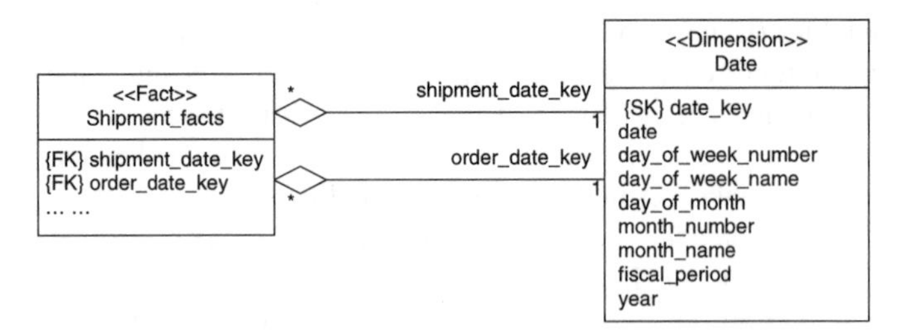

Figure 8: Modeling multiple foreign key relationships.

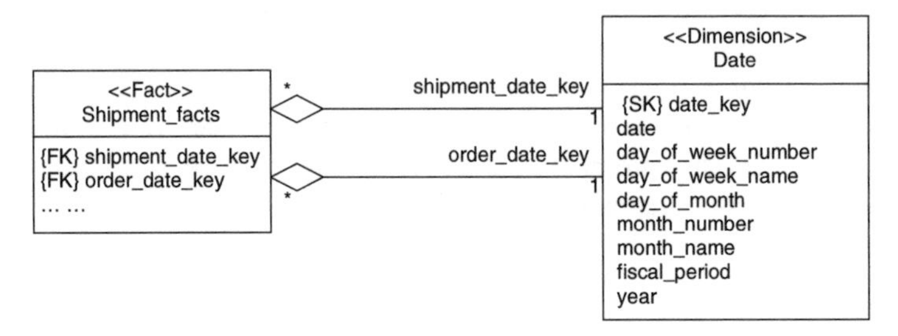

Figure 9: Modeling many-to-many relationship.

the bridge table refers to one customer in the `Customer` table. Since *account_key* in the bridge table does not uniquely identify a record (it is part of the primary key), each record in the fact table can refer more than one customer in the `Customer` dimension.

- Model aggregate fact table. A common convention in data warehouse modeling is to pre-store aggregate data in the data warehouse (for performance improvement). Figure 10 gives an example of using the aggregate fact table. First, a new dimension called `Category` is created as a shrunken table of dimension `Product`. It only contains the product key (surrogate key), category, and department. The surrogate key here is only a subset of the product keys in dimension `Product` with each representing a category product. We subsequently create an aggregate class called `Sales_by_Category`. In this class, each row contains data such as time, store, product category, as well as some measures such as dollars sold, quantity sold, and dollars cost along the time, store and product category dimensions. The directed dashed lines represent dependency relationships. This is due to the fact that `Category` is a subset of `Product`, and `Sales_by_Category` is calculated from Sales.

4 Conclusion

Modeling a data warehouse is a complex task due to a number of complex factors, including the huge size thereof, the constraints imposed by operational data

Management Information Systems, C. A. Brebbia (Editor)

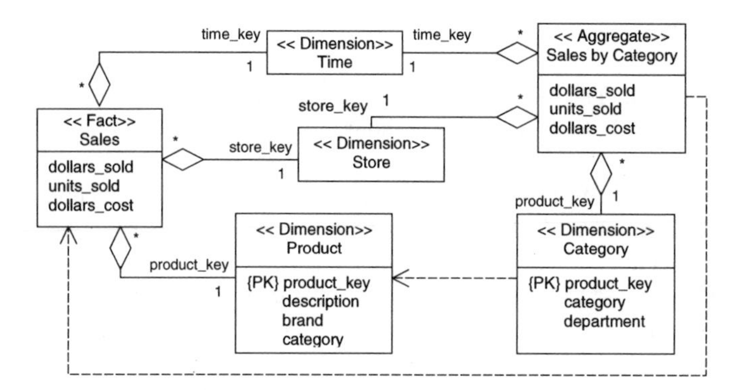

Figure 10: Using the aggregate table.

sources and the complex business requirements as obtained from numerous users, amongst others. Dimensional modeling separates business concerns and addresses one business need at a time. The purpose of dimensional modeling is to bridge the communication gap between domain users and data warehouse developers. However, a traditional dimensional model is a data-driven model. It contains little semantics for notations and it is therefore difficult to embed the inherent meaning of the data and data relationships therein. Data warehouse developers therefore have difficulty in understanding the subtle meanings of conformed dimensions and facts, which may lead to poorly designed data warehouse systems.

This paper proposes an approach to represent the dimensional model in a semantically rich manner. As a standard modeling language, UML provides rich and extensible semantics to a model. With the UML as the foundation, this paper proposes a metamodel for dimensional modeling, in which we define a vocabulary to model basic data warehouse concepts using UML notations. The metamodel extends the UML metamodel. It defines the object-oriented dimensional model that we propose. Based on this metamodel, we model the business processes and requirements, and the dimensional model with data warehouse bus matrix and data marts. The modeling diagrams are easy to understand without losing explicit semantics.

This paper shows how dimensional modeling can be extended using object-oriented modeling techniques, thus providing a foundation for modeling the logical core of a data warehouse system. Future work will focus on extending the model to capture the entire business life cycle of a data warehouse, including its behavioral modeling and physical modeling. In addition, the extension of UML diagrams for modeling misuse of the data warehouse will be further investigated. That is, the investigation of approaches to design our system is such a way to anticipate and model the illegal use of the data warehouse is an exciting new research direction, especially within the domain of so-called data webhouses.

Management Information Systems, C. A. Brebbia (Editor)
© 2004 WIT Press, www.witpress.com, ISBN 1-85312-728-0

References

[1] Kimball, R., The Data Warehouse Toolkit. John Wiley & Sons Inc, 1998.

[2] Kimball et al., R., The Data Warehouse Lifecycle Toolkit. John Wiley & Sons Inc, 1998.

[3] Trujillo et al., J., Designing Data Warehouses with OO Conceptual Models. IEEE Computer, V 34, No 12, pp. 66–75, 2001.

[4] Abello et al., A., Understanding Analysis Dimensions in a Multidimensional Object-oriented Model. Proceedings of the international workshop on Design and Management of Data warehousing (DMDW' 2001), Interlaken, Switzerland, pp. 16–19, 2001.

[5] Abello et al., A., Understanding Facts in a Multidimensional Object-oriented Model. Proceedings of the fourth ACM international workshop on Data warehousing and OLAP, 2001.

[6] Bruckner et al., R.M., Developing Requirements for Data Warehouse Systems with Use Cases. The Seventh Americas Conference on Information Systems, 2001.

[7] Lujan-Mora et al., S., Extending the UML for Multidimensional Modeling. Proceedings of UML 2002 - The Unified Modeling Language Model Engineering, Languages, Concepts, and Tools 5th International Conference, Dresden, Germany, 2002.

[8] Lujan-Mora et al., S., Multidimensional Modeling with UML Package Diagrams. The 21st International Conference on Conceptual Modeling (ER 2002), Tampere, Finland, 2002.

[9] Ambler, S.W., A UML Profile for Data Modeling. wwwagiledataorg, 2002.

[10] Blaha, M. & Premeriani, W., Using UML to Design Database Applications. http://wwwumlchinacom/Indepth/usingumlhtm, 2002.

[11] Gornik, D., UML Data Modeling Profile. Rational Software White Paper (TP162), 2002.

[12] OMG, Common Warehouse Metamodel. Object Management Group (OMG), http://wwwomgorg/cwm/, 2000.

[13] OMG, MetaObject Facility (MOF) Specification. Object Management Group (OMG), http://cgiomgorg/docs/formal/00-04-03pdf, 2000.

[14] Johnson, R., Dynamic Object Model. Object-Oriented Programming Systems, Languages & Applications (OOPSLA' 1997), Atlanta, Georgia, 1997.

[15] Johnson, R. & Woolf, B., Type Object, Pattern Languages of Program Design 3. Addison Wesley, 1998.

[16] Adamson, C. & Venerable, M., Data Warehouse Design Solutions. John Wiley & Sons Inc, 1998.

[17] OMG, Unified Modeling Language (UML) Specification (version 1.4). Object Management Group (OMG), 2001.

[18] Booch, G., Jacobson, I. & Rumbaugh, J., The Unified Modeling Language User Guide. Addison-Wesley, 1999.

Management Information Systems, C. A. Brebbia (Editor)
© 2004 WIT Press, www.witpress.com, ISBN 1-85312-728-0

Using common data acquisition in a database for optimal location and distribution problems

J. Marasović & M. Čić
Faculty of Electrical Engineering, Mechanical Engineering and Naval Architecture, University of Split, Croatia

Abstract

Finding optimal solutions in heuristic problems is based on intuitive and empirical rules that intelligently move one solution point to another, according to the model criteria. It is well known that each heuristic problem is specific and optimal solution achievements must be treated, from the very beginning, independently. In this paper we propose the original idea of common input data acquisition. Common databases can be easily transformed and used in more than one heuristic problem (location and distribution problems). We present the application of our own software, developed for creating such a database. Our idea and our contribution are based on work concerning different heuristic problems where all the efforts are motivated for optimum time, resource and money savings.
Keywords: heuristic problems, location and distribution problems, input database.

1 Introduction

Development of modern decision theory is confronted with many different and very large tasks, created, for example in economy, biology, communications, and automatic control. The problems are viewed as different systems with common mathematical characteristics to make optimal solution finding easier and faster. Different problems from human life can be commonly modelled with costs, capacities and specific positions in the chosen area. Models, in which all relevant variables are not quantifiable and are not related by appropriate mathematical functions, can be solved only with heuristic methods.

The heuristic methods are based on intuitive and empirical rules. They are searching procedures that intelligently move one solution point to another with

Management Information Systems, C. A. Brebbia (Editor)

the objective of improving the value of the model criterion (*branch and bound rules*). Mostly, development of optimal heuristic problems solving methods is confronted with the lack of useful general relations. So, any idea that can find any common parts in different heuristic problems is a good idea.

Situations where a physical object has to be optimally located are of very frequent occurrence. Optimal location problems can be categorized in several ways [2], [3]. The sites which object may occupy will be limited to a set that may be *discrete or continuous.* The number of objects to be sited may be *one (1-dimensional case)* or *several (multi-dimensional case).* In multi-dimensional cases a distinction is made between *private* and *public sector.* Private sector problems are concerned with minimization of cost or maximization of profit and such problems are mathematically called *simple location problems,* [6], [7], [8]. In public sector decisions, called *emergency problems* (hospitals, police stations, placing of communication base stations), the appropriate constraint "maximal covering location problem" was added.

Location problems in which the facilities within a transportation system are assumed to be fixed and given, but the question is an efficient distribution with constrained vehicle number and their capacity, are called *distribution problems,* [2], [3], [9]. Distribution problems can be seen as the special cases, called *salesman* problems, when the number of vehicles and their capacity are not constrained. Distribution problems become important even in *motion planning and robot control* when the robot vehicles are planned for transportation, [2], [3], [4].

Given problems are all heuristic ones, different, with specific model criterion, constraints and one or more solving algorithms developed particularly. All known solutions for described problems start each time from the very beginning, particularly with input data collection. In this paper the original idea of common data acquisition, useful for all given problems, is presented. That means, the first step for different heuristic problems, presented here, with useful organization can be done only once (and stored), saving time and computational effort. This can be done with the application of our own developed software. The problems are mathematically established and the idea of general input data is introduced in Chapter 2. The main elements of software are presented and the way of its use is illustrated in Chapter 3; the conclusions are given in Chapter 4.

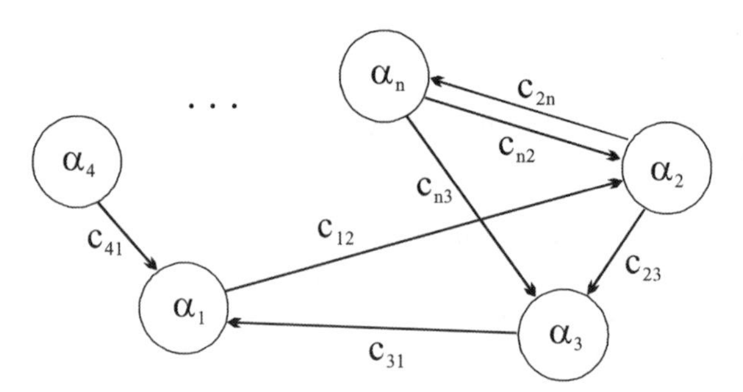

Figure 1: Sites distribution on global plan.

2 Common characteristics in location and distribution problem mathematical models

Let's say that N is the set of n sites distributed on a given plan, illustrated in figure 1.

According to figure 1. the sites are indexed as presented and it is commonly supposed that:

$$
\begin{array}{lll}
i \in M, & i = 1, \dots, n & \text{rows} \\
j \in F, & j = 1, \dots, n & \text{columns} \\
M = F = N, & k = i = j &
\end{array}
\qquad (1)
$$

and

$\{c_{ij}\}$ is the set of constants denoting all possible distances (costs) between sites, for all $i \neq j$. It is supposed that is $c_{ij} \equiv c_{ji}, \forall i, j$.

$\{\alpha_k\}$ is the set of specific sites constants.

From figure 1 and relation (1) it is possible to create two significant input matrices given bellow, where X denotes impossible distances (c_{11}, c_{22},…., c_{nn}).

$$
\begin{bmatrix} \alpha_1 \\ \alpha_2 \\ \dots\dots \\ \alpha_n \end{bmatrix} \quad \text{and} \quad \begin{bmatrix} X & c_{12} & \dots\dots & c_{1n} \\ c_{21} & X & \dots\dots & c_{2n} \\ \dots\dots & \dots\dots & \dots\dots & \dots\dots \\ c_{n1} & c_{n2} & \dots\dots & X \end{bmatrix}
$$

Can this global plan and relations be a common start (input data) for different heuristic problems, transformable easily to any particular one?

All problems presented are modelled with input data only. Detailed modelling includes set of specific constraints for each problem whose particular explanation is not the aim of this paper.

2.1 Transportation problem

The total cost minimization problem of transporting a commodity from a number of sources to several destinations is called the transportation problem. The supply at each source and the demand at each destination are known and number of sources and destinations is fixed.

Let's say that the number of sources is m and the number of destinations is (n-m). Optimality is defined with function φ, with equation (2), [1].

$$\text{minimize } \varphi = \sum_{i=1}^{m} \sum_{j=m+1}^{n} c_{ij} x_{ij} \qquad (2)$$

Management Information Systems, C. A. Brebbia (Editor)

where: x_{ij} is *unknown transporting quantity from i to j, which becomes optimal after minimization*

c_{ij} is the constant cost of supplying from i to j

If the sites on global plan (figure 1.), indexed 1 to m, are placed on source positions, and the others on destinations positions, relation (1) can be transformed into:

$$\begin{array}{lll} i \in M, & i = 1, \dots, m & \text{rows} \\ j \in F, & j = m+1, \dots, n & \text{columns} \\ N = M \cup F, & k = i + j & \end{array} \tag{3}$$

If the supplying cost c_{ij} can be measured with distance between the sites and the set of specific site constants can be connected with supply capacities (relation (4), global model can be transformed in useful transportation problem input data.

$$\{\alpha_k\} \equiv \{S_k\}, \tag{4}$$

$\{S_i\}$ is the set of source supply capacities,

$\{S_j\}$ is the set of destination supply capacities.

The transportation problem input data for planned area, are given in table 1, where sites 1 and 2 are chosen to be the sources and sites 3,4 and 5 are chosen to be the destinations. The number of sources (m) and the sites supply capacities must be added as specific input parameters.

Table 1: Necessary costs as part of input data (m =2, n= 5) for transportation problem.

Sites	1	2	3	4	5
1	x		c_{13}	c_{14}	c_{15}
2		x	c_{23}	c_{24}	c_{25}
3			x		
4				x	
5					x

2.2 Multi dimensional simple location problem

Let's say that the number of sources is m and the number of possible object location is (n-m). The problem, called simple location problem, is how to *choose p optimal locations* ($p \leq (n-m)$) among given ones to satisfy demands of the sources according to the equation (5), [2], [3].

$$\text{minimize } \varphi = \sum_{i=1}^{m} \sum_{j=m+1}^{n} c_{ij} x_{ij} + \sum_{j=m+1}^{n} CE_j y_j \tag{5}$$

Management Information Systems, C. A. Brebbia (Editor)
© 2004 WIT Press, www.witpress.com, ISBN 1-85312-728-0

where: $x_{ij} = \begin{cases} 1 & \text{if the object at j supplies total demand from i} \\ 0 & \text{otherwise} \end{cases}$

$y_j = \begin{cases} 1 & \text{if the object is located at j} \\ 0 & \text{otherwise} \end{cases}$

$\{c_{ij}\}$ is the set of constant costs of supplying location j from i, measured with distances,
$\{CE_j\}$ is the set of constant costs of establishing an object at j.

The given problem can be modelled from global relations (figure 1. and relation (1)) transforming them into relations (6), (7) according to table 1. The number of sources (m) and the costs of establishing object at chosen location must be added as a specific input parameters.

$$\begin{array}{lll} i \in M, & i = 1, \dots, m & \text{rows} \\ j \in F, & j = m+1, \dots, n & \text{columns} \\ N = M \cup F, & k = j & \end{array} \quad (6)$$

$$\{\alpha_k\} \equiv \{CE_k\} \equiv \{CE_j\} \quad (7)$$

Table 2: Necessary costs as part of input data (n=5) for distribution problems. Shaded elements present symmetric constants.

Sites	1	2	3	4	5
1	x	c_{12}	c_{13}	c_{14}	c_{15}
2		x	c_{23}	c_{24}	c_{25}
3			x	c_{34}	c_{35}
4				x	c_{45}
5					x

2.3 Distribution problems

The simplest form of distribution problem concerns supplying (n-1) destinations from a single source by one vehicle. Each destination has a fixed requirement Q_k that must be met. The objective is to determine *optimal route for the vehicle* according to the equation (8), [2], [3], when:

a) a vehicle has constrained capacity which must not be exceeded, or
b) a vehicle has unconstrained capacity, known as a salesman problem.

$$\text{minimize } \varphi = \sum_{i=1}^{n} \sum_{j=1}^{n} c_{ij} x_{ij} \quad (8)$$

Management Information Systems, C. A. Brebbia (Editor)
© 2004 WIT Press, www.witpress.com, ISBN 1-85312-728-0

where: $x_{ij} = \begin{cases} 1 & \text{if the vehicle connects locations i and j} \\ 0 & \text{otherwise} \end{cases}$

$\{c_{ij}\}$ is the set of constant costs of supplying location j from i, measured with distance,

$\{Q_k\}$ is the set of destinations requirements.

Given problem can be modelled from global relations (figure 1, relation (1)) transforming them into relations (9), (10) according to the table 2. For the salesman problem there are no other input parameters, otherwise a vehicle capacity must be added.

$$i \in M, \quad i = 1, \dots, n \quad \text{rows}$$
$$j \in F, \quad j = 1, \dots, n \quad \text{columns}$$
$$N = M = F, \quad k = i = j \tag{9}$$

$$\{\alpha_k\} \equiv \{Q_k\} \tag{10}$$

3 Programming

According to the idea of common data acquisition from a given plan (geographic map, for example), we developed the software in which sites can be marked easily, on a screen (with a mouse) or on a touch screen (with a finger). Connection with a geographic map is presented in figure 2 and data acquisition program interface is presented in figure 3.

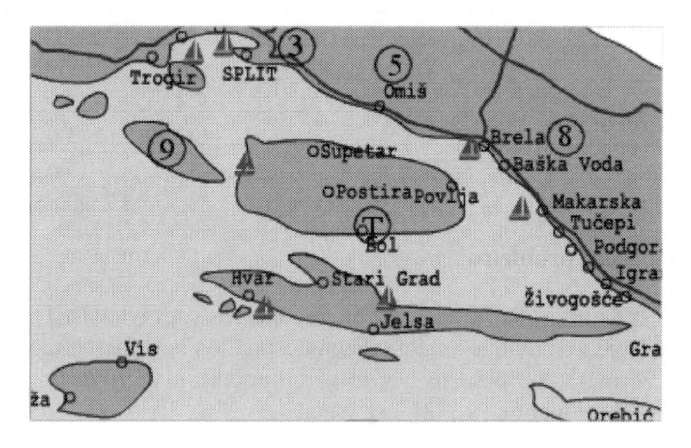

Figure 2: Geographic map as data input form.

The software is developed for Croatian language. Main interface messages are: (Obriši zadnje skladište = *Delete last site*, Obriši podatke = *Delete all data,*

Management Information Systems, C. A. Brebbia (Editor)

Matrični unos = *Matrix input*, Proračun = *Calculate*!, Izlaz = *Exit*, Optimizacija = *Optimisation*, Legenda = *Legend*, Mreža = *Grid)*

The program interface presented in figure 3. is connected with distribution problems, so it has two specific textual messages: (Unesite kapacitet skladišta = *Enter supply capacity*, Kapacitet vozila = *Vehicle capacity*). Distances are automatically measured for all sites (costs ≡ distances).

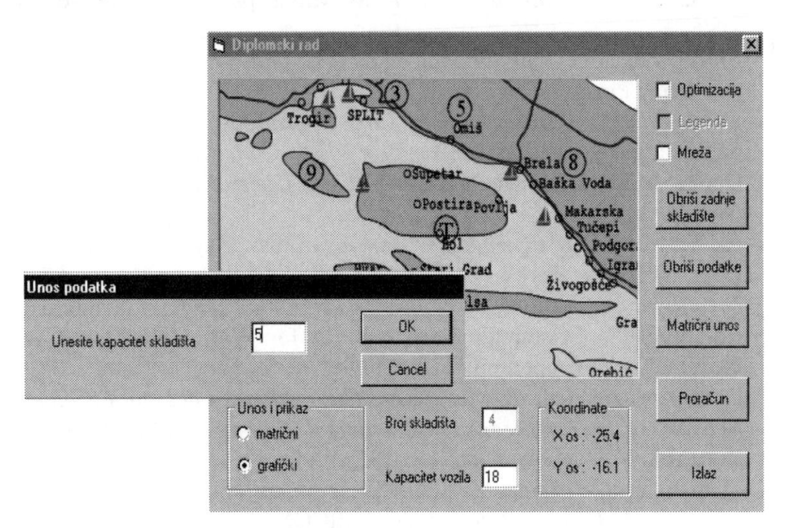

Figure 3: Data acquisition program interface.

The results create complete distance database; independent and useful to any problem that begins with a site plan.

4 Conclusion

In this paper the original idea of common input data acquisition for more than one heuristic problem is introduced. The idea is developed according to the heuristic models for specific problems that deal and begin with similar planar area demands (geographic map). For the transportation problem, for multidimensional location problem and for distribution problems (constrained and unconstrained vehicle capacity) input data can be transformed easily from given common input matrices.

All problems presented are modelled with input data only. Detailed modelling includes a set of specific constraints for each problem whose particular use is not the aim of this paper.

The software developed performs easy input database acquisition and a database can be created for a given area independently. The advantage of the proposed idea and the software is that it can be used for different types of problems and it can permit suitable algorithms to start without wasting input

Management Information Systems, C. A. Brebbia (Editor)
© 2004 WIT Press, www.witpress.com, ISBN 1-85312-728-0

time. The database can also be used for connections among dislocated program files and Internet communications.

References

[1] Taha, A.H., *Operations research*, McMillan publishing co., New York, 1976.
[2] Morin, T.L., *Computational Advances in Dynamic Programming*, Academic Press, 1978.
[3] Boffey, T.B., *Graph Theory in Operations Research*, McMillan publishing co., London, 1982.
[4] Gupta, M.M. & Sinha, N.K., *Intelligent Control Systems*, IEEE Press, New York, 1996.
[5] Pavlić, M., *Razvoj informacijskih sustava*, Znak, Zagreb, 1996.
[6] Uber, J.G. & Tryby, M.E., *Optimal Location and Scheduling of Disinfectant Additions in Water Distributions Networks*, International Conference on Computing and Control for Water Industry CCWI'99, Exter, UK, September, pp.13-15, 1999.
[7] De Los Mozos, C.L. & Mesa, J.A., *The Variance Location Problem on a Network with Continuously Distributed Demand*, RAIRO Operations Research, Vol.34, No. 2, pp. 155-182, 1999.
[8] Marasović, J., Božanić, M. & Jurić, J., *Optimal Location Problems Solved with Graph Theory*, Proc. of the SoftCOM 2000, Vol.1, pp. 475-484, October, 2000.
[9] Fruhwirth, T. & Brisset, P., *Placing Base Stations in Wireless Indoor Communication Networks*, IEEE Intelligent Systems, Vol. 15, No. 1, January/February, 2000.

Management Information Systems, C. A. Brebbia (Editor)

Time dependent spatial price equilibrium problems

M. Milasi & C. Vitanza
Dipartimento di Matematica, Universitá di Messina, Italia

Abstract

We consider a time dependent economic market in order to show the existence of the dynamical market equilibrium. Moreover, the Generalized Lagrange Multipliers and the Lagrangian theories are studied and, as an interesting consequence, we obtain the Lagrangian variables. This theory plays an extraordinary role in the estimation of the dynamical market equilibrium. In our opinion this study makes these problems more general and realistic than classical static spatial price equilibrium problems.
Keywords: supply excess, demand excess, Lagrangian multipliers, subgradient method.

1 Introduction

The aim of this paper is to consider a dynamic economic model. We see that this problem can be incorporated directly in to a Variational Inequality models and moreover we characterize the equilibrium solution by means of Lagrangian multipliers applying the duality theory in the case of infinite dimensional space. In the static case the spatial price equilibrium problem has been formulated in terms of a Variational Inequality by Nagurney and Zhao in [8] and A. Nagurney in [7]. Subsequently P. Daniele in [1] has been concerned with the spatial price equilibrium problem in the case of the quantity formulation under the assumption that the data evolve in the time. In [6] the authors extended the result of [1] considering a model with supply and demand excesses and with capacity constraints on prices and on transportation costs. In the last years some papers have been devoted to the study of the influence of the time on the equilibrium problems (see [2], [3] and [4]). In fact, we cannot avoid to consider that each phenomenon of our economic and physical world is not stable with respect to the time and that our static models of equilibria are a first useful abstract approach.

Management Information Systems, C. A. Brebbia (Editor)
 ISBN 1-85312-728-0

2 Evolutionary market model

Let us consider n supply markets $P_i, \quad i = 1,2,..,n$, and m demand markets $Q_j, \; j = 1,2,..,m$ involved in the production and in the consumption respectively of a commodity during a period of time $[0,T], \; T > 0$. Let $g_i(t), \; t \in [0,T], \; i = 1,2,..,n$ denote the supply of the commodity associated with supply market i at the time $t \in [0,T]$ and let $p_i(t), \; t \in [0,T], \; i = 1,2,..,n$ denote the supply price of the commodity associated with supply market i at the time $t \in [0,T]$. A fixed minimal and maximum supply price $\underline{p}_i(t), \; \overline{p}_i(t) \geq 0$, respectively, for each supply market, are given. Let $f_j(t), \; t \in [0,T], \; j = 1,2,..,m$ denote the demand associated with the demand market j at the time $t \in [0,T]$ and let $q_j(t), \; t \in [0,T], \; j = 1,2,..,m$, denote the demand price associated with the demand market j at the time $t \in [0,T]$. Let $\underline{q}_j(t), \; \overline{q}_j(t) \geq 0$, for each demand market, be the fixed minimal and maximum demand price respectively. Since the markets are spatially separated, let $x_{ij}(t), \; t \in [0,T], \; i = 1,2,..,n, \; j = 1,2,...,m$ denote the nonnegative commodity shipment transported from supply market P_i to demand market Q_j at the same time $t \in [0,T]$. Let $c_{ij}(t), \; t \in [0,T], \; i = 1,2,..,n, \; j = 1,2,..,m$ denote the nonnegative unit transportation cost associated with trading the commodity between (P_i, Q_j) at the same time $t \in [0,T]$. Let we suppose that we are in presence of excesses on the supply and on the demand. Let $s_i(t), \; t \in [0,T], \; i = 1,2,...,n$ denote the supply excess for the supply market P_i at the time $t \in [0,T]$. Let $\tau_j(t), \; t \in [0,T], \; j = 1,2,..,m$ denote the demand excess for the demand market Q_j at the time $t \in [0,T]$. We assume that the following feasibility conditions must hold for every $i = 1,2,..,n$ and $j = 1,2,..,m$ a. e. in $[0,T]$:

$$g_i(t) = \sum_{j=1}^{m} x_{ij}(t) + s_i(t), \tag{1}$$

$$f_j(t) = \sum_{i=1}^{n} x_{ij}(t) + \tau_j(t). \tag{2}$$

Grouping the introduced quantities in vectors, we have the total supply vector $g(t) \in L^2([0,T], \mathbb{R}^n)$ and the total demand vector $f(t) \in L^2([0,T], \mathbb{R}^m)$. Furthermore in order to precise the quantity formulation, we assume that two mappings $p(g(t))$ and $q(f(t))$ are given:

$$p : L^2([0,T], \mathbb{R}^n) \to L^2([0,T], \mathbb{R}^n),$$

$$q : L^2([0,T], \mathbb{R}^m) \to L^2([0,T], \mathbb{R}^m).$$

The mapping p assigns for each supply $g(t)$ the supply price $p(g(t))$ and the mapping q assigns for each demand $f(t)$ the demand price $q(f(t))$. Analogously $x(t) \in L^2([0,T], \mathbb{R}^{nm})$ is the vector of commodity shipment and the mapping

$$c : L^2([0,T], \mathbb{R}^{nm}) \to L^2([0,T], \mathbb{R}^{nm})$$

assigns for each commodity shipment $x(t)$ the transportation cost $c(x(t))$. Moreover let $s(t) \in L^2([0,T], \mathbb{R}^n), \; \tau(t) \in L^2([0,T], \mathbb{R}^m)$ be the vectors of supply and demand excess. Denoting by $w(t) = (g(t), f(t), x(t), s(t), \tau(t))$, we set

$$\widetilde{L} = \{w(t) = (g(t), f(t), x(t), s(t), \tau(t)) : w(t) \in L^2([0,T], \mathbb{R}^n)$$
$$\times L^2([0,T], \mathbb{R}^m) \times L^2([0,T], \mathbb{R}^{nm}) \times L^2([0,T], \mathbb{R}^n) \times L^2([0,T], \mathbb{R}^m)\},$$

and

$$\|w(t)\|_{\widetilde{L}} = \Big(\|g(t)\|^2_{L^2([0,T],\mathbb{R}^n)} + \|f(t)\|^2_{L^2([0,T],\mathbb{R}^m)} + \|x(t)\|^2_{L^2([0,T],\mathbb{R}^{nm})}$$
$$+ \|s(t)\|^2_{L^2([0,T],\mathbb{R}^n)} + \|\tau(t)\|^2_{L^2([0,T],\mathbb{R}^m)} \Big)^{\frac{1}{2}}.$$

Furthermore we assume that the feasible vector
$w(t) = (g(t), f(t), x(t), s(t), \tau(t))$ satisfies the condition

$$w(t) \geq 0 \quad \text{a. e. in } [0,T] \tag{3}$$

Taking into account conditions (1), (2) and (3), we denote with $\widetilde{K}$ the set of feasible vectors. $\widetilde{K}$ is a convex, closed, not bounded subset of the Hilbert space $\widetilde{L}$.
Finally the presence of the capacity constraints on $p,\ q,\ c$ can be expressed in the following way, for each $w(t) = (g(t), f(t), x(t), s(t), \tau(t)) \in \widetilde{K}$:

$$\underline{p}(t) \leq p(g(t)) \leq \overline{p}(t), \quad \underline{q}(t) \leq q(f(t)) \leq \overline{q}(t), \quad \underline{c}(t) \leq c(x(t)) \leq \overline{c}(t).$$

Then the time-dependent market equilibrium condition in the case of the quantity formulation takes the following form:

Definition 2.1. Let $w^*(t) = (g^*(t), f^*(t), x^*(t), s^*(t), \tau^*(t))$ in $\widetilde{K}$. $w^*(t)$ is a market equilibrium if and only if for each $i = 1, 2, .., n$ and $j = 1, 2, .., m$ the following conditions hold a. e. in $[0,T]$:

$$\begin{cases} \text{if } \ s_i^*(t) > 0 \Rightarrow p_i(g^*(t)) = \underline{p}_i(t) \\ \\ \text{if } \ \underline{p}_i(t) < p_i(g^*(t)) \Rightarrow s_i^*(t) = 0 \end{cases} \tag{4}$$

$$\begin{cases} \text{if } \ \tau_j^*(t) > 0 \Rightarrow q_j(f^*(t)) = \overline{q}_j(t) \\ \\ \text{if } \ q_j(f^*(t)) < \overline{q}_j(t) \Rightarrow \tau_j^*(t) = 0 \end{cases} \tag{5}$$

$$\begin{cases} \text{if } \ x_{ij}^*(t) > 0 \Rightarrow p_i(g^*(t)) + c_{ij}(x^*(t)) = q_j(f^*(t)) \\ \\ \text{if } \ p_i(g^*(t)) + c_{ij}(x^*(t)) > q_j(f^*(t)) \Rightarrow \quad x_{ij}^*(t) = 0 \end{cases} \tag{6}$$

Moreover we are able to characterize time dependent market equilibrium as a solution to a Variational Inequality; in fact the following result holds:

Management Information Systems, C. A. Brebbia (Editor)

Theorem 2.1. *$u^* = (g^*, f^* x^*, s^*, \tau^*) \in \widetilde{K}$ is a time dependent market equilibrium if and only if u^* is a solution to Variational Inequality:*

$$\int_0^T \Big(p(g^*(t))(g(t)-g^*(t)) - q(f^*(t))(f(t)-f^*(t)) + c(x^*(t))(x(t)-x^*(t)) +$$

$$-\underline{p}(t)(s(t)-s^*(t)) + \overline{q}(t)(\tau(t)-\tau^*(t))\Big)dt \geq 0$$

$$\forall \;\; w(t) = (g(t), f(t), x(t), s(t), \tau(t)) \in \widetilde{K};$$

or, equivalently, the Variational Inequality:

$$\int_0^T \Big\{\big(p(x^*(t), s^*(t)) - q(x^*(t), \tau^*(t)) + c(x^*(t))\big)\big(x(t) - x^*(t)\big)$$

$$+(p(x^*(t), s^*(t)) - \underline{p}(t))(s(t) - s^*(t)) - (q(x^*(t), \tau^*(t))$$

$$-\overline{q}(t))(\tau(t) - \tau^*(t))\Big\}dt \geq 0$$

$$\forall \;\; u(t) = (x(t), s(t), \tau(t)) \in K." \tag{7}$$

where: $L = L^2([0,T], \mathbb{R}^{nm}) \times L^2([0,T], \mathbb{R}^n) \times L^2([0,T], \mathbb{R}^m)$ and $K = \{u(t) = (x(t), s(t), \tau(t)) \in L \; : u(t) \geq 0\}$.

Proof. See e. g. [6]. □

If we consider the function $v : L \to L$ defined setting for each $u \in L$:

$$v(u) = \Big(p(x,s) - q(x,\tau) + c(x), p(x,s) - \underline{p}, \overline{q} - q(x,\tau)\Big),$$

Variational Inequality (7) can be rewrite as:

"Find $u^* \in K$ such that:

$$< v(u^*), u - u^* > \geq 0, \quad \forall u \in K." \tag{8}$$

Now we can observe that it is possible to obtain the following existence result of equilibria adapting a classical existence theorem for solution of a variational inequality to our problem (for more details see e. g. [6]).

Theorem 2.2. *Each of the following conditions is sufficient to ensure the existence of the solution of (8):*

1. *$v(u) = v(x(t),\, s(t),\, \tau(t))$ is hemicontinuous with respect to the strong topology and there exist $A \subseteq \mathbb{K}$ compact and there exist $B \subseteq \mathbb{K}$ compact, convex with respect to the strong topology such that*

$$\forall u_1 \in \mathbb{K} \setminus A \quad \exists u_2 \in B : \; \langle v(u_1), u_2 - u_1 \rangle < 0;$$

Management Information Systems, C. A. Brebbia (Editor)

2. *v is pseudomonotone, v is hemicontinuous along line segments and there exist $A \subseteq \mathbb{K}$ compact and $B \subseteq \mathbb{K}$ compact, convex with respect to the weak topology such that*
$$\forall p \in \mathbb{K} \setminus A \quad \exists \tilde{p} \in B : \ \langle v(p), \tilde{p} - p \rangle < 0;$$
3. *v is hemicontinuous on $\mathbb{K}$ with respect to the weak topology, $\exists A \subseteq \mathbb{K}$ compact, $\exists B \subseteq \mathbb{K}$ compact, convex with respect to the weak topology such that*
$$\forall p \in \mathbb{K} \setminus A \quad \exists \tilde{p} \in B : \ \langle v(p), \tilde{p} - p \rangle < 0.$$

3 Example

The spatial price equilibrium model before presented is now illustrated with a simple example consisting of 2 supply markets, P_1, P_2 and 1 demand market, Q_1. It follows from theorem 2.2 that we have existence of the time depending market equilibrium. The supply price, transportation cost and demand price functions are, respectively:

$$p_1(g_1(t), g_2(t)) = 2g_1(t) + g_2(t) + 5, \quad p_2(g_1(t), g_2(t)) = g_2(t) + 10.$$
$$c_{11}(x(t)) = 5x_{11}(t) + x_{21}(t) + 9, \quad c_{21}(x(t)) = 3x_{21}(t) + 2x_{11}(t) + 19.$$
$$q_1(f_1(t)) = -f_1(t) + 80.$$

The capacity constraints on price are:

$$p_1(t) \geq 21, \quad p_2(t) \geq 16, \quad q_1(t) \leq 60.$$

From the equilibrium conditions we have the following:

$$\begin{cases} p_1(g_1(t), g_2(t)) + c_{11}(x(t)) = q_1(f_1(t)) \\ \\ p_2(g_1(t), g_2(t)) + c_{21}(x(t)) = q_1(f_1(t)) \\ \\ p_1(g_1(t), g_2(t)) = \underline{p}_1(t) \\ \\ p_2(g_1(t), g_2(t)) = \underline{p}_2(t) \\ \\ q_1(f_1(t)) = \overline{q}_1(t) \end{cases}$$

The time depending equilibrium is then given by:

$$g_1^*(t) = 5, \quad g_2^*(t) = 6, \quad f_1^*(t) = 20, \quad x_{11}^*(t) = 5, \quad x_{21}^*(t) = 5,$$
$$s_1^*(t) = 0, \quad s_2^*(t) = 1, \quad \tau_1^*(t) = 10,$$

with equilibrium supply prices, costs and demand prices:

$$p_1(t) = 21, \quad p_2(t) = 16, \quad c_{11}(t) = 39, \quad c_{21}(t) = 44, \quad q_1(t) = 60.$$

Management Information Systems, C. A. Brebbia (Editor)

4 Lagrangian theory

It is possible to apply the Lagrangian and the duality theory which plays an extraordinary role in economic theory in order to characterize the evolutionary market equilibrium conditions in terms of the Lagrangian multipliers.
We observe that if $u^* \in K$ is a solution to problem (8) then

$$\min_{u \in K} < v(u^*), u - u^* >= 0.$$

Let us introduce the functional

$$\psi_{u^*}(u) =< v(u^*), u - u^* >, \quad \forall u \in K,$$

and let us observe that

$$\psi_{u^*}(u) \geq 0,$$

$$\min_{u \in K} \psi_{u^*}(u) = 0.$$

We associate to Variational Inequality (8) the following Lagrangian function: $\forall u \in L,\ l = (\alpha, \beta, \gamma) \in C^*$

$$\mathcal{L}(u, l) = \psi_{u^*}(u) - \Big(\int_0^T \alpha(t)x(t)dt + \int_0^T \beta(t)s(t)dt + \int_0^T \gamma(t)\tau(t)dt \Big)$$

where

$$C^* = \{l(t) = (\alpha(t), \beta(t), \gamma(t)) \in L : \ \alpha(t),\ \beta(t),\ \gamma(t) \geq 0 \ \ a.\ e.\ in\ [0, T]\}$$

is the dual cone of L. We observe that the dual cone of L, from Riesz theorem, is equal to the ordering cone of $L : \quad C^* = C$. Then, using the infinite dimensional Lagrangian and duality theory (see chapter 5 and 6 in [5]), we can obtain the following:

Theorem 4.1. *Let $u^* \in K$ be a solution to problem (8). Then there exist three functions $\alpha^* \in L^2([0, T], \mathbb{R}^{nm})$, $\beta^* \in L^2([0, T], \mathbb{R}^n)$, $\gamma^* \in L^2([0, T], \mathbb{R}^m)$ such that:*
i) $\alpha^*(t), \beta^*(t), \gamma^*(t) \geq 0 \quad a.e.\ in\ [0, T]$;
ii) $\alpha^* \cdot x^* = 0, \quad \beta^* \cdot s^* = 0, \quad \gamma^* \cdot \tau^* = 0$;
iii)

$$\begin{cases} p(x^*, s^*) - q(x^*, \tau^*) + c(x^*) = \alpha^*, \\ \\ p(x^*, s^*) - \underline{p} = \beta^*, \\ \\ \overline{q} - q(x^*, \tau^*) = \gamma^*. \end{cases}$$

Management Information Systems, C. A. Brebbia (Editor)

Proof. See e. g. [6].

Remark 4.1. *Moreover it is easy to show that if* $u \in K$ *and there exist* $\alpha^*,\ \beta^*,\ \gamma^*$ *as in theorem (4.1) such that conditions* $i),\ ii)$ *and* $iii)$ *are fulfilled, then* u *verifies the Variational Inequality (8).*

5 Calculus of the solution

We start to observe that in our assumptions on the capacity constraint we can assume, in natural way, that u is upper limited by a function $\overline{u} \in L\ \overline{u}(t) \geq 0$ a. e. in $[0, T]$; then we get:

$$K = \{u(t) \in L : \ 0 \leq u(t) \leq \overline{u}\,(t)\}.$$

Now, we suppose that the pseudomonotonicity and hemicontinuity conditions hold in order to ensure the existence of the solution to our variational inequality. By pseudo-monotonicity hypothesi on v,

$$< v(u) - v(w), u - w > \geq 0 \qquad \forall u,\ w \in K :$$

Variational Inequality (8) is equivalent to the following Minty Variational Inequality:

"*Find* $u^*(t) \in K$ *such that:*

$$< v(u), u - u^* \geq 0 \quad \forall u \in K. \tag{9}$$

Let us set, for all $u \in K$:

$$\psi(u) = \max_{w \in K} < v(w), u - w > .$$

ψ is well defined, because the operator

$$w \to < v(w), u - w >$$

is weakly upper semicontinuous, K is a bounded convex subset of L and hence weakly compact set. Moreover $\psi(u)$, being the maximum of a family of continuous affine functions, is convex and weakly lower semicontinuous. $\psi(u)$ is also:

$$\psi(u) \geq 0, \ \ \psi(u) = 0 \ \ \Leftrightarrow \ \ u \ \text{ is a solution to VI (8).}$$

We can consider the subdifferential of $\psi(u)$:

$$\partial\psi(u) = \{\tau \in L : \ \psi(w) - \psi(u) \geq < \tau, w - u >, \ \forall w \in K\}.$$

We will show that $\psi(u)$ has a nonempty subdifferential for all $u \in K$. The subgradient method for finding a solution of Variational Inequality (8) runs as follows:

Management Information Systems, C. A. Brebbia (Editor)

choose $u_0 \in K$ arbitrary. Given $u_n \in K,\ \psi(u_n) > 0$, let

$$u_{n+1} = \text{Proj}_K(u_n - \rho_n \tau_n),$$

where

$$\tau_n \in \partial\psi(u_n) \left(\tau_n \neq 0 \text{ because } \psi(u_n) > 0\right), \qquad \rho_n = \frac{\psi(u_n)}{||\tau_n||^2}.$$

We shall see below that τ_n can be chosen in such a way that $||\tau_n||$ remains bounded. If τ_n remains bounded and $\psi(u_n) > 0$ for all n, then we have the following result:

Theorem 5.1. *There holds that* $\lim_{n\to\infty} \psi(u_n) = 0$. *The sequence* $\{u_n\}_{n\in\mathbb{N}}$ *has weak cluster points and in every weak cluster point* ψ *is equal to zero.*

Proof. Let us w^* a solution to Variational Inequality (8); by inequality:

$$< \tau_n, w^* - u_n > \leq \psi(w^*) - \psi(u_n) = -\psi(u_n);$$

and by nonexpansivity of the projection mapping, we have:

$$||u_{n+1} - w^*||^2 = ||Proj(u_n - \rho_n\tau_n) - Proj\, w^*||^2 \leq ||u_n - \rho_n\tau_n - w^*||^2$$

$$= ||u_n - w^*||^2 + ||\rho_n\tau_n||^2 + 2\rho_n < \tau_n, w^* - u_n >$$

$$\leq ||u_n - w^*||^2 + \rho_n^2||\tau_n||^2 - 2\rho_n\psi(u_n)$$

$$= ||u_n - w^*||^2 + \rho_n^2||\tau_n||^2 - 2\rho_n^2||\tau_n||^2 = ||u_n - w^*||^2 - \rho_n^2||\tau_n||^2.$$

Since the sequence $\{||u_n - w^*||\}$ is decreasing and bounded, and since $||\tau_n||$ are bounded from above, it follows that:

$$\lim_{n\to\infty} \psi(u_n) = 0,$$

and in particular

$$||u_n - w^*|| \leq ||u_0 - w^*||, \qquad \forall n \in \mathbb{N}.$$

This show that $\{u_n\}$ is bounded and therefore has a weak cluster point. Since $\psi(u_n) \to 0$ and ψ is weakly lower semicontinuous, it follows that, if u is a weak cluster point:

$$0 = \liminf_{k\to\infty} \psi(u_{n_k}) \geq \psi(u) \geq 0,$$

namely $\psi(u) = 0$. □

Management Information Systems, C. A. Brebbia (Editor)

We show that $\psi(u)$ has a nonempty subdifferential for all $u \in K$: we call u^* the element of K such that:

$$\max_{w \in K} < v(w), u - w > = < v(u^*), u - u^* > = \psi(u),$$

and we pose $\tau = v(u^*)$. We prove that $\tau = v(u^*) \in \partial\psi(u)$, that is:

$$\psi(v) - \psi(u) \geq < v(u^*), v - u >, \qquad \forall v \in K.$$

$$< v(u^*), v - u > = < v(u^*), v - u^* + u^* - u > = < v(u^*), v - u^* >$$
$$+ < v(u^*), u^* - u > \leq \max_{w \in K} < v(w), v - w >$$
$$- < v(u^*), u - u^* > = \psi(v) - \psi(u).$$

Then:

$$\tau = v(u^*) \in \partial\psi(u) \Rightarrow \qquad \partial\psi(u) \neq \emptyset.$$

Let us $u_n^* \in K$ such that:

$$\max_{w \in K} < v(w), u_n - w > = < v(u_n^*), u_n - u^* > = \psi(u_n),$$

and we choose $\tau_n = v(u_n^*) \in \partial\psi(u_n) \quad \forall n \in \mathbb{N}$. Because v is bounded, then $||\tau_n||$ remains bounded. Finally the approximating sequence is so defined:

$$u_{n+1} = Proj_K\Big(u_n - \frac{\max_{w \in K} < v(w), u_n - w >}{||v(u_n^*)||^2} v(u_n)\Big)$$
$$= Proj_K\Big(u_n - \frac{< v_n^*, u_n - u_n^* >}{||v(u_n^*)||^2} v(u_n^*)\Big)$$

6 Conclusions

The model we are concerned with is the spatial price equilibrium model in the presence of excess of the supplies and of demands. We stress that it is surprising to see that this kind of network problem is incorporated into Variational Inequality model as in the static case. An existence theorem for this Variational Inequality is provided and the same Variational Inequality formulation is used in order to provide a computation of the equilibrium patterns and to study the generalized Lagrange multipliers. The existence of the equilibrium solution of the time depending model seems to be important because it allows us to follow the evolution in time of prices and commodities. Moreover the importance of the functions α^*, β^*, γ^*, studied in section [4], derives from the fact that they are able to describe the behavior of the evolutionary market. In fact the set $A_+^{ij} = \{t \in [0,T] : \ \alpha_{ij}^*(t) > 0\}$ indicates the time when there is not trade between the supply marker i and the demand market j. Analogously $B_+^j = \{t \in [0,T] : \ \beta_i^*(t) > 0\}$ indicates the time when there is a zero supply excess of the market i. The same holds for γ_j^* which indicates when the demand market j has not demand excess.

Management Information Systems, C. A. Brebbia (Editor)

References

[1] P. Daniele, Time-Dependent spatial Price Equilibrium Problem: Existence and Stability results for the Quantity Formulation Model, *Journal of Global Optimization*, to Appear.

[2] P. Daniele and A. Maugeri, On Dynamical Equilibrium Problems and Variational Inequalities *Equilibrium Problems: Nonsmooth Optimization and Variational Inequality Models*, Kluwer Academic Publishers, F. Giannessi- A. Maugeri- P. Pardos Eds.,59-69 (2001).

[3] P. Daniele, A. Maugeri and W. Oettli, Variational Inequalities and time-dependent traffic equilibria, *C. R. Acad. Sci. Paris* t.326, serie I, 1059-1062 (1998).

[4] P. Daniele, A. Maugeri and W. Oettli, Time Dependent traffic Equilibria, *Jou. Optim. Th. Appl.*, Vol. 103, No 3, 543-555 (1999).

[5] J. Jahn, *Introduction to the theory of nonlinear Optimization*, Springer (1996).

[6] M. Milasi and C. Vitanza, Variational Inequality and evolutionary market disequilibria: the case of quantity formulation, in *Variational Analysis and Applications*, F. Giannessi and A. Maugeri Editors, Kluwer Academic Publishers.

[7] A. Nagurney, *Network Economics - A Variational Inequality Approach*, Kluwer Academic Publishers, 1993.

[8] A. Nagurney and L. Zhao, Disequilibrium and Variational Inequalities, *I Comput. Appl. Math.* 33 (1990), 181-198.

Management Information Systems, C. A. Brebbia (Editor)
 ISBN 1-85312-728-0

Section 5
Hydro and
Geo informatics

Satellite derived impervious surface area as an indicator for water resource impacts in a semi-arid environment, Utah, USA

P. Rieke Arentsen, R. R. Gillies & N. Mesner
Department of Aquatic, Watershed, and Earth Resources,
Utah State University, Utah, USA

Abstract

Impervious surface area (*ISA*) has recently emerged as an important ecological indicator of cumulative water resource impacts to urban watersheds. An *ISA* study was conducted in Cache County, Utah, to examine the potential of *ISA* as an indicator of water resource impacts in the region. Two LANDSAT 7 images from the spring of 2000 and, the summer of 2002 were analysed to compute *ISA*. Water chemistry and discharge data were collected from canals during five stormwater runoff events. Variables measured included dissolved metals, nutrients, sediment, oil and grease, fecal coliforms, and field parameters (e.g., water temperature). Percent *ISA* and loading coefficients were calculated for each of the contributing basins for a total of eight sampling locations. Results of linear regressions correlating mean site concentrations and percentage *ISA* were positive and strong ($p<0.03$) for total phosphorus ($R^2=0.86$), total nitrogen ($R^2=0.94$), orthophosphate ($R^2=0.94$), total suspended solids ($R^2=0.67$), boron ($R^2=0.82$), copper ($R^2=0.98$), lead ($R^2=0.96$), zinc ($R^2=0.97$), and turbidity ($R^2=0.67$). Mean site loading coefficients were similarly correlated. Additionally, potential water quality impacts were evaluated by comparing mean site concentrations and percentage *ISA* with Utah and European water quality standards or pollution indicators. Those that exceeded the Utah standards or pollutant indicator values were total phosphorus, total suspended solids, lead, aluminium, and fecal coliform. Concentrations above the European standards were aluminium and iron. The Cache County study indicates that *ISA* is an applicable parameter as an indicator for cumulative water resource impacts and, further, suggests its application as an additional resource tool for the assessment and subsequent planning for water resource protection in urban developments.
Keywords: impervious surface area, water quality, remote sensing, nonpoint source pollution, runoff, urbanisation, planning, watershed management, nutrients, metals.

Management Information Systems, C. A. Brebbia (Editor)

1 Introduction

Water resource planners and managers have recognized impervious surface area (*ISA*) as an important ecological indicator of cumulative impacts to urban aquatic systems. However, cost and technical issues have precluded its widespread use in environmental management. Historically, methods such as mapping from aerial photography have been employed to calculate percent *ISA* (Ragan and Jackson [1]). The cost of gathering and analysing the photos is both time-consuming and expensive. In recent years, remote sensing techniques have been developed to derive *ISA* in a cost and time efficient manner. Moreover, remote sensing data covers a significantly larger temporal and spatial domain over traditional methods (Carlson and Arthur [2]).

ISA is defined as anything that water cannot penetrate including rock outcrops, parking lots, rooftops, and roads (Leopold [3]). During storm events, runoff is often routed from impervious surfaces to storm drains and eventually to receiving water bodies, usually rivers and lakes. The impacts to these water bodies can be physical, chemical and biological in nature (Leopold [3]; Klein [4]; Booth and Jackson [5]). Increases in percent *ISA* have been correlated to increased water quality degradation in urban watersheds (Schueler [6]; Arnold and Gibbons [7]). Schueler [6], for example, undertook a review of multiple studies on the correlation between percent *ISA* and watershed health and developed the following relationship: watersheds with less than 10% *ISA* are healthy, those between 10% and 25% *ISA* indicate impairment to aquatic systems, and those greater than 25% indicate water quality degradation which is potentially irreversible.

Recently, researchers have mapped percent *ISA* using remotely sensed data and have related these measurements of *ISA* to water resource impacts. For example, Gillies *et al* [8] performed a time-series analysis of 3 Landsat images to track changes in percent *ISA* over 18 years in an Atlanta, Georgia, U.S.A. metropolitan watershed. They correlated increases in percent *ISA* with mussel population degradation, and subsequently, a decrease in water quality. A threshold between 50-70% *ISA* was found to correlate with a noticeable decline in water quality. Roy *et al* [9] determined percentage of land cover in another Georgia watershed using Landsat TM data. They were able to relate the percentage of land use at >15-20% urban land cover to a decline in macroinvertebrate assemblages. Clausen *et al* [10] researched the relationship between percent *ISA* and water quality in the state of Connecticut, U.S.A. Landsat TM and ETM data were used to determine percent *ISA*. They found that percent *ISA* was significantly related to all of the water quality variables tested.

Prior research on the relationship between percent *ISA* and watershed functions has been performed in the northwest, midwest, and eastern parts of the United States. However, Schueler's [6] relationship has not yet been applied in a semi-arid climate region (Booth and Jackson [5]). The goal of this research was to examine the relationship between percent *ISA* (as derived from Landsat TM satellite imagery) and water quality in a semi-arid region in northeastern Utah, U.S.A. Of particular concern are the nonpoint source pollution (NPS) impacts on

Management Information Systems, C. A. Brebbia (Editor)

irrigation, aquatic life, and recreational uses of downstream waters. In addition, much of the water that is not allocated for agricultural or municipal use eventually infiltrates to groundwater or flows into the Bear River, where NPS impacts could affect aquatic life and other beneficial uses. Moreover, an established relationship between percent *ISA* and water quality for the western location could be useful in regional planning of water resources. Therefore, the research objectives were to: 1) Calculate percent *ISA* using Landsat TM data, 2) monitor stormwater runoff pollutant concentrations and loads from contributing land areas, and 3) determine if a relationship exists between both mean site loading coefficients and mean site concentrations with percent *ISA*.

2 Methods

2.1 Study area

The Logan urban area (latitude: 41.8° N, longitude: 111.8°W) is the region of interest for this study, located in Cache County in the northeast corner of Utah (ref., Figure 1). The population is approximately 59,000 in an area of approximately 94 square kilometres (km^2).

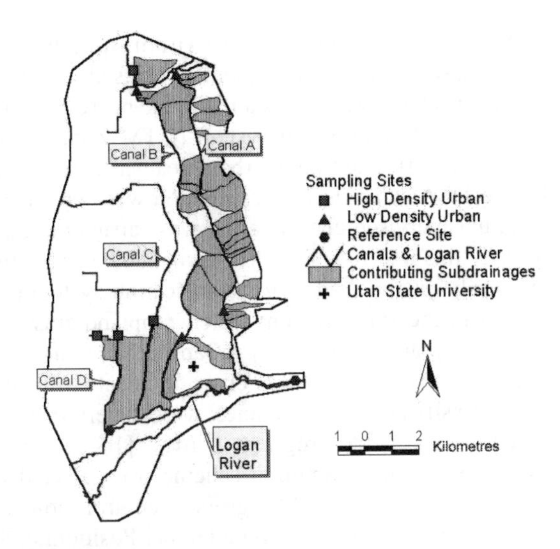

Figure 1: Canal system, sampling sites, and their respective contributing subdrainages within the Logan urban area, as outlined. Note the area designated as Utah State University does not contribute since its drainage is self-contained.

Logan is bordered to the east and west by the Bear River Mountain and the Wellsville Mountain Ranges, respectively. Land uses in the Logan area consist primarily of agricultural and urban and suburban land uses.

Management Information Systems, C. A. Brebbia (Editor)

The Logan River is a tributary to the Bear River, the main-stem tributary in the county, and its watershed drains approximately 65,964 hectares (ha) (UDWQ [11]). Water in the Logan River is diverted at varying locations into four main canals for agricultural and municipal irrigation. The uppermost diversion occurs in Logan Canyon and the subsequent three diversions occur downstream at locations throughout the city of Logan, as noted in Figure 1. In each of the canals, water flows north through the municipalities of Logan, North Logan, Hyde Park and Smithfield, and unincorporated areas. Water is generally diverted into the canals starting the beginning of May until early autumn. The canals run parallel to the hillslope and so provide a unique opportunity to observe stormwater runoff, as they intercept stormwater runoff from up-gradient land areas.

2.2 *ISA* calculation

The following methods describe the steps taken in calculating percent *ISA*. The first step involved classifying the land cover types within the satellite images. Next, the fractional vegetation cover per pixel was calculated, followed by the percent of *ISA* for each pixel.

2.2.1 Classification

Two Landsat 7 TM 30 metre resolution images (from April 26, 2000 and July 5, 2002) were first classified and subsequently utilised in the percent *ISA* computation. The images were georectified using the nearest neighbor technique and referenced to the UTM coordinate system (UTM Zone 12, spheroid GRS 1980, datum NAD 83 North). The Logan urban area was delineated for use in the analysis (ref., Figure 1). A supervised classification was performed on the April 2000 image and an unsupervised classification was carried out on the July 2002 image using ERDAS Imagine 8.5. For the April 2000 image, five land cover classes, based primarily on the Anderson classification system (Anderson *et al* [12]) were identified in the study region: water, cropland/grassland, bare soils, low density urban (LDU), and high density urban (HDU) (Yang and Lo [13]). A forest land cover class was an extra class delineated in the July 2002 classification. The classifications' accuracies were determined using ERDAS Imagine's accuracy assessment program (Erdas [14]). For the accuracy assessments, a stratified random sampling scheme was selected (Yang and Lo [13]). For the April 2000 image, 209 ground control points (GCP) were randomly selected and ground truthed using a Global Positioning System (GPS). A relatively smaller region of the July 2002 Landsat image was assessed for accuracy and, consequently, 43 GCP's were ground truthed.

2.2.2 Fractional vegetation cover

The next step was to establish a relationship between the Normalized Difference Vegetation Index (NDVI) and fractional vegetation cover (Fr), as established by Gillies and Carlson [15]. Calculations were performed using Imagine 8.5's geospatial modeling function. First, the images were converted to apparent reflectance (NASA [16]). The NDVI was then calculated as:

Management Information Systems, C. A. Brebbia (Editor)
© 2004 WIT Press, www.witpress.com, ISBN 1-85312-728-0

$$NDVI = \frac{NIR - RED}{NIR + RED}, \tag{1}$$

where *NIR* is the near-infrared and *RED* is the red apparent reflectances, respectively. Following *NDVI*, N^* was computed as:

$$N^* = \frac{NDVI - NDVI_0}{NDVI_S - NDVI_0}, \tag{2}$$

where $NDVI_o$ is the bare soil value and $NDVI_s$ is the dense vegetation value (Gillies and Carlson [15]).

Fr was finally computed as:

$$Fr = N^{*2}. \tag{3}$$

2.2.3 *ISA*

Based on the observation that *ISA* is inversely related to vegetation cover in an urban setting (Ridd [17]), the assumption is made (Carlson and Arthur [2]) that the percent *ISA* of the pixel was calculated as:

$$ISA = (1 - Fr)_{HDU / LDU}, \tag{4}$$

where the subscript, *HDU / LDU*, is included in the equation to indicate that percent *ISA* is calculated only for *HDU / LDU* classes Carlson and Arthur [2]. All other classes were assigned zero percent *ISA*.

2.3 Contributing subdrainage delineation

Contributing subdrainages for each sampling location were determined from USGS topographic (1:24,000 scale) quadrangles (ref., Figure 1). The polygons were then digitized and area computed using ArcView 3.3. The delineated contributing subdrainages were imported into ERDAS Imagine 8.5 and converted to areas of interest (AOI). The AOI's were then used to compute the *ISA* statistics of the contributing subdrainages. The resulting subset pixels for each contributing subdrainage were then converted to ASCII files. The ASCII files were imported into Microsoft's Excel where mean pixel values, including zero values, were calculated to determine the percent of *ISA* for each contributing subdrainage. Percent *ISA* for five of the eight contributing subdrainages was determined from the April 2000 image. The July 2002 image was used to compute percent *ISA* for the three remaining sites. This was justified since considerable development had occurred between image periods.

2.4 Water quality analysis

As stated previously, the canals originating from the Logan River were the subjects for the water quality analysis. During the spring, summer, and autumn of 2003, five stormwater runoff events were sampled.

Management Information Systems, C. A. Brebbia (Editor)

2.4.1 Sampling design and data collected

A stratified random sampling scheme was used to characterize NPS impacts from *HDU* and *LDU* sites. Ten locations (ref., Figure 1), including four *HDU*, four *LDU* and two reference sites, were sampled during storm events of greater than 2.5 mm total precipitation (Brezonik and Stadelmann [18]).

Water samples were analysed for the following: Dissolved metals (arsenic, barium, boron, copper, lead, iron, zinc, and aluminium), total nitrogen (TN), total phosphorus (TP), nitrate (NO_3-N), orthophosphate (PO_4-P), total suspended solids (TSS), hydrocarbons (HC), and fecal coliforms (FC). Field variables measured include: pH, temperature, dissolved oxygen (DO), specific conductivity, and turbidity. Most samples were collected and analysed according to Standard Methods' [19] guidelines.

2.4.2 Discharge and precipitation data

Flow was measured with the aid of a Flo-Mate portable flowmeter (Model 2000, Marsh-McBirney, Inc.) at each site at varying water stages. A rating curve was subsequently developed for each location to help expedite flow measurements during storm events. Hourly precipitation was collected and, the rainfall intensity was calculated as total precipitation divided by the duration (Hornberger *et al* [20]).

2.4.3 Loading analysis

Loading coefficient calculations were as follows for concentrations [conc] in mg/L as given by eqn 5:

$$\frac{[conc]\left(\frac{mg}{L}\right) \times discharge(cfs) \times 2.4468\ (conversion\ factor)}{area(ha)} = \frac{kg}{ha \cdot day} \tag{5}$$

Pollutant concentrations and loads during the five runoff events were subsequently averaged for each site.

3 Results

The following sections detail the results for the elements of *ISA* determination along with those comparisons of pollutant measurements.

3.1 Classifications and percent *ISA*

The accuracy assessment for the April 26, 2000 Landsat image had an overall classification accuracy of 81% and an overall kappa index of agreement of 0.69. Similarly, the July 5, 2002 image's overall classification accuracy was 77% and the overall kappa index of agreement was 0.67.

Computed percent *ISA* for the contributing subdrainages ranged from 9% to 58%. The *ISA* maps for the Logan urban area are shown in Figure 2:

Management Information Systems, C. A. Brebbia (Editor)

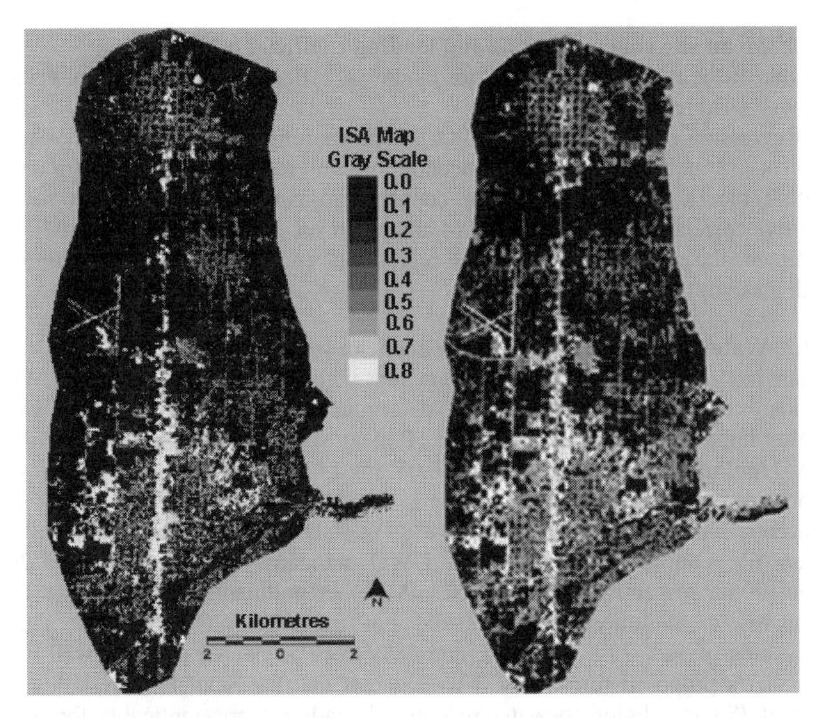

Figure 2: *ISA* maps of the Logan urban area. Left to right: April 2000 and July 2002.

3.2 Stormwater variables and percent *ISA*

Linear regressions were performed using SAS JMP version 4 statistical software (SAS Institute, Inc.) to determine if relationships existed between averaged storm event loading coefficients and concentrations with percent *ISA*.

Table 1: Regression results of both mean site loading coefficients and concentrations with percent *ISA*.

Mean Site Concentrations			*Mean Site Loading Coefficients*		
Variable	*R^2 Value*	*P-Value*	*Variable*	*R^2 Value*	*P-Value*
TN	0.94	<0.0001	*TN*	0.71	0.008
TP	0.86	0.0008	*TP*	0.62	0.021
NO_3-N	0.44	0.073	*PO_4-P*	0.50	0.052
PO_4-P	0.94	0.0001			
TSS	0.67	0.013			
Arsenic	0.71	0.074			
Boron	0.82	0.036			
Copper	0.98	0.001			
Lead	0.96	0.004			
Zinc	0.97	0.002			
Turbidity	0.67	0.013			

Management Information Systems, C. A. Brebbia (Editor)

3.2.1 Mean site concentrations and loading coefficients

Results of the regression analysis for loading coefficients and concentrations are shown in Table 1.

Regression results for TP, TN, PO_4-P, show strong positive correlations with percent *ISA* and both mean site concentrations and loading coefficients ($p<0.05$) (ref., Table 1). TSS, NO_3-N, boron, copper, lead, zinc, and turbidity also reveal positive correlations with percent *ISA* and mean site concentrations ($p<0.07$). As expected, the graphed results of the regressions (not shown) for TSS coincide with those of turbidity ($p<0.01$).

3.2.2 Water quality and Utah and European water quality standards

Mean site concentrations were compared with Utah water quality (UWQ) standards or pollutant indicators to determine the degree of potential water quality impact. Findings revealed that TP mean site concentrations exceeded the UWQ pollution indicator level of 0.05 mg/L at 12% *ISA*, lead mean site concentrations exceeded 0.0032 mg/L at 42% *ISA*, and TSS mean site concentrations exceeded 90 mg/L at 48% *ISA* (UDAR [21]). Aluminium and FC mean site concentrations exceeded UWQ standards (0.087 mg/L and 200 conc/100ml, respectively) in all cases. All other pollutants that have specified standards or pollution indicators did not exceed UWQ standards. These thresholds of 12% *ISA*, 42% *ISA*, and 48% *ISA* correlate reasonably well with Schueler's proposed thresholds. This suggests that the Schueler thresholds for percent *ISA* in relation to water resource degradation are applicable for some water quality variables in this semi-arid region.

European water quality standards (EWQS) (European Commission Water Quality Legislation [22]) were also compared to mean site concentrations. Aluminium was found to exceed the EWQS of 0.2 mg/L at 23% *ISA* and iron exceeds 0.2 mg/L in all cases. All other pollutants that have specified standards did not exceed the EWQS.

4 Discussion and conclusions

The main goal of this study was to examine the nature of the relationship between percent *ISA* (as computed from a remote sensing technique) and water quality (as indicated by water quality variables) in a semi-arid climate. The regression statistics indicate that *ISA* is a useful measure for assessing cumulative water resource impacts in such a climatic regime. In particular, high correlation coefficients are shown to exist between certain dependant variables and percent *ISA* (the independent variable) for such a region/regime. This suggests that *ISA* as derived, may be used as a surrogate to infer the potential loadings to be expected and, moreover those concentrations likely to exceed water quality standards in such rural-urban regions, such as those examined here. Furthermore, given the relative ease of determining *ISA* from remotely sensed sources coupled with the advantages of the spatial and temporal extent of such data are persuasive arguments that *ISA*, as computed, can serve as an important planning tool for protecting current and potential impacts to sensitive waters.

Management Information Systems, C. A. Brebbia (Editor)
© 2004 WIT Press, www.witpress.com, ISBN 1-85312-728-0

An important aspect of this analysis lies in the reliance of classifying the image into land cover types. While there exist multiple methods, with increasing complexity, to perform classifications to arguably better levels of accuracy, their implementation precludes their use for planners and natural resource managers. The method used here to classify land cover types and subsequently compute *ISA* is a relatively straightforward methodology that may be implemented effectively by planners and natural resource managers (Carlson *et al* [23]).

References

[1] Ragan, Robert M., M.ASCE and Jackson, Thomas J., A.M. ASCE, Use of satellite data in urban hydrologic models. *Journal of the Hydraulics Division*, pp. 1469-1475, 1975.

[2] Carlson, Toby N. and Arthur, S. Traci, The impact of land use-land cover changes due to urbanization on surface microclimate and hydrology: a satellite perspective. *Global and Planetary Change*, **25**, pp. 49-65, 2000.

[3] Leopold, L.B, *Hydrology for Urban Land Use Planning: A Guidebook on the Hydrologic Effects of Urban Land Use*, US Geological Survey Circular 554, 1968.

[4] Klein, Richard D., Urbanization and stream quality impairment. *Water Resources Bulletin: American Water Resources Association*, **15(4)**, pp. 948-963, 1979.

[5] Booth, Derek B. and Jackson, C. Rhett, Urbanization of aquatic systems-degradation thresholds, stormwater detention, and the limits of mitigation. *Journal of the American Water Resources Association*, **22(5)**, pp. 1-19, 1997.

[6] Schueler, Thomas, The importance of imperviousness. *Watershed Protection Techniques*, **1(3)**, pp. 1-12, 1994.

[7] Arnold, Chester L., and Gibbons, C. James, Impervious surface coverage: the emergence of a key environmental indicator. *Journal of American Planning Association*, **62(2)**, pp. 243-258, 1996.

[8] Gillies, Robert R., Brim-Box, Jayne, Symanzik, Jurgen, and Rodemaker, Eli J., Effects of urbanization on the aquatic fauna of the Line Creek Watershed, Atlanta- a satellite perspective. *Remote Sensing of the Environment*, **86**, pp. 411-422, 2003.

[9] Roy, A.H., Rosemond, A.D., Paul M.J., Leigh, D.S., and Wallace, J.B., Stream macroinvertebrate response to catchment urbanization (Georgia, U.S.A.). *Freshwater Biology*, **48**, pp. 329-346, 2003.

[10] Clausen, John C., Warner, Glenn, Civco, Dan, and Hood, Mark. Nonpoint education for municipal officials impervious surface research. www.nemo.uconn.edu/publications/research_reports/clausen_is-wq.pdf

[11] Bear River Watershed Description; Utah Division of Water Quality (UDWQ).www.eq.state.ut.us/EQWQ/watersheds/bear/watershed_description.htm#Physiography_20_20Geology

Management Information Systems, C. A. Brebbia (Editor)

[12] Anderson, R.J., Hardy, E.E., Roach, J.T., and Witmer, R.E., *A Land Use and Land Cover Classification System for Use with Remote Sensor Data*, US Geological Survey Professional Paper 964, 1976.
[13] Yang, X. and Lo, C.P., Using a time series of satellite imagery to detect land use and land cover changes in the Atlanta, Georgia metropolitan area. *International Journal of Remote Sensing*, **23(9)**, pp. 1775-1798, 2002.
[14] Erdas Imagine 8.4 *Tour Guides*. Erdas, Inc. Atlanta, GA., pp. 429-477, 1999.
[15] Gillies, R.R. and Carlson, Toby N., Thermal remote sensing of surface soil water content with partial vegetation cover for incorporation into climate models. *Journal of Applied Meteorology*, **34**, pp. 745-756, 1995.
[16] Science Data User's Handbook; NASA. ltpwww.gsfc.nasa.gov/IAS/handbook/handbook_toc.html
[17] Ridd, M.K., Exploring a v-i-s (vegetation-impervious surface-soil) model for urban ecosystem analysis through remote sensing: comparative anatomy for cities. *International Journal of Remote Sensing*, **16**, pp. 2165-2185, 1995.
[18] Brezonik, Patrick L. and Stadelmann, Teresa H., Analysis and predictive models of stormwater runoff volumes, loads, and pollutant concentrations from watersheds in the Twin Cities Metropolitan Area, Minnesota, USA. *Water Research*, **36**, pp. 1743-1757, 2002.
[19] Greenburg, Arnold E., Trussell, R. Rhodes, and Clesceri, Lenore S., (eds). *Standard Methods for the Examination of Water and Wastewater, 16th Ed.*, American Public Health Association: Washington, D.C., 1985.
[20] Hornberger, George M., Raffensperger, Jeffrey P., Wiberg, Patricia L., and Eshleman Keith N., *Elements of Physical Hydrology*, The John Hopkins University Press: Baltimore and London, pp. 25 & 102, 1998.
[21] Utah Division of Administrative Rules (UDAR), 2004. Rule R317-2 Standards of Quality for Waters of the State, www.rules.utah.gov/publicat/ code/ r317/r317-002.htm
[22] European Commission Water Quality Legislation (ECWQ), www.italocorotondo.it/tequila/module4/legislation/drink_water_directive_98.htm#ANNEX%20I
[23] Carlson, Toby N., Arthur, Traci, Ripley, David A., and Lembeck, Stanford, *Urban Planning and Satellite Remote Sensing: A Pilot Training Program for Planners*. The Pennsylvania State University: University Park, PA, 1999.

Management Information Systems, C. A. Brebbia (Editor)

Farm types as an alternative to detailed models in evaluation of agricultural practise in a certain area

I. T. Kristensen & I. S. Kristensen
Department of Agroecology, Danish Institute of Agricultural Sciences, Denmark

Abstract

Agriculture has a significant impact on the environment in Denmark due to nutrient emissions. To reduce the impact, it is necessary to have improved nutrient management both at national and regional scales. Our knowledge about important model parameters to describe nutrient balances has increased, but at a local level there is often a lack of the necessary data. Therefore, the Danish Institute of Agricultural Sciences has tried to find alternative data sources to describe nutrient management. One approach is to use data from databases of land and fertiliser use (e.g. registers established to support the administration of the EU Community agricultural support) combined with very detailed knowledge about a limited number of farms from a random sample of farm accounts and study farms.

A typology based on main production and stock density is established from a set of representative data sets. Models of production and nutrient flows for each farm type are developed based on detailed information from case studies etc. Based on register data (including geo-references for each farm), the farms are classified according to the typology and the total N-surplus in the region is estimated based on the amount of areas within each farm type.

The first test of the method has been on a regional scale in a vulnerable watershed. The preliminary results look promising and the work is now extending the model to analyse possible improvements in selected farm type factors on e.g. nutrient loss in the watershed.

Management Information Systems, C. A. Brebbia (Editor)

1 Introduction

The local effect of different initiatives to reduce nutrient emissions will often depend on the farm structure in the area, e.g. demands for specific handling of organic manure only effect farms that actually handle organic manure. General improvement in nutrient utilization takes place at different farm types. When knowledge about local effects of an initiative is needed, it is therefore important to know the local farm structure.

To some extent, it is possible to use a relative simple model to estimate the potential nutrient surplus in a given region, based on average N-surplus per ha for different types of farms and knowledge about the composition of farms in the region. The nutrient losses differ a lot depending on climate, soil, stocking, feeding, and handling of organic manure etc. The variation between farms is less within the same production branch and soil type, but still there will be differences between farms within the same type.

The process involves four main steps:

1. Defining farm type
2. Implementing farm types
3. Calculating N surplus fore each farm type
4. Calculating N-surplus in a region

2 Defining farm types

All Danish farms are obliged to keep detailed records of purchases and sales for tax purposes and the yearly accounts are made with professional help. A representative set of these accounts, are reported by the advisors to the Danish Research Institute of Food Economics and constitute the basic empirical input to the farm types presented here. Besides the economical data, information on the land use, livestock numbers and amounts produced are included in the data set by the advisors.

Farms are classified into 31 different farm types based on the parameters shown in Table 1. To secure representativity within each established typology only farm types that could be described by at least 14 accounts from the sample were allowed.

2.1 Example: dairy farms

The present dairy farm types are based on eight sub samples (Table 2). Together, they represent all Danish dairy farms with a maximum of 10% of Gross Margin from pig production. The total milk production on these types accounts for 85% of the total milk produced in Denmark.

The farms have been divided into groups to represent dairy production on sandy and loamy soil types respectively and with different stocking rates (number of standard livestock units per hectare (LSU)). Two separate types represent organic dairy farms that by definition have a stocking rate < 1.4 LSU. Farms with low or medium stocking rates usually produce 1-3 secondary products, which may differ from farm to farm. The resulting farm type thus

Management Information Systems, C. A. Brebbia (Editor)

represents an average of these secondary enterprises, but the number of small enterprises is not typical for a single farm.

Table 1: Parameters to define farm types.

	Percent of income from:	Percent of income from:
Sandy or loamy soil Calculated working hours per year Organic farming Cultivated area Stocking rate	- Horticulture crops - Fruit berry or tree - Seed crops - Sugar beet - Potatoes	- Pigs - Cattle (divided into dairy cattle and beef cattle) - Poultry (divided into chick and egg production)

Table 2: Dairy farm types.

Soil type	Loamy soil (clay)				Sandy soil			
Stocking rate	<1,4	1,4-2,3	>2,3	Organic farms	<1,4	1,4-2,3	>2,3	Organic farms
Number of accounts	23	32	14	24	83	182	16	127
Pct of total Danish milk production	4	7	3	1	15	43	4	9

Table 3: Parameters used in implementation.

Sandy or loamy soil Calculated working hours per year Percent of area with organic crops Cultivated area Stocking rate	Percent of area with: - Horticulture crops - Fruit berry or tree - Seed crops - Sugar beet - Potatoes	Percent of livestock unit from: - Pigs - Cattle (divided into dairy cattle and beef cattle) - Poultry (divided into chicken and egg production)

3 Implementing farm types

Data from three registers; animal, area subsidy and fertiliser accounts are chained together to form a picture of each farm. Then the amount of animals and farm size can be determined [1]. The parameters used differ in some cases from the parameters used within the representative data set. Table 3 shows the parameters used in classification based on register data.

In the basic formation of farm types, the production-value and standard working hours is used to determine primary production and part-time farms. That is not possible with register data. Instead of production value, the amount of livestock units are calculated and used in determining primary animal production. The production value per LSU is nearly the same for milk and pig production. Therefore, the same criteria can be used on LSU as well as production value. Cash crop production is determined by percent of area with certain cash crops.

The representative economic data, dos not include very small farms, therefore a hobby farm type is added to the types defined. A part-time farm is then defined as having less than 220-832 working hours and a hobby farm as having less than 220 hours. To calculate working hours a simple method is used which defines 1 cattle LSU = 46 hours and 1 ha = 22 hours.

The farm types are divided into groups depending on soil-type. This is determined by a GIS analysis. The farms are geo-referenced by their address using common address coordinates and the main soil type determined based on the Danish soil map. A more precise determination could be done using the placement of each farm fields by linking these to the Danish Field Block map [2].

To extend the information on organic farming, this farm type has been subdivided. The resulting 38 farm types are shown in Figure 1.

When the farms have been classified according to the typology, it is possible to calculate the area of each farm type within each field block. In Figure 2 the area belonging to farms with dairy cattle production and cash crop production, respectively are shown. The area within each field block has been aggregated to one km^2 grid.

4 Calculating N surplus for each farm type

Nutrient-balances in agriculture can be set up at several levels (scales). Balances can be set up for the single field, all fields of a farm, stable management, the whole farm or a region. No matter which level, the principle is the same. All products and sources of input and output are identified, and the amounts of nutrient in the products are calculated. The difference between nutrient input and nutrient output makes up the nutrient surplus of the particular system. The surplus consists of the sum of all losses from the system, changes in the nutrient amount in the system and the sum of uncertainty. The principle is shown in Figure 3 [3].

Farm N surplus is an expression of the long-term potential losses. The nutrient surplus is a good indicator for the total amount of losses to the surroundings. However, the nutrient surplus is not an unambiguous expression for loss to the environment. The surplus also includes changes in the amount of nutrient in the system. A part of the N-surplus will consist of an increase or decrease in the N-content of the soil caused by changing the amount of organic material in the soil. In many cases, it is desirable to know how the loss divides on different accounts as ammonia-evaporation, denitrification and N-leaching, e.g. when estimating nutrient leaching in a watershed. In this case, the N-leaching to the recipient will be crucial in estimating the demands for reduction.

In modelling the technical processes, data are used from: the advisory services (feeding and grazing practices), the Directorate for Food, Fisheries and

Management Information Systems, C. A. Brebbia (Editor)

Agri-business and Statistic Denmark (countrywide use of fertilizer and concentrates, partition of land use on different crops and their total yields).

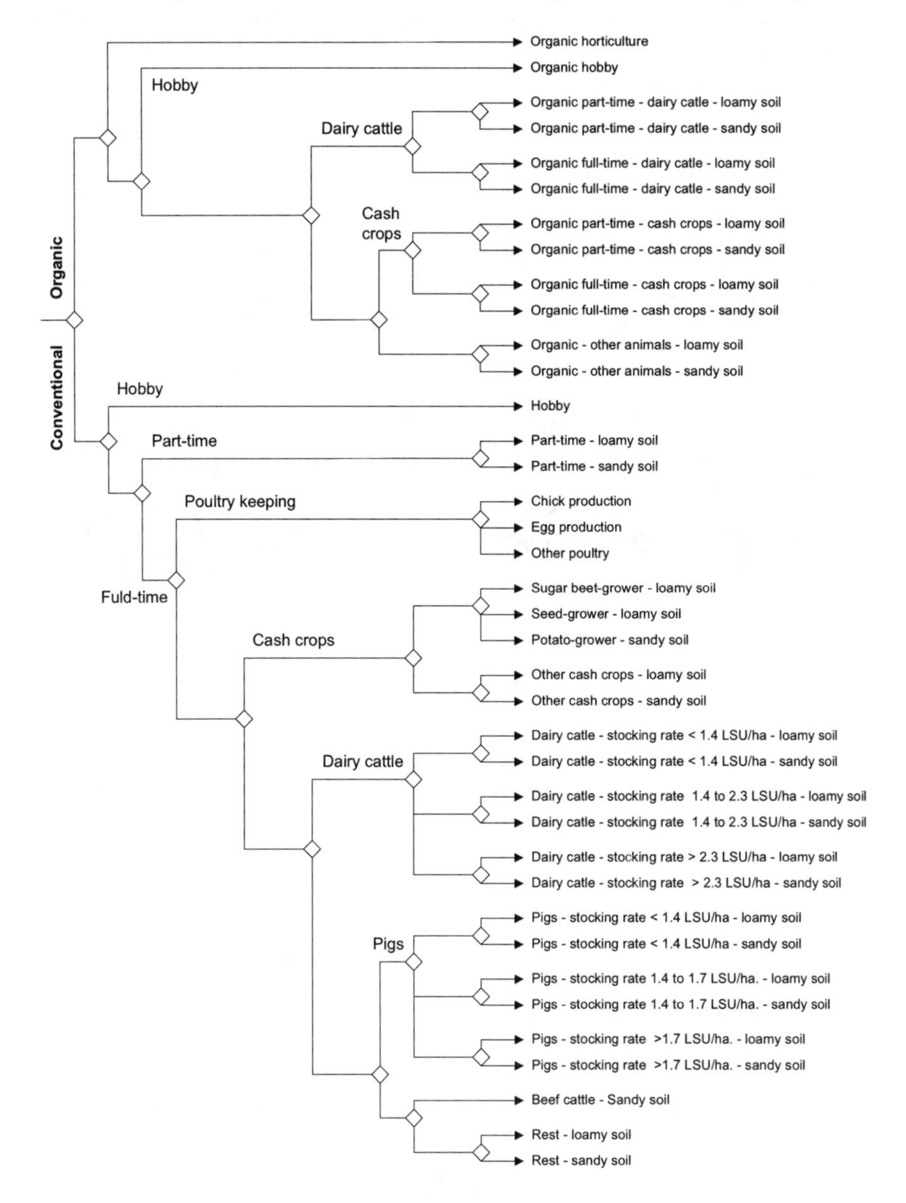

Figure 1: Resulting farm types.

N-balances are set up for each farm type. Average data from each farm type are used to model typical farms in terms of land use, herd size and production. The inputs and outputs (products and emissions) from the production processes

Management Information Systems, C. A. Brebbia (Editor)
© 2004 WIT Press, www.witpress.com, ISBN 1-85312-728-0

are quantified using the farm level as the basic unit. All the single enterprises have been described so that they fit coherently into the overall farm balances (e.g. crop production must fit the sum of home-grown feed used and exported). Thus, inputs of fertilizer, feeds and minerals are calculated to match the livestock and cash crop production after correction for home-grown feed.

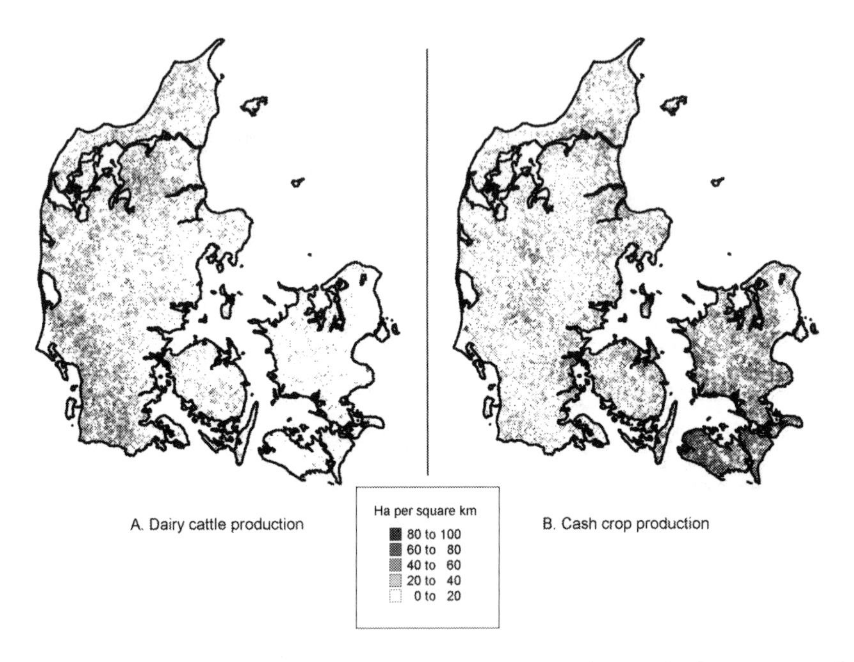

Figure 2: Distribution of area belonging to: A. Dairy cattle farms B. Cash crop production farms.

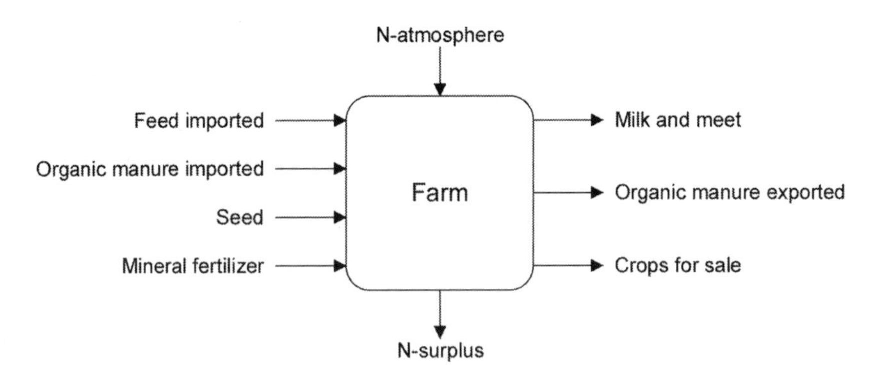

Figure 3: Elements in calculating the farm nitrogen turnover.

The nutrient turnover on the farm is calculated by multiplying the physical turnover of inputs and products with standard N- and P contents.

4.1 Distributing N-surpluses between losses

Emissions of ammonia, methane and nitrous oxide (N_2O) from the livestock, stables, manure storage and handling and from crop residues and soil are calculated using standard coefficients on the amounts of nutrients and feed dry matter [4]. The field N-balance was calculated by deducting the ammonia loss in animal housing and manure storage from the farm surplus. The ammonia losses after spreading and denitrification were calculated as a percent of the manure N applied. The remaining surplus includes the soil-N changes and leaching.

Partitioning this remaining field surplus between changes in soil-N and leaching is difficult. A preliminary attempt at partitioning the remaining field N-surplus between changes in soil-N and leaching is shown in Table 4.

Estimating leaching as the difference between farm gate surplus and gaseous emissions means that any errors in the surplus or gaseous emissions will be rolled up into that estimate. Therefore alternative methods are needed.

Table 4: Example: dairy cattle N-surplus per ha divided into accounts.

Soil type	Loamy soil (clay)				Sandy soil			
Stocking rate	<1,4	1,4-2,3	>2,3	Organic farms	<1,4	1,4-2,3	>2,3	Organic farms
N-surplus	116	188	227	37	128	177	217	101
Ammonia evaporation	27	48	62	18	29	44	60	24
Denitrification	32	41	36	30	13	17	20	15
Increase in soil N	10	29	33	18	23	41	45	37
N-Leaching	46	69	96	-30	64	75	93	25

4.2 Representativity of the farm accounts

The representativity of the farm accounts has been checked using standard methodology. The resource use and production on the farms have been validated at two levels: Internal coherence within each farm type and overall coherence between the sum of farm types and national level input use and production.

On the farm level the quantification of each type has been validated primarily by checking the coherence between land uses, crop yields and livestock production. At a higher hierarchical level the land use has been validated by comparing the sum of each crop acreage over all types with national statistics for the same year, e.g. checking that the total wheat area and total wheat yield does not differ more than a few percent from the national statistics.

Likewise, the total estimated use of inputs like fertilizer and concentrated feeds across all farm types have been checked against statistical information on national level. In case of differences that could not be ascribed to an error in a specific type, a general correction factor was multiplied into all types for the relevant input item.

Management Information Systems, C. A. Brebbia (Editor)
© 2004 WIT Press, www.witpress.com, ISBN 1-85312-728-0

5 Calculating N-surplus in a region

The method has been tested in a vulnerable watershed in northern Jutland. The farms in the watershed are identified by GIS analysis. Then the total N-surplus in the area is calculated by multiplying the amount of area belonging to each farm type with the national N-surplus per ha. Figure 4 shows farms in the area by their main types and sizes, and Figure 5 shows the distribution of area belonging to dairy cattle farms [5].

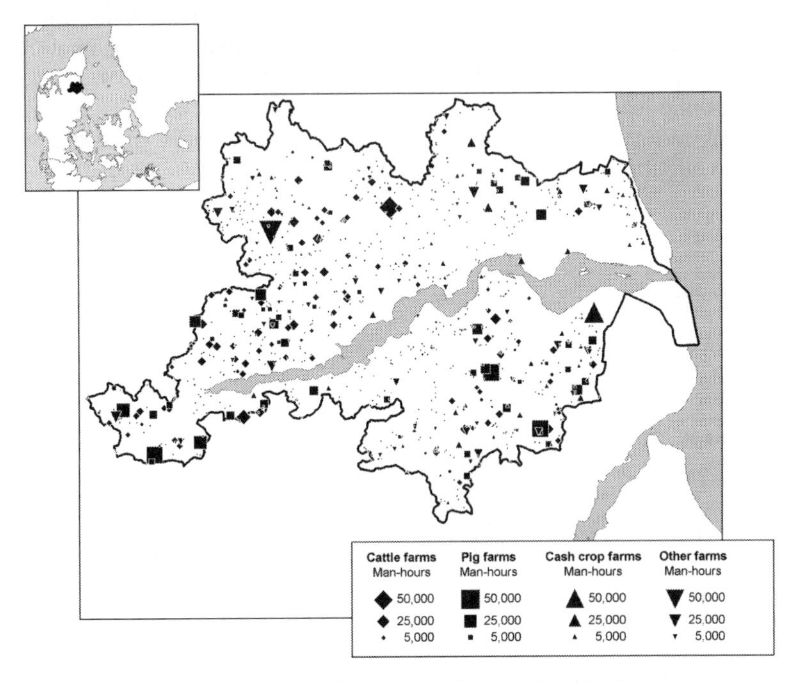

Figure 4: Farm main types and farm sizes in the area.

As shown in Figure 6 about three quarters of the estimated N-surplus comes from animal production farms, whereas cash crop and part-time only contribute with one quarter despite that they manage half the area. In case the result should be used to decide on the priority of initiatives to reduce N-leaching in the area it is also interesting that part-time farms seems to contribute with a considerable share. The ammonia loss mainly comes from cattle and pig farms. The ammonia loss from cash crop production mainly comes from import of organic manure.

How well the calculated N-surplus fits in a smaller region depends on how well the farms in the region, the climate etc. fits the national means that form the base of the typologies. The most important farm types in Mariager have a little higher stocking rate than the same farm types at national scale. This means that the estimated N-surplus in the area probably is 3-5% too low. This is supposed to have less importance for evaluating whether the loss is a problem and in case in which sub-area and for which farm type.

Management Information Systems, C. A. Brebbia (Editor)

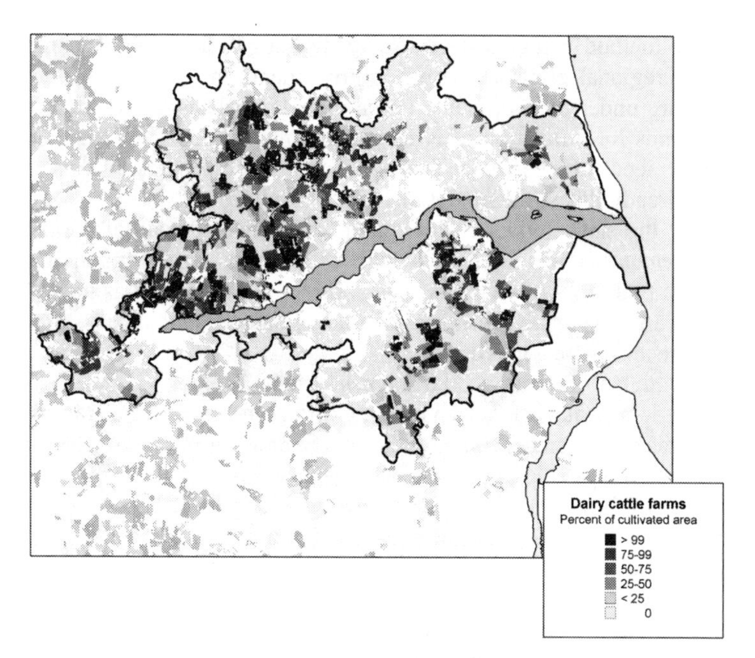

Figure 5: Percent of the cultivated area belonging to dairy cattle farms.

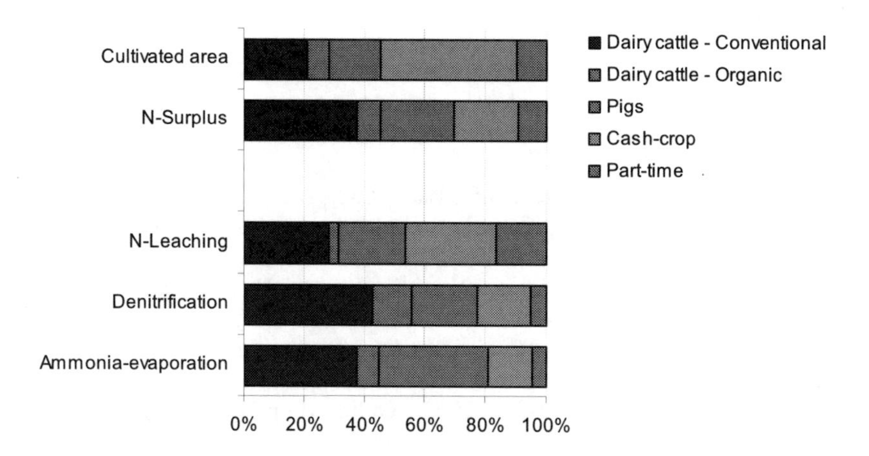

Figure 6: Distribution of N-surplus and loss on main farm types.

6 Evaluation

The method used in estimation of N-turnover at national level is considered sturdy. However, both ammonia loss and denitrification are subjects to some uncertainty and the used calculation of N-leaching as a residual is not satisfactory in the long term. Calculations of N-leaching as a difference must be treated with caution.

Management Information Systems, C. A. Brebbia (Editor)

Now, the method gives an overview over status in a region. Techniques to estimate the regional effect of partial improvements in N-utilization on single farm types are under development. These include analysis of the affect of e.g. higher demands for utilization of N in organic manure or exchanges of organic manure and feedstuff between farms. Turnover of other nutrients can be modelled. Already today, the model includes phosphorous.

However, the static model is not suited for analysis of partial regulation of field management as increased area with catch crops, straw incorporation or changed crop rotation; these factors affect both N in soil as well as the N-loss. If these factors are to be analysed a dynamic model that can calculate N-turnover in soil and plants has to be applied.

To obtain higher precision the farm types have to be extended. The farm types in the model do not take account to climatic differences across the country. Even though Denmark is a small country, the precipitation differs enough to have a major effect on the partitioning of the N-surplus on losses and changes in soil-N. The influence of these aspects will be evaluated by sensitivity analysis. Also, the amount of manure at cash crops or part-time farms will differ depending on the overall amount of animal per ha in the area. Areas with a high number of animals will be under-estimated and areas with few animals over-estimated. Extending some of the types with climatic regions and dividing the non-animal types depending on the amount of organic manure used, will improve regional results.

References

[1] Kristensen, I.T., Larsen, P.E., Utilisation of data created in control and administration in physical planning: *Management Information Systems 2000, GIS and Remote Sensing*. Eds: Brebbia, C.A., Pascolo, P, pp. 213-224, 2002.

[2] Sørensen, M.B., Pedersen, B.F., Bach, E.O., Larsen, P.E. Field maps and CABS revolutionise agricultural Science, *Remote Sensing controls 1999 - Final Technical Meeting*, Stresa, Italy, 1999.

[3] Kristensen, I. S., Halberg, N., Nielsen, A. H., Dalgaard, R., and Hutchings, N. N-turnover on Danish mixed dairy farms. *Workshop: Nutrient management on farm scale: how to attain European and national policy objectives in regions with intensive dairy farming.* Quimper, France. Part 2, pp 1-21. 2003. Online http://web.agrsci.dk/jbs/isk/DK_country_report_partII.pdf.

[4] IPCC, 2000. Intergovernmental Panel on Climate Change. Good Practice or Guidance and Uncertainty Management in Greenhouse Gas Inventories.

[5] Kristensen, I.T. & Rasmussen, B.M., Determining the geographical variation in Danish farming, *Third International Conference on Management Information Systems Incorporating GIS & Remote Sensing.* Eds: Brebbia, C.A., Pascolo, P, Haidiki, Greece, pp. 35-42, 2002.

Management Information Systems, C. A. Brebbia (Editor)

Expressing the configuration of the equipment and its characteristics in discharge lines of pumping wells

J. S. Santana & E. Pedroza-González
Instituto Mexicano de Tecnología del Agua, México

Abstract

The problem of information management of pumping stations is discussed. As a step towards its solution, a graphical language aimed at expressing the configuration of discharge lines is introduced in the context of an information system for water administration, and it is formally described by its syntactic and semantic rules. Its incorporation into the information system by means of an adequate user interface is shown, and it is concluded that tools like this are very useful in the gathering of large amounts of pumping stations' information.
Keywords: information systems, water administration, graphical languages, discharge lines, pumping stations.

1 Introduction

A project for improving water flow measurements in more than four thousand agricultural wells along the country, was entrusted to the Mexican Institute for Water Technology by the Mexican Government in 2001 [1]. The information about well's discharge lines was summarized with the aid of sketches (Fig. 1). Although their resulting quantity was enough for the verification of the current status of the measurement equipment and the generation of a recommendation for improving the measurements, it did not suffice as a foundation for recording the operation conditions of the discharge lines. The work presented here is aimed at the most accurate record of each discharge line information.

Management Information Systems, C. A. Brebbia (Editor)

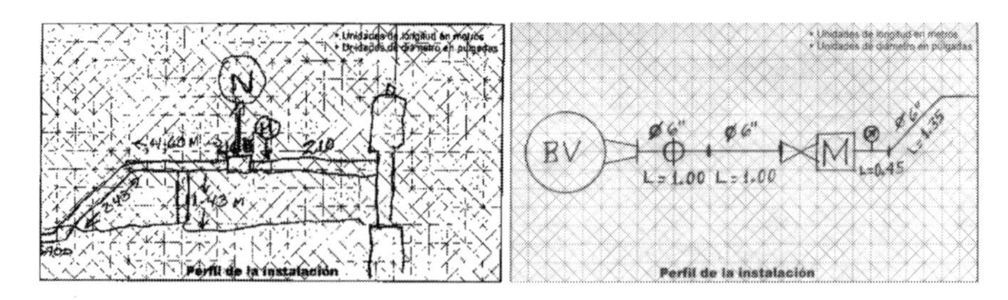

Figure 1: Two discharge line sketches.

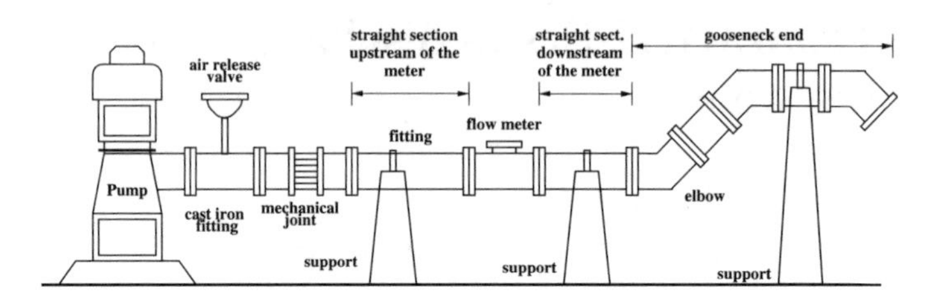

Figure 2: A typical discharge line.

2 The information of pumping wells discharge lines

Fig. 2 depicts a typical discharge line. We developed a database to capture arrangements like this [2]. Its tables are: **(1) WellForm**: Information of the well, **(2) Equipment**: Information of the discharge line equipment, **(3) Characteristic**: Characteristics of each piece of equipment, **(4) EqDict**: Interpretation of the **Equipment** and **Characteristic** tables, **(5) States** and **(6) Counties**: Names of the states and counties of the Country (Fig. 3). Table 1 shows the database tables **Equipment**, and **Characteristic**, with a GLang expression in its first row.

3 GLang: the graphical objects language

The syntax is given according to a formalism related to the BNF and the DCG formalisms [3, 4, 5]. Each syntactic rule defines a way to build the symbol in its *Left Hand Side* (LHS) and has the form: ***n* LHS→RHS.**, where, ***n*** is an integer. **LHS** is a single *non terminal* symbol or *syntactic category*, written in normal up roman characters. **RHS** is a sequence of *non terminal,* and *terminal* symbols, written in *italics* or enclosed in apostrophes, denoting the characters enclosed in them. When in italics, *Terminal* symbols denote the conventional expansion of the symbol, e.g., *GNumber* denotes all real numbers, and *ANSeq,* all alphanumeric sequences.

3.1 Language syntax: GLang grammar rules

1 SLSeq → SLine SLSeq .
2 SLSeq → SLine .

11 NSeq → *Number* ',' NSeq .
12 NSeq → *Number* .

Management Information Systems, C. A. Brebbia (Editor)

Table 1: **Equipment** and **Characteristic** tables example.

Folio	Eq id	EqName
...	...	...
81	0	GLang
81	1	Pmp
81	2	Pipe
81	3	Met
81	4	Pipe
81	5	Elbow45
81	6	Fitting
81	7	Elbow45
81	8	End
...	...	...
122	3	Vlv

Folio	Eq id	Chr id	ChrType	ChrDescr
81	0	Expr	TTxt	0:Pmp/1,Pipe/2,Met/3,Pipe/4,Elbow45/5, Fitting/6,Elbow45/7,End/8 .
81	1	Type	TTxt	Vertical
81	1	Diam	TNum	10
81	1	Pow	TNum	60
...	...	...	...	...
81	3	Type	TTxt	Propeller
81	3	Brand	TTxt	Micrometer
81	3	Diam	TNum	10

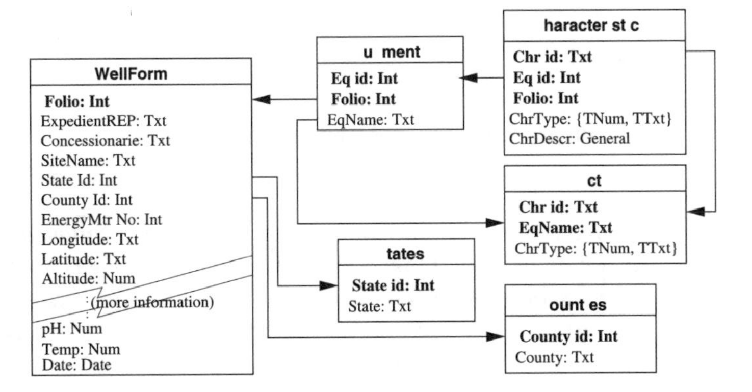

Figure 3: Database schema.

3 SLine → IdLine SimSeq '.' .
4 IdLine → *Number* ':' .
5 SimSeq → Sblock ',' SimSeq .
6 SimSeq → Sblock .
7 Sblock → SimId Args .
8 Sblock → SimId .
9 Args → '(' SimId ')' .
10 Args → '(' NSeq ')' .
13 SimId → SName '/' *Number* .
14 SName → *ANSeq* '{' ArgsList '}' .
15 SName → *ANSeq* .
16 ArgsList → Arg ',' ArgsList .
17 ArgsList → Arg .
18 Arg → *GNumber* .
19 Arg → *Number* .
20 Arg → SName .

The following is a valid expression of the language:
0: Pmp/1, Pipe/2, Met/3, Pipe/4, Elbow45/5, Fitting/6, Elbow45b/7, End/8 .

Fig. 4 depicts its graphical interpretation, and Fig. 5 shows a fragment of its phrase structure analysis, corresponding to the segment: "0: Pmp/1, Pipe/2, ...".

The phrase structure analysis of GLang expressions is done with a top-down chart parser we developed according to the techniques proposed elsewhere [5, 6, 7]. This structure, is described here by means of the following notation:

Notational convention 1 : SYNTACTIC TREES

1. *The empty syntactic tree is represented by* $\lceil\rceil$.
2. $\alpha\lceil\beta_1, \beta_2, ..., \beta_n\rceil$ *is a tree with root* α, *and branches* $\beta_1, \beta_2, ..., \beta_n$.

Management Information Systems, C. A. Brebbia (Editor)
© 2004 WIT Press, www.witpress.com, ISBN 1-85312-728-0

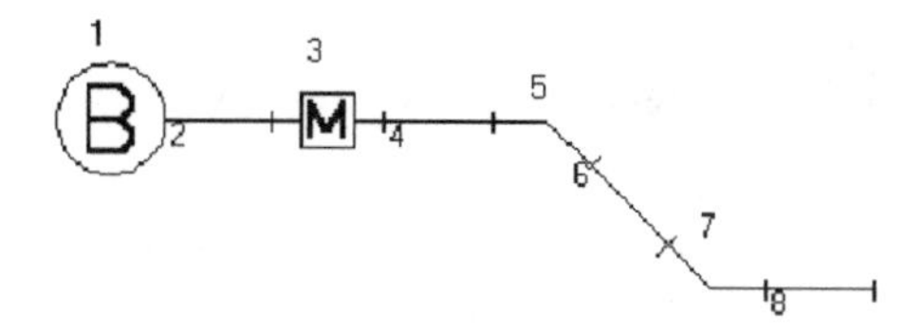

Figure 4: Graphical interpretation of a GLang expression.

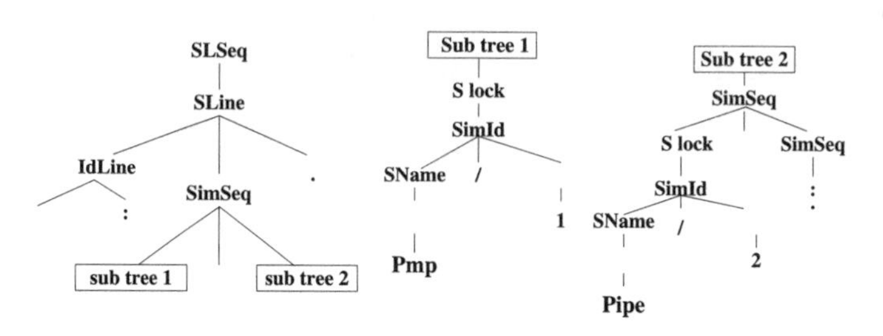

Figure 5: Example of a phrase structure tree.

3. *$\alpha\lceil ...\rceil$ is a tree with root α, and whose branches are not mentioned.*
4. *The linearisation of a tree T, denoted by $\ell(T)$, consists of the concatenation of all the leaves found in a depth-first order traversal of the tree.*
5. *The phrase structure tree given by the syntactic analysis of a phrase p, according to category* Cat, *is denoted by* $\Lambda(\text{Cat})(p)$.

Sub-tree 1, shown in Fig. 5, results from the syntactic analysis of the phrase "Pmp/1" according to syntactic category SBlock. That is, Λ(SBlock)(Pmp / 1) = sub-tree 1 = SBlock⌈SimId⌈SName⌈*ANSeq*⌈**Pmp**⌉⌉, '/' , *Number*⌈**1**⌉⌉⌉, and its linearisation is ℓ(sub-tree 1) = "Pmp/1". ℓ and Λ are inverse, i.e., $\ell(\Lambda(\text{Cat})(p)) = p$.

3.2 Language semantics

The description of GLang semantics is based on Montague semantics [8, 9, mainly ch. 3], and on denotational semantics [10, 11]. Some notation is introduced below.

STANDARD SETS: $\mathbb{R}$ is the set of real numbers, $\mathbb{N}$ is the set of positive integer numbers, $\mathbb{T}$ = {**false**, **true**}, and $\mathbb{U}$ ={**undef**}.

Notational convention 2 : SEQUENCES

Let D be any domain, then D^* denotes the domain of all finite sequences of elements of D, that is: $D^* = \{()\} \cup D \cup [D \times D] \cup [D \times D \times D] \cup \ldots$.

Definition 1 : SEQUENCES

If $d = (d_0, d_1, \ldots, d_i, \ldots, d_n) \in D^*$, then the function $\varepsilon_i : D^* \to D$ is defined as $\varepsilon_i(d) = d_i$. This function can also be expressed as: $d\,[i] = d_i$.

Notational convention 3 : ASSIGNMENTS

Let $g : U \to D$, be a function from domain U into domain D. $g^{u/d}$ indicates the value assignment exactly like g except that it assigns the element $d \in D$ to element $u \in U$. Similarly, $g' = g^{u_0/d_0, u_1/d_1, \ldots, u_n/d_n}$ indicates a value assignment exactly

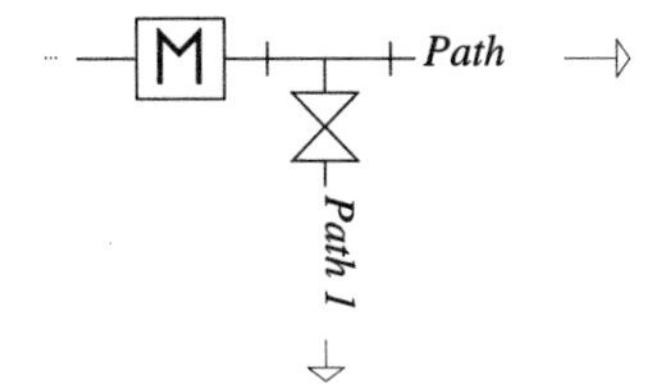

Figure 6: Insertion paths within a discharge line sketch.

like g, except that: $g'(u_0) = d_0, g'(u_1) = d_1, \ldots, g'(u_n) = d_n$.

Definition 2 : CONDITIONAL

Let **cond**:[D×D]→ $\mathbb{T}$ →D*, where* D *is any domain, be the function defined by* **cond**$(d_1, d_2)\, t = d_1$ *if* $t =$**true**, *and* **cond**$(d_1, d_2)\, t = d_2$ *if* $t =$**false**

This function can also be expressed as: $t \to d_1, d_2$*, meaning* **cond**$(d_1, d_2)\, t$.

Definition 3 : SEQUENCING OPERATOR $\star$

Let $f : [\mathbb{T} \times (D_1 \cup \mathbb{U})] \to [\mathbb{T} \times (D_2 \cup \mathbb{U})]$, *and* $g : [\mathbb{T} \times (D_2 \cup \mathbb{U})] \to [\mathbb{T} \times (D_3 \cup \mathbb{U})]$ *be two functions, where* D_1, D_2, and D_3 *are arbitrary domains, then* $f \star g : [\mathbb{T} \times (D_1 \cup \mathbb{U})] \to [\mathbb{T} \times (D_3 \cup \mathbb{U})]$ *is a function given by* $f \star g = \lambda(t, x)_\bullet \varepsilon_0(f(t, x)) \to g(f(t, x)), (\mathsf{false}, \mathsf{undef})$. *Here, functions are defined using standard lambda calculus* [12, 10, 8]. *The set* $\mathbb{U}$ *is introduced here to give the functions the possibility of assigning an undefined value.*

Definition 4 : CONCATENATION

Let α *and* β *be any two grammar symbols, then* $\alpha||\beta$ *is another grammar symbol built from the concatenation of* α *and* β*, that is:* $\alpha||\beta = \alpha\beta$.

3.2.1 State of interpretation

A *state of interpretation*, a set of variables describing a relevant state-of-affairs of *the geometrical world*, provides the way for placing graphical objects into a bi-dimensional space, where sketches describing discharge lines are realised. These objects are placed in the so called *insertion paths*, as shown in Fig. 6.

The state of interpretation variables are the following:

1. **PDot[]** $\in (\mathbb{R}^2)^*$ and **Angle[]** $\in \mathbb{R}^*$ are two arrays of as many elements as insertion paths there are in the discharge line. Each dot in **PDot[]** indicates the insertion place of any coming graphical object into the corresponding path, and each number in **Angle[]** indicates its insertion orientation angle.
2. **InPth** $\in \mathbb{N}$. This integer identifies the insertion path to be used.
3. **nLabel** $\in \mathbb{N}$. This integer is used as the label of the symbol being inserted.
4. **IDef** $\in \mathbb{N}$. This identifies elements within arrays **PDot[]** and **Angle[]**.
5. **OnInserted** $\in \mathbb{T}$. This marks if the graphical object is being inserted in the insertion path (**false**), or as a part of another graphical object (**true**).
6. **LastSymb** $\in$ Ⓢ. Identifier of the class of the last inserted graphical object.
7. **Sketch** $\subset \mathbb{R}^2$. This contains the graphical description of the equipment, abstracted from the physical space to get an insight into its actual arrangement.

Let the set of these variables identifiers, **StVars**, be defined as: **StVars** =
{**PDot[], Angle[], InPth, nLabel, IDef, OnInserted, LastSymb, Sketch**}.

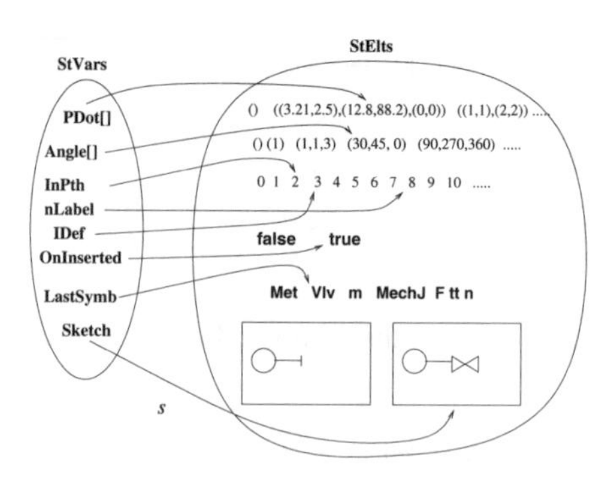

Figure 7: A state of interpretation s, viewed as function.

Then, the states of interpretation space is: **IntSt** = $(\mathbb{R}^2)^* \times \mathbb{R}^* \times \mathbb{N} \times \mathbb{N} \times \mathbb{N} \times \mathbb{T} \times$ Ⓢ $\times P(\mathbb{R}^2) + \mathbb{U}$, where Ⓢ is a set containing the identifiers of graphical object of symbol classes: Ⓢ = {Met, FCross, Elbow45, Elbow45b, Reduc, Suprt, Pmp, ...}. A^* is the set of all finite sequences of elements of any set A, and $P(A)$ is its power set: the set of all its subsets. $\mathbb{U}$ is used to represent any undefined state.

If **StElts** = $(\mathbb{R}^2)^* + \mathbb{R}^* + \mathbb{N} + \mathbb{T} +$ Ⓢ $+ P(\mathbb{R}^2)$, where the symbol $+$ represents the disjoint union of sets, then a state s ∈**IntSt**, can be viewed as a dot in the space **IntSt**, or as an assignment function, s :**StVars** → **StElts**, as illustrated in Fig. 7.

Some identifiers, e.g. **PDot[]**, are associated with sequences (see Definition 1). The identifiers of their individuals are generated according to:

Notational convention 4

If **X[]** *or* **X** *is an identifier which a state of interpretation function, s, associates with a sequence of elements* $d = (d_0, d_1, \ldots, d_i, \ldots, d_n)$*, then* **X[i]** *is the identifier that the same function, s, associates with the i-th sequence element* d_i*. Thus, using Definition 1,* s(**X**[i]) = $\varepsilon_i(s(\mathbf{X}))$ = $s(\mathbf{X})[i]$.

Let function s be defined as shown in Fig. 7, then s(**OnInserted**) = **true**, and s(**LastSymb**) = **Vlv** , but $s^{\text{LastSymb/JMec}}$(**LastSymb**) = **MechJ**, though $s^{\text{LastSymb/JMec}}$(**OnInserted**) = **true**.

Let **Mem** $\subseteq P(\mathbb{R}^2)$ be a set containing the graphical descriptions of symbol classes as well as their geometrical transformations. These graphical descriptions can be picked out by means of function *mem*: Ⓢ → **Mem.** Practically, this function is defined with the aid of a catalogue like the one presented in Code 1. Some of these graphical descriptions are shown in Fig. 8.

The catalogue registers the symbol identifier, a series of *relevant* dots and angles (with format 'D(X,Y),α'), and graphical primitives describing the symbol. Among others, these graphical primitives can be: a sequence of lines (L), a polyline (p), a closed polygon (P), or a character (f). The two functions defined below are used to pick out individual relevant dots or angles of a given symbol.

Definition 5

Dot:[**Mem**$\times\mathbb{N}$] $\rightarrow \mathbb{R}^2$ *is a function that assigns to each pair composed of a graphical object,* ***sg****, and an integer,* ***i****, the* ***i****-th relevant dot of the graphical object* ***sg****.*

Management Information Systems, C. A. Brebbia (Editor)
© 2004 WIT Press, www.witpress.com, ISBN 1-85312-728-0

Code 1 A symbols catalogue fragment.

```
Met
  D(10,10)
  D(0,5)
  D(10,5)
  t(3,9.5)
  L{(0,4),(0,6),(0,5),(2.5,5)}
  P{(2.5,2.5),(7.5,2.5),(7.5,7.5),(2.5,7.5)}
  w{2}
  f{M,(3.5,3.5),(3,3)}
  w{1}
  L{(7.5,5),(10,5),(10,4),(10,6)}
.
```

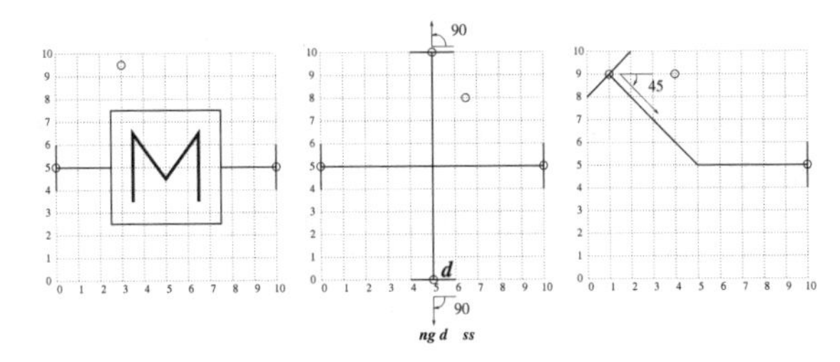

Figure 8: Graphical depictions of some symbols.

Table 2: Some graphical symbols.

Identifier	Symbol	Meaning
Enlr		Enlargement
Sprt		Support
Pmp		Pump
Fitting		Fitting

Identifier	Symbol	Meaning
Elbow45		45 deg elbow
ToGrnd		To ground
MechJ		Mechanical Joint
Met	M	Meter

That is, **Dot(*sg*, *i*)** *is the* ***i****-th relevant dot of the graphical object* ***sg****.*
Ang:[**Mem** $\times \mathbb{N}$] $\rightarrow \mathbb{R}$ *is a function that assigns to each pair composed of a graphical object,* ***sg****, and an integer,* ***i****, the* ***i****-th relevant angle of* ***sg****.*

Thus, according to Fig. 8(*a*), *mem*(Met) = M . Table 2 shows a fragment of this function, with arguments in column "Identifier" and function values in column "Symbol". The graphical symbols or objects are described in a normalized virtual space. In order to be incorporated into a sketch in a given state of interpretation, they will require to be affinely transformed [13]. For a given symbol, whose identifier is ***id***, the resulting transformed graphical symbol required by a given state of interpretation, ***s***, is τ(***s***,***id***), where ***s***∈**IntSt**, and ***id***∈ Ⓢ.

Management Information Systems, C. A. Brebbia (Editor)
© 2004 WIT Press, www.witpress.com, ISBN 1-85312-728-0

Table 3: Meaning of some lexical constants.

Category (α)	Description	Example s (β)	Denotation Δ[β] (v,s)
GNumber	Real numbers	3.1416 000.032	(3.1416, *s*) (0.032, *s*)
Number	Positive integer num.	0 88	(0, *s*) (88, *s*)
ANSeq	Alphanumeric sequences	Elbow45 FCross	(Elbow45, *s*) (FCross, *s*)

3.2.2 Semantic rules

Let us call the semantic value of each syntactic rule given in section 3.1, its *denotation*, and let us introduce a function Δ, which associates each valid GLang expression with its denotation as: Δ: **Exp** → [[**Val** × **IntSt**] → [**Val** × **IntSt**]], where, **Exp** is the set of all valid GLang expressions, and **Val** is the set of possible values: **Val** = $\mathbb{T} \cup \mathbb{R} \cup \mathbb{N}$. Then Δ is a *semantic* function such that $\Delta[e]\,(v,s) = (v_0, s_0)$ is the denotation of expression e with an input *value* v in state of interpretation s.

Let us define here a function to generate a new state from another state:

Definition 6 : STATE TRANSITION

The state transition function, φ : (**StVars**×**StElts**)* → [[**Val**×**IntSt**] → [**Val**×**IntSt**]], *is defined as* $\varphi = \lambda((i_0,a_0),(i_1,a_1),\ldots,(i_n,a_n))_\bullet[\lambda(v,s)_\bullet(v \neq$ **false**$) \rightarrow (v, s^{i_0/a_0, i_1/a_1, \ldots, i_n/a_n,})$,(**false**,**undef**)].

Then, $\varphi(i,a)(v,s) = (v, s^{i/a})$ if $v \neq$ **false**. That is, given s, i, and a, φ generates a new state, s_0, exactly like s, except that now $s_0(i) = \alpha$.

Function Δ is defined by semantic rules, corresponding to the syntactic ones:

Rule 0 (Terminal symbols or **lexical elements)**: If a subtree $\gamma = \alpha\,\lceil\beta\rceil$ is introduced by a lexical category α, e.g. *GNumber*, and β is a lexical constant of category α, then $\Delta\,[\gamma]\,(v,s) = \Delta\,[\beta]\,(v,s)$. Table 3 shows the meaning of some lexical constants.

Rule 1: If α is SLine$\lceil\ldots\rceil$, β is SLSeq$\lceil\ldots\rceil$, and γ is SLSeq$\lceil\alpha,\beta\rceil$, then $\Delta[\gamma](v,s) = (\Delta\,[\alpha] \star \Delta\,[\beta])\,(v,s)$, or, simplifying, $\Delta[\gamma] = (\Delta\,[\alpha] \star \Delta\,[\beta])$.

Rule 2: If α is SLine$\lceil\ldots\rceil$, and γ is SLSeq$\lceil\alpha\rceil$, then $\Delta[\gamma](v,s) = (\Delta\,[\alpha])\,(v,s)$.

Rule 3: If α is IdLine$\lceil\ldots\rceil$, β is SimSeq$\lceil\ldots\rceil$, and γ is SLine$\lceil\,\alpha,\beta,\,'.'\,\rceil$, then $\Delta[\gamma](v,s) = (\Delta\,[\alpha] \star \Delta\,[\beta])\,(v,s)$, or $\Delta[\gamma] = (\Delta\,[\alpha] \star \Delta\,[\beta])$.

Rule 4: If α is *Number*$\lceil\ldots\rceil$, and γ is IdLine$\lceil\alpha,':'\rceil$, then $\Delta[\gamma](v,s) = (v_0, s_0)$, where v_0= **true**, and $s_0 = s^{\text{IIdPth}/\varepsilon_0(\Delta[\alpha](v,s))}$. Note that if v =**false**, then (v_0, s_0) =(**false**, **undef**), whatever the value of α is. The application of this rule implies a state transition from s to s_0, where $\varepsilon_0(\Delta\,[\alpha]\,(v,s))$ is assigned to **InPth**.

Rule 5: If α is SBlock$\lceil\ldots\rceil$, β is SimSeq$\lceil\ldots\rceil$, and γ is SimSeq$\lceil\alpha,',',\beta\rceil$, then $\Delta[\gamma](v,s) = (\Delta\,[\alpha] \star \Delta\,[\beta])\,(v,s)$, or $\Delta[\gamma] = (\Delta\,[\alpha] \star \Delta\,[\beta])$.

Rule 6: If α is SBlock$\lceil\ldots\rceil$, and γ is SimSeq$\lceil\alpha\rceil$, then $\Delta[\gamma](v,s) = (\Delta\,[\alpha])\,(v,s)$.

Rule 7: If α is SimId$\lceil\ldots\rceil$, β is Args$\lceil\ldots\rceil$, and γ is Sblock$\lceil\alpha,\beta\rceil$, then $\Delta[\gamma](v,s) = (\Delta\,[\alpha] \star \Delta\,[\beta])\,(v,s)$, or $\Delta[\gamma] = (\Delta\,[\alpha] \star \Delta\,[\beta])$.

Management Information Systems, C. A. Brebbia (Editor)

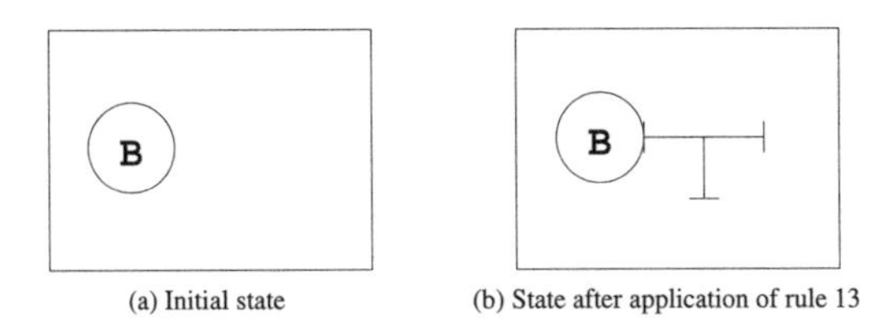

(a) Initial state (b) State after application of rule 13

Figure 9: Interpretation of semantic rule 13.

Rule 8: If α is SimId$\lceil\ldots\rceil$, and γ is Sblock$\lceil\alpha\rceil$, then $\Delta[\gamma](v,s) = (\Delta[\alpha])(v,s)$.
Rule 9: If α is SimId$\lceil\ldots\rceil$, and γ is Args$\lceil'(',\alpha,')'\rceil$, then $\Delta[\gamma](v,s) = (\varphi(\textbf{OnInserted}, \textbf{true})\star\Delta[\alpha]\star\varphi(\textbf{OnInserted}, \textbf{false}))(v,s)$ (see Definition 6).

This rule manages the insertion of a symbol α as a part of another graphical symbol and not as a part of the main insertion path.
Rule 10: If α is NSeq$\lceil\ldots\rceil$, and γ is Args$\lceil'(',\alpha,')'\rceil$, then $\Delta[\gamma](v,s) = (\varphi(\textbf{IDef},2)\star\Delta[\alpha])(v,s)$, or $\Delta[\gamma] = (\varphi(\textbf{IDef},2)\star\Delta[\alpha])$.
Rule 11: If α is *Number*$\lceil\ldots\rceil$, β is NSeq$\lceil\ldots\rceil$, and γ is NSeq$\lceil\alpha,',',\beta\rceil$, then $\Delta[\gamma](v,s) = (\varphi((\textbf{PDot}[\varepsilon_0(\Delta[\alpha](v,s))], \textbf{Dot}(\tau(s,\textbf{LastSymb}), s(\textbf{IDef}))), (\textbf{Angle}[\varepsilon_0(\Delta[\alpha](v,s))], \textbf{Ang}(\tau(s,\textbf{LastSymb}), s(\textbf{IDef}))), (\textbf{IDef}, s(\textbf{IDef})+1))\star\Delta[\beta])(v,s)$.
Rule 12: If α is *Number*$\lceil\ldots\rceil$, and γ is NSeq$\lceil\alpha\rceil$, then $\Delta[\gamma](v,s) = \varphi((\textbf{PDot}[\varepsilon_0(\Delta[\alpha](v,s))], \textbf{Dot}(\tau(s,\textbf{LastSymb}), s(\textbf{IDef}))), (\textbf{Angle}[\varepsilon_0(\Delta[\alpha](v,s))], \textbf{Ang}(\tau(s,\textbf{LastSymb}), s(\textbf{IDef}))), (\textbf{IDef}, s(\textbf{IDef})+1))(v,s)$.
Rule 13: If α is SName$\lceil\ldots\rceil$, β is *Number*$\lceil\ldots\rceil$, and γ is SimId$\lceil\alpha,'/',\beta\rceil$, then $\Delta[\gamma](v,s) = \varphi((\textbf{nLabel}, \varepsilon_0(\Delta[\beta](v,s))), (\textbf{Sketch}, s(\textbf{Sketch}) \cup \tau(s, \varepsilon_0(\Delta[\alpha](v,s)))), (\textbf{PDot}[s(\textbf{InPth})], \textbf{Dot}(\tau(s, \varepsilon_0(\Delta[\alpha](v,s))), \textbf{1})), (\textbf{Angle}[s(\textbf{InPth})], \textbf{Ang}(\tau(s, \varepsilon_0(\Delta[\alpha](v,s))), \textbf{1})), (\textbf{LastSymb}, \Delta[\alpha](v,s)))(v,s)$.

This rule inserts a new graphical symbol into the sketch. Fig. 9(a) shows an initial state of $s(\textbf{Sketch})$, and Fig. 9(b) shows the representation of $s(\textbf{Sketch}) \cup \tau(s, \varepsilon_0(\Delta[\Lambda(\text{SName})(\text{“FTee”})](v,s)))$, where a flanged Tee is inserted.
Rule 14: If α is *ANSeq*$\lceil\ldots\rceil$, β is ArgsList$\lceil\ldots\rceil$, and γ is SName$\lceil\alpha,'\{',\beta,'\}'\rceil$, then $\Delta[\gamma](v,s) = ((\varepsilon_0(\Delta[\alpha](v,s)) \,||\, '\{' \,||\, \varepsilon_0(\Delta[\beta](v,s)) \,||\, '\}'), s)$.

This rule enables the use of arguments for symbol names. Their interpretation is done when consulting the symbols catalogue (Code 1), in a process similar to the *morphological analysis* of the natural languages processing [14].
Rule 15: If α is *ANSeq*$\lceil\ldots\rceil$ and γ is SName$\lceil\alpha\rceil$, $\Delta[\gamma](v,s) = (\varepsilon_0(\Delta[\alpha](v,s)), s)$.
Rule 16:If α is Arg$\lceil\ldots\rceil$, β is ArgsList$\lceil\ldots\rceil$, and γ is ArgsList$\lceil\alpha,',',\beta\rceil$, then $\Delta[\gamma](v,s) = ((\varepsilon_0(\Delta[\alpha](v,s)) \,||\, ',' \,||\, \varepsilon_0(\Delta[\beta](v,s))), s)$.
Rule 17: If α is Arg$\lceil\ldots\rceil$ and γ is ArgsList$\lceil\alpha\rceil$, $\Delta[\gamma](v,s) = (\varepsilon_0(\Delta[\alpha](v,s)), s)$.
Rule 18: If α is *GNumber*$\lceil\ldots\rceil$, and γ is Arg$\lceil\alpha\rceil$, then $\Delta[\gamma](v,s) = (\alpha, s)$.
Rule 19: If α is *Number*$\lceil\ldots\rceil$, and γ is Arg$\lceil\alpha\rceil$, then $\Delta[\gamma](v,s) = (\alpha, s)$.
Rule 20: If α is SName$\lceil\ldots\rceil$, and γ is Arg$\lceil\alpha\rceil$, then $\Delta[\gamma](v,s) = \Delta[\alpha](v,s)$.
EXAMPLE. The analysis of expression “FTee/2” gives tree γ =SimId$\lceil$ Λ(SName) (“FTee”), '/', *Number*$\lceil 2\rceil\,\rceil$, whose denotation, according to semantic rule 13, is:

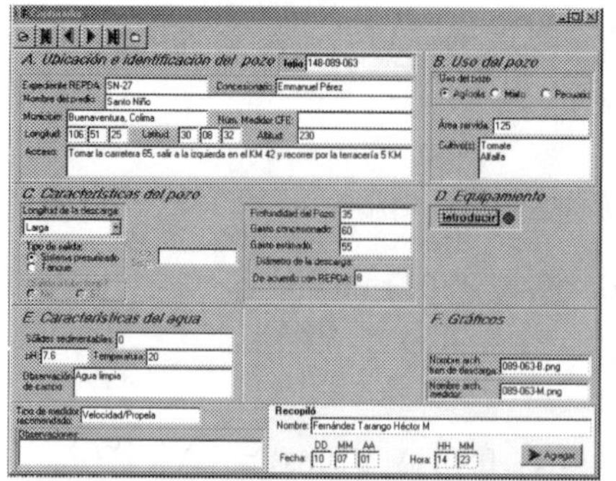

(a) Screen to introduce well's information

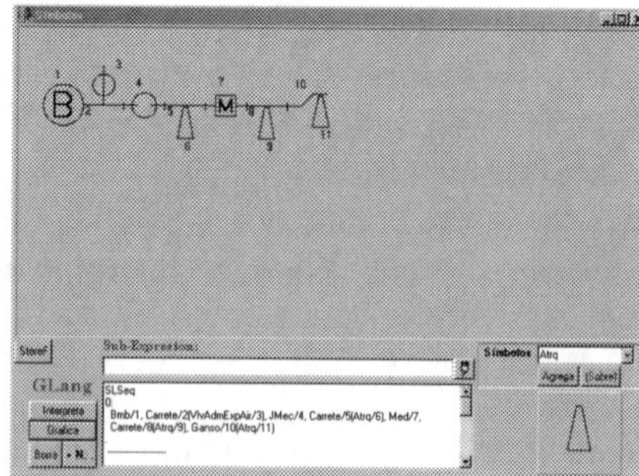

(b) Screen to introduce GLang expressions

Figure 10: Two aspects of the user interface.

$\Delta[\gamma](v,s) = \varphi((\textbf{nLabel}, \varepsilon_0(\Delta[\textit{Number}\lceil 2\rceil]\,(v,s))), (\textbf{Sketch}, s(\textbf{Sketch}) \cup \tau(s, \varepsilon_0(\Delta[\Lambda(\text{SName})(\text{“FTee”})]\,(v,s)))), (\textbf{PDot}[s(\textbf{InPth})], \textbf{Dot}(\tau(s, \varepsilon_0(\Delta[\Lambda(\text{SName})(\text{“FTee”})]\,(v,s))), \mathbf{1})), (\textbf{Angle}[s(\textbf{InPth})], \textbf{Ang}(\tau(s, \varepsilon_0(\Delta[\Lambda(\text{SName})(\text{“FTee”})]\,(v,s))), \mathbf{1})), (\textbf{LastSymb}, \Delta[\Lambda(\text{SName})(\text{“FTee”})]\,(v,s)))\,(v,s)$.
The graphical result of this example has been shown in Fig. 9.

4 User interface

The user interface has several levels to capture and display information:
Physical information of a well. This form (Fig. 10(a)) is to capture each well information stated in Table **WellForm** (Fig. 3). The button "Introducir" displays the next interface level.
Discharge line equipment information. This form (Fig. 10(b)) allows the user the introduction of the equipment arrangement, either, by writing directly a GLang expression, or, by selecting orderly the pieces of equipment from the *Combo-Box* and the graphical panel in the right-down corner of the form. In this case, the system will produce the GLang expression. The GLang expression graphical depiction can be viewed by depressing the button labelled "Grafica".
Characteristics of each piece of equipment. Each time a symbol is selected to be included in the discharge line (Fig. 10(b)), the system will display a new form, similar to that shown in Fig. 10(a), to introduce the characteristics of the piece of equipment represented by this symbol.

5 Conclusion

A method to standardise the description of equipment installed in discharge lines of pumping wells has been presented. The kernel of this method is the GLang

language, formally described by its syntactic and semantic rules. A tool like this will enable the capturing of information of a large amount of pumping wells in an easy and ambiguities-free manner. Furthermore, it will allow, in a simpler way, the automatic issuing of recommendations for enhancing the discharge lines of pumping wells whose information has been captured with this tool.

References

[1] Santana, J.S. & Pedroza González, E., Un sistema de información para el mejoramiento de la medición en pozos agrícolas. *Memorias del XVII Congreso Nacional de Hdráulica*, eds. A.I. Ramírez, F.J. Aparicio, F. Arreguín, G. Paz Soldán, B. de León, H. Marengo, G. Sotelo, N.H. García, J. Hidalgo, E. Gutiérrez & G. Ortiz, Monterrey, N.L. México, pp. 481–486, 2002.

[2] Stones, R. & Matthew, N., *Beginning databases with PostgreSQL*. Wrox Press Ltd: Arden House, 1102 Warwick Road, Acocks Green, Birmingham, B27 6BH, UK, 1st edition, 2001.

[3] Aho, A. & Ullman, J., *Principles of Compiler Design*. Addison-Wesley: Reading, MA, U.S.A., 1977.

[4] Naur, P., Revised report on the algorithmic language Algol 60. *Communications of the ACM*, **6(1)**, pp. 1–17, 1963.

[5] Gazdar, G. & Mellish, C., *Natural language processing in PROLOG*. Addison-Wesley: U.S.A., 1989.

[6] Allen, J., *Natural language understanding*. The Benjamin/Cummings Publishing Company, Inc.: 390 Bridge Parkway, Redwood City, CA 94065, U.S.A., 2nd edition, 1995.

[7] Kay, M., Algorithm schemata and data structures in syntactic processing. *Readings in natural language processing*, eds. B.J. Grosz, K.S. Jones & B.L. Webber, Morgan Kaufmann: Los Altos, U.S.A., pp. 35–70, 1980/1986.

[8] Dowty, D.R., Wall, R.E. & Peters, S., *Introduction to Montague semantics*, volume 11 of *Studies in Linguistics and Philosophy*. D. Reidel Publishing Company: P.O.Box 17, 3300 AA Dordrecht, Holland, 1981.

[9] Santana, J.S., *The Generation of Coordinated Natural Language and Graphical Explanations in Design Environments*. Ph.D. thesis, University of Salford, Salford M5 4WT UK, 1999.

[10] Gordon, M.J.C., *The Denotational Description of Programming Languages*. Springer-Verlag: New York, NY, U.S.A., 1979.

[11] Tennent, R.D., *Semantics of Programming Languages*. Prentice-Hall, 1991.

[12] Church, A., A formulation of a simple theory of types. *Journal of Symbolic Logic*, **5**, pp. 56–68, 1940.

[13] Arttzy, R., *Linear geometry*. Mathematics, Addison-Wesley Publishing Co. Inc.: Reading, MA, U.S.A., 1965.

[14] Sproat, R., Lexical analysis. *Handbook of natural language processing*, eds. R. Dale, H. Moisl & H. Somers, Marcel Dekker, Inc.: New York, U.S.A., pp. 37–57, 2000.

Management Information Systems, C. A. Brebbia (Editor)

Tourism information based on visualisation of multimedia geodata – ReGeo

M. Mosch[1], I. Frech[1], C. Schill[1], B. Koch[1], H. Stelzl[2], A. Almer[2] & T. Schnabel[2]
[1]*Department Remote Sensing and Landscape Information Systems, Albert-Ludwigs-University Freiburg i.Brsg., Germany*
[2]*JOANNEUM RESEARCH, Institute of Digital Image Processing, Graz, Austria*

Abstract

Tourist regions in rural areas are particularly interested in using advanced presentation technologies to raise their attractiveness. Tourism information systems, especially on the World Wide Web, can be enhanced through the use of interactive maps [1] that represent information in either two or three dimensions [2]. An information system has been developed for the ReGeo project that is based on a virtual geo-multimedia database [3] and serves a variety of different applications. The system will be accessible to offline and online devices, such as the Internet, palmtop computers, info-terminals and PCs. The visualisation component of this project demonstrates how attractive and useful 2D- and 3D-visualisations can be integrated into information systems. The central object of this system is a 2D-map that is able to display information on demand and in a range of theme combinations (i.e. accommodation + hiking tours + sightseeing). The user is able to choose between either a topographic map or a remote sensing image as a background. The latter offers an impression of the region's landscape, which is even better represented through the interactive 3D-map. This 3D-map allows the user to navigate through a virtual model of the chosen destination and to display information mentioned in the 2D-map upon demand. Panoramic views and virtual flights over the landscape complete the representation possibilities.
Keywords: tourist information system, 2D and 3D visualisation, geodata, interactive map, 3D object extraction.

Management Information Systems, C. A. Brebbia (Editor)

1 Introduction

The ReGeo project has produced a tourism information system that visualises geo-referenced tourist information on maps and in 3D-views. The system is based on a Virtual Geo-Multimedia Database and Exchange Framework, which was developed by our partner Joanneum Research (IIS) and enables the customer to store, retrieve and share all types of geo-referenced multimedia content. Customers have to ability to easily access geographic and multimedia objects via a standard Web service interface for any purpose - from marketing to pure administration. Geo-referenced objects, together with their associated multimedia presentations, are integrated into maps as points of interest such as hotels, restaurants and sights; or as tours, such as hiking, biking or cross-country skiing tours. Multimedia presentations include text, pictures, audio, video or 3D models. Each ReGeo application has been especially designed to conform to a specific purpose or device. The basic design of the 2D-maps and 3D-views is used by both online and offline applications. The web application, which has been developed by our partner Geokonzept, is the most visible part of Regeo and is based on an integration of the Open Source Web Application Server ZOPE and the open source "Minnesota mapserver". The application requests data from the above-mentioned database via the WFS protocol, and creates an internal representation for the Zope content management system by "reading" the GML code retrieved. This information can then be presented in HTML format over the Internet. The content management system contains 2D-maps, as well as a 3D-viewer, which requires a Shockwave plug-in.

Taxus, another partner in the ReGeo project, is currently working on mobile applications for PDAs and CD ROMs. Both versions are technically offline in nature. While the CD ROM application delivers a unique multimedia presentation of a region with a somewhat "frozen" information content, the mobile PDA component also has the capacity to update its information via an Internet connected PC or from terminal stations in the test sites. Because information stored in the handheld device can be regularly updated, additional multimedia content, such as virtual flyovers and 3D models can also be integrated. If the PDA is used in conjunction with a GPS module, the system will even display the precise position of the user on the current map.

A large amount of time and effort has been invested in developing the unique visualisation techniques used to present geographical data in these applications.

2 Interactive 2D maps

Two-dimensional maps are the central application for visualising information within the ReGeo tourist information system. The information represented in prototype 2 covers events and objects for the themes accommodation, infrastructure, hiking, biking, sightseeing and nature. Each of the objects has been geo-referenced (with the help of field measurements or using the ReGeo management tool), and can be displayed on a map background upon demand.

Management Information Systems, C. A. Brebbia (Editor)
© 2004 WIT Press, www.witpress.com, ISBN 1-85312-728-0

2.1 Geo data and map background

The maps, 3D-scenes and 'flyover movies' displayed in the ReGeo system are based on a range of different data sources. This information is retrieved from the virtual database by the various applications as mentioned above, and may vary according to the specific requirements of each test site. Clients have the ability to provide descriptive data for objects within each theme, including details on accommodation, hiking or sightseeing. The geo-referencing of these objects, and the composition of background maps for each of the test sites have been completed during the project.

Because the project concentrates on tourism in rural areas, the visualisation of local landscapes is most important for local suppliers. In order to offer enhanced visualisation to users, two map background versions have been designed for each of the test sites. In addition to the familiar topographic map style, ReGeo provides the user with the ability to switch to a remote sensing based map.

The remote sensing based map for the Thuringian forest test site, for example, was created from a combination of Landsat TM 7, IRS-1c panchromatic and aerial photographs. The Landsat image is used for an overall view (30 metre resolution), while aerial photographs are used for close-up views (1 metre resolution). For the middle scale range, a panchromatic IRS-1c scene was combined with the Landsat TM 7 image, using the AIF image fusion [4] at a resolution of 5 metres.

The topographic map background for this test site was created using classification results from Landsat and IRS-1c images; which differentiated water, forest, agricultural areas and settlements. The classification results were shaded on the base of a digital elevation model to highlight the relief of the region. Additional information layers were also added, which included place names, streets and railroad lines. As well with the background image, the information contained in these additional layers is displayed according to three scales: up to 1:300 000, 1:150 000 and 1:25 000.

2.2 2D map presentation

ReGeo's two-dimensional maps are integrated into both offline and online applications, and perform the same functions according to the actions of the user. The user is able to choose which thematic layers are displayed on top of the background maps, and thus has the ability to create a thematic map that is customised to his/her specific requirements. Icons on the map represent single objects corresponding to a particular theme (i.e. a hotel), and can be identified by scrolling over with the mouse to display the name of the object (i.e. Schieferhof Hotel). Clicking the mouse provides a direct link to descriptive text or various multimedia presentations. The features displayed on a map can be specified according to the results of a textual search, whereby only the objects resulting from a search (i.e. for certain hotels), are displayed on the map.

The user also has the ability to switch between map backgrounds or zoom to different scale levels to assist orientation, and may save the result to his/her 'private' favourites for use at a later stage.

Management Information Systems, C. A. Brebbia (Editor)
© 2004 WIT Press, www.witpress.com, ISBN 1-85312-728-0

3 3D viewer and flyover movies

One of the main objectives of the REGEO project is to present geographically-oriented tourist information. In order to fulfil user expectations, such presentations must be made in an interesting and appealing manner. ReGeo presentations therefore include the latest 3D visualisation techniques to present the tourist information in an appropriate way.

3.1 Real time 3D viewer

The Institute of Digital Image Processing has developed a real-time landscape visualisation interface according to the requirements of the ReGeo project. As it was necessary to choose an appropriate 3D software technology, a number of technologies were evaluated, including Java3D, VRML, MTS, QuickTime VR and Shockwave3D. The latter was eventually chosen due to its superior performance, flexibility, simple programming and ease of data integration. Further advantageous features include the compatibility with Mac and PC, as well as support for DirectX and OpenGL. In addition, the development of an online and offline version is identical except for access to the databases, which consequently lowers development costs concerning porting between online and offline systems. Fig 1 shows the integration of the 3D-viewer into the existing online presentation developed by the German partner company Geokonzept. The ability to navigate the map, as well as the visualisation of additional objects, is all controlled by means of JavaScript based on HTML. This solution facilitates a clear distinction of functionality and design.

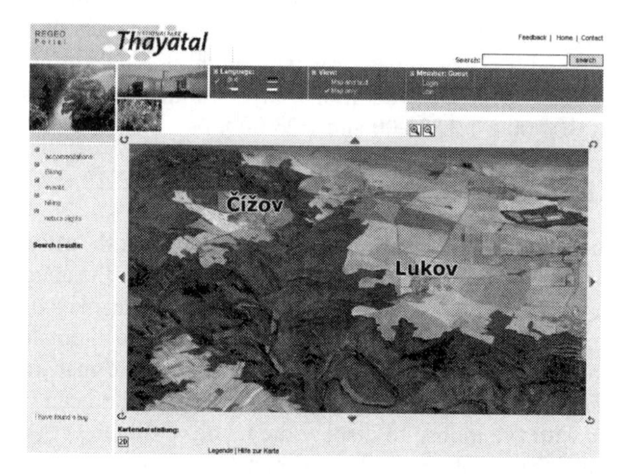

Figure 1: Integration of the 3D-Viewer into Geokonzept's CSA.

3.1.1 Conditions and past experience

In order to create the 3D images, a wireframe model was developed from the digital landscape model, which was cut out and resampled in advance. In order to use the model in Director, the DEM had to be reduced to an 8-bit format. Digital landscape models are often available as 16-bit pictures, and contain far more colours than can be shown in the final product. In our case, a scale using 256 gradations between the highest and the lowest point of the region is sufficient.

Another important aspect concerning the preparation of digital landscape models is the number of pixels to display. The larger the landscape model is increases the number of triangles required to generate the wireframe model. The number of triangles created may exceed 40,000, without the likelihood of performance loss, when using the 3D acceleration features inherent in modern video cards through DirectX and OpenGL. The basic DEM may therefore total 200x200 pixels. Without the necessary hardware supported 3D-acceleration, visualisation performance may be fairly low; however this is also highly dependent on the speed of the user's processor. According to the experience of many developers, an area measuring 10km x 10km is easy to visualise with approximately 20,000 triangles. The texture used here, with a resolution of 10 metres, is perfectly adequate for this purpose. When selecting the grid space, and thus the number of triangles, it is essential to consider the target system, the landscape model and the accuracy of the texture. More triangles are required for a terrain displaying a very high resolution or for a landscape showing very cragged rocks and peaks. When fewer triangles are used there is a danger that a sharp mountaintop may be shown with a rounded peak instead. In addition, it is crucial to pay attention to the memory of the graphic card, as this memory is a basic criterion for the size of the used texture. It is theoretically possible to use textures of 2048x2048 pixels or more in Shockwave3D, however, this depends very much on the capacity of the graphic card. It is necessary to bear in mind that this texture limit applies for a single texture. Using several models simultaneously facilitates the representation of an entire landscape in high resolution, as every model can be assigned its own texture. However, as mentioned above, this is limited by the memory of the graphic card. In the case of online-applications, the bandwidth available to the user provides a further limit to the number of textures displayed. In order to calculate the memory requirements of a particular image, it is necessary to assume that the uncompressed picture must be available in the memory of the graphic card. At a resolution of 2048x2048 pixels this equals approx. 12.3 MB. This corresponds to approximately 1.6 MB for the transference JPEG compressed images via a network [5].

3.2 Virtual flights

Virtual flyovers can be produced from artificial landscape models in order to present landscapes in an interesting way. These models are built from digital elevation models and artificial textures such as computer-generated patterns, vegetation-images or 3D-objects.

In order to generate a virtual flight, many single images must be rendered and combined to form an animation file (e.g. mpg-1-file). It is vital to define a flight path and to determine the variable view direction along this flight path, as well as the equally variable flight height. Starting from a standardised frame rate of 25

Management Information Systems, C. A. Brebbia (Editor)
© 2004 WIT Press, www.witpress.com, ISBN 1-85312-728-0

images per second, 1,000 separate images must be calculated for a 40 second animation.

The virtual flyover makes use of the "keyframe-animation" technique, in which all the necessary parameters, such as flight height, view direction, banking, sun light, etc are fixed for a small number of points (keyframes). The values for the frames in between are calculated from interpolation between the keyframes, which is normally done by a cubic spline interpolation. This ensures that the flying path is smooth and realistic [2].

Fig.2 displays three singe frames of a virtual flight over Thayatal National Park using artificial texture:

Figure 2: Frames of a flight over the national park using artificial texture.

Anti-aliasing techniques have been applied to the render process in order to avoid flickering in the final animations. These single images have been rendered using a frame size of 352 x 288 pixels (1/2 PAL resolution).

For the generation of animation files, the frames have been coded into a movie format. Different compression rates have been used for online and offline application in order to produce a useful data size. The movie formats currently being are mpeg-1 and mpeg-4.

4 3D building extraction from high-resolution satellite images of rural areas

The visualisation of 3D spatial data is becoming increasingly important in the fields of tourism and spatial planning [6]. Descriptiveness and ease of orientation are the basic requirements of tourism and planning applications. The high resolution, multi-spectral data offered by today's remote sensing satellites, such as IKONOS or Quickbird, offers the potential to visualise small-scale objects appealing and understandable at a regional scale. 3D building models can be generated from satellite imagery and integrated into 3D landscape models to represent the region in an interesting manner and to enhance the amount detail presented. An increase in content and detail may have not only a corresponding influence on the viewers' impressions of the visualisation, but also on the ability to recognise landmarks. Furthermore, incorporating 3D building models into modern small to medium scale landscape visualisations can be a valuable way of

Management Information Systems, C. A. Brebbia (Editor)
 ISBN 1-85312-728-0

applying colourness and simplicity in an appealing way. The complex tourist information system provided by the ReGeo project is contained in 3D landscape models at small scales. A method for extracting buildings from high-resolution satellite imagery has been developed within the framework of the project. The multi-spectral analysis of high-resolution satellite images offers the ability to recognise and extract buildings according to certain abstraction. In addition to get height relevant data related attributes are exported and converted in the visualisation process to concrete 3D object data.

4.1 The basis of data

The multi-spectral data used for evaluation has a resolution of 2.4 metres and includes the blue, green, red and near-infrared spectral bands. The image represents a densely settled suburban area. The multi-spectral data is supplemented by the corresponding panchromatic imagine from the same sensor with a resolution of 0.60 metres and a stereo pair of airborne RMK Top15 images degraded to pixel size of 0.60 square metres. After reconstruction the stereo pairs' photogrammetric bundle of rays, the corresponding surface model (DSM) was calculated and used for data evaluation.

Figure 3: Quickbird ms 2.4 m.

Figure 4: Quickbird fusion image 0.6 m.

During the data pre-processing, spatially high-resolution panchromatic channel was fused with multi-spectral data of higher radiometric density. An ideal fusion method for evaluating the spectral values is to use the grey values of the high-resolution material to "sharpen" the small spatial resolution objects [7]. The part pixel, part object-based fusion method Adaptive Image Fusion (AIF) used by Freiburg University fulfils these crucial demands [8]. In order to supplement the processed data record with high frequency information, further fusion is necessary by way of principle component analysis and the substitution of the first principle component with image of higher spatial resolution.

4.2 Procedure

The current generation of satellites not only produce high-resolution spatial information, but also offer the ability to receive several reflection spectra on separate channels; thus opening new possibilities to automate the analysis of spatial data. In the past it was often necessary to work with panchromatic data to detect buildings by variances in contrast; which depended largely on the material

Management Information Systems, C. A. Brebbia (Editor)
 ISBN 1-85312-728-0

available. Nowadays, it is possible to build n-dimensional featurespaces for extractions.

In a similar manner to human orientation, object detection begins with a general survey of the area, before accurately focusing on structured shapes. These various levels of accuracy can be achieved by means of multi-resolution segmentation. If the segmented objects are recognised as buildings, it is possible to detect the particular roof style by analysing the grey-scale values of the corresponding digital surface model and correlating them with the vectorised building outlines as a separate parameter. The above-mentioned DSM is then used to add a third dimension as another parameter of the building outline polygons, before the data is exported as a thematic dataset.

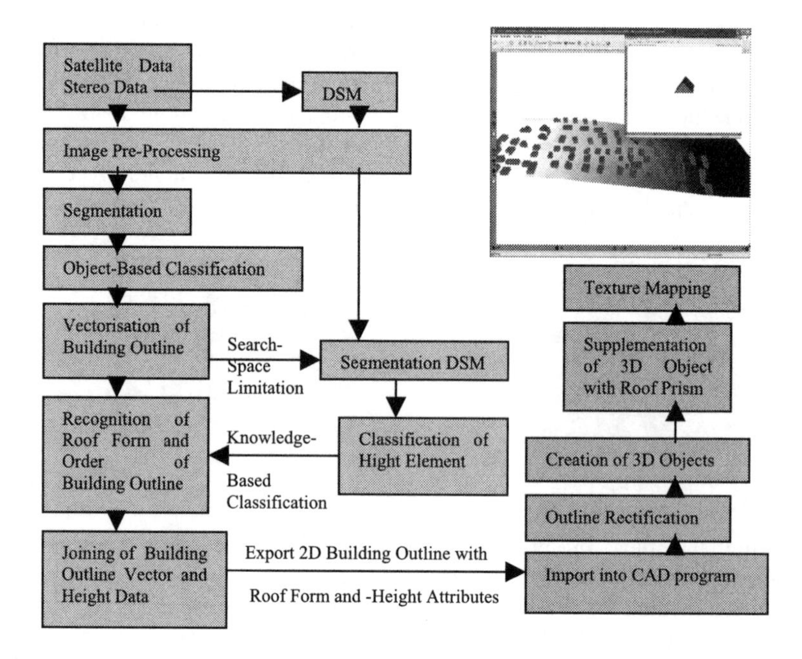

Figure 5: Process model: input data -> Data extraction through sample recognition -> 3D building reconstruction and visualisation.

4.2.1 Segmentation

Various object features are represented across a range of scales. The use of varying segmentation levels results in the creation of different objects at different abstractions levels. As a consequence, certain features become more clearly outlined and can therefore be systematically and effectively analysed.

When segmenting an image, the grey-values are compared across the entire model according to their variance, and a particular geometric generalisation value is then assigned.

Management Information Systems, C. A. Brebbia (Editor)
 ISBN 1-85312-728-0

4.2.2 2D building outline extraction

An object-oriented classification system based on an abstraction set was applied in order to obtain 2D building vectors derived from the satellite images. An n-dimensional featurespace was used to categorise vectors from clusters of multispectral, geometrical and relational features for the class description. The discrimination of individual classes was assessed by a fuzzy analysis of the distances of position vectors (appropriate minimum distance classifier).

The grey-values of the images of sun-reflection, or their corresponding synthetically-generated derivatives (such as channel ratios, principle components and Normalised Difference Vegetation Index), have been compared. Furthermore, the geometrical properties of an object are very useful for the process of building extraction. Artificial objects can be separated according to regular shapes and typical characteristics.

In this way, discrimination according to spatial properties such as object size or object compactness and the search for rectangular objects appears to be promising. In addition, grading with relational attributes linked to a classification key seems logical. This method would not only accommodate an analysis of neighbouring values of objects of the same abstraction level, but also a comparison in vertical direction with objects of a following abstraction or subobjects of a lower abstraction level.

The classification rules compiled also cover hierarchical structures and logical operators. Features can be combined with Boolean operators and weighted according to appropriate functions, which consequently combines a functional interpretation key is set together. The extracted building outlines have been vectorised, and during this process generalised in a required order.

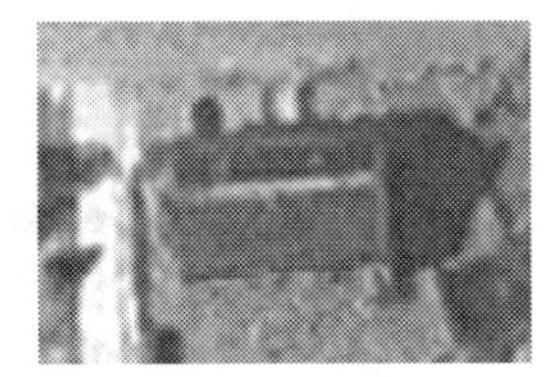

Figure 6: Building.

Figure 7: Classified and vectored object.

4.2.3 Analysis of roof shapes

Classifying of the grey-values (coded height information) of the DSM is a common method of estimating the roof shape through measuring data or indirect measures of a derived DSM. Once the building has been extracted, the DSM within the building outline is analysed. The segments of equal heights of one building object are then compared with each others in order to assign a particular roof shape to a building polygon.

Flat and saddle roofs have been initialised as separate classes. Because many roofs are built with a inclined plane, the height objects have resulted in many parallel segments, if the slope is sufficient. In order to adequately distinguish between the various roof shape classes, height differences in form of object

Management Information Systems, C. A. Brebbia (Editor)

quantities and the segment parallelism (which typically characterise saddle roofs) must be taken into consideration.

The detected classes comprise potential absolute heights that can be used for subsequent visualisation.

In order to "define" the location of the ridge in the final visualisation, the saddle roof class is classified according to a fuzzy analysis of the relationship between the roof segmentation and the building object by evaluating the object mean directions of both. The resulting class membership value offers information about how parallel the roof segments are to the longest side of the building. The roof prism is then reconstructed in corresponding location for the purpose of visualisation.

Figure 8: Roofs are typically peaked or flat.

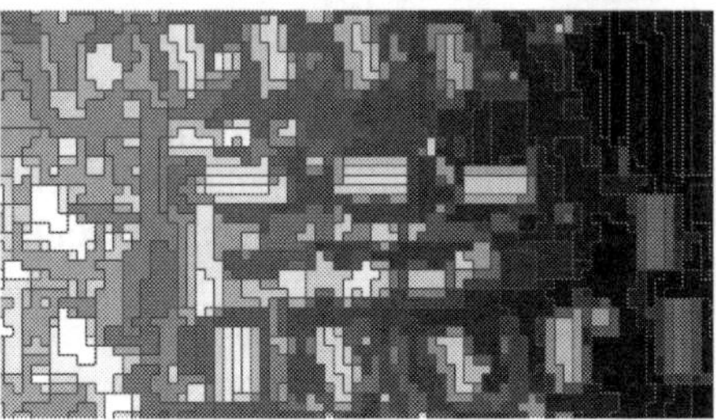

Figure 9: Segmented DSM.

5 Conclusions

The solutions described above are currently in practise. A prototype of the multimedia tourist information system is ready and functions well as a basic package available over Internet. The application comprises the map presentation and is complemented by thematic information and 3D landscape visualisations. Data requests are processed by ZOPE, WSF resp., as mentioned above, while the geo-coded data is stored in the virtual database. A future improvement will be the application of 3D building extraction, as specified in this paper. This enhancement will bring more realism to the computed scenery.

References

[1] Faby, H. (2004): Individuelle Reisevorbereitung mit Internetkarten – Status quo und Potenziale. Kartographische Nachrichten, Fachzeitschrift für Geoinformation und Visualisierung, Vol. 1, 2004, Kirschbaum Verlag, pp. 3-9.

[2] Almer, A.; Stelzl, H. (2002): Multimedia Visualisation of Geoinformation for Tourism Regions based on Remote Sensing Data. ISPRS - Technical Commission IV/6, ISPRS Congress Ottawa, 8-12 July 2002.

Management Information Systems, C. A. Brebbia (Editor)

[3] Zeiner, H, Almer, A., Stelzl, H., Frech, I., Haas, W. (2003): Virtual Environmental Scenarios Applied in a Geo-Multimedia System. EnviroInfo, 17th International Conference: Informatics for Environmental Protection, Cottbus, 24.-26.9.2003.
[4] Steinocher, K. (1999): Adaptive fusion of multisource raster data applying filter techniques. International Archives of Photogrammetry a Remote Sensing, Vol. 32, Part 7-4-3 W6, Valladolid, Spain, 3-4 June, 1999.
[5] Schnabel T., Almer A., Stelzl H., Nischelwitzer A. (2003): Multimediale Echtzeit 3D Visualisierung von Geo-Daten. CORP 2003. Geo Multimedia 2003. Wien, 25.02. – 01.03.2003.
[6] Fischer K. (2004): "3D-Visualisierung von Waldstrukturen und Waldstrukturentwicklungen- Instrument für die waldbezogene Umweltbildung sowie für partizipative Planungsansätze", University of Freiburg, Germany.
[7] Wald L. (2002): Data Fusion, Ecol des Mines de Paris.
[8] Smith D.M. (1996): Speckle reduction and segmentation of Synthetic Aperture Radar images. *Int. J. Remote Sensing,* vol. 17, no. 11, pp. 2043-2057.

Management Information Systems, C. A. Brebbia (Editor)
© 2004 WIT Press, www.witpress.com, ISBN 1-85312-728-0

Visual elements inventory in urban landscapes: collecting data for a visual environmental G.I.S.

A. Tsouchlaraki[1] & G. Achilleos[2]
[1]*Hellenic Open University, Greece*
[2]*National Technical University of Athens, Greece*

Abstract

The environment that we perceive, record and analyze as "urban landscape" is the most complex of all landscapes, due to the multiple interactions between its built components and the dynamic social, economical and cultural standards of the society. Each urban landscape has its own history. Visual resource analysis can contribute to the management of urban environments by assisting in decision making or actions, such as determining which geographic areas are suitable for urbanization or deciding what to keep or what to develop at the district or neighborhood level because of its inherent or associated visual quality. There are many elements that can be recorded for an integrated analysis of the urban environment. In this paper, emphasis is given to those elements, methods and techniques that describe, analyze and evaluate the visual elements of a city. An application of visual element inventory and analysis in the Zografos area is presented and these data are entered into a spatial database, within a visual environmental Geographical Information System. Furthermore, the common problems of the Greek urban landscape are briefly discussed in this paper.
Keywords: urban landscape, visual elements, spatial database, GIS.

1 Introduction

The environment, as it is perceived, registered and analysed in terms of "urban landscape" is the most complicated system among those that are considered to be environmentally available. The complexity of the urban environment lies in the multiple interactions of its constituent elements.

Management Information Systems, C. A. Brebbia (Editor)

"The urban landscape is a familiar and codified area that has been subjected to uses, habits and arrangements and, in this sense, it is neither a free nor a void space. History is recorded upon it. It is a dynamic space, resulting from the relations between economic activity, social structures, cultural values and their ever-changing natural background" [1].

Analysing an urban landscape requires a methodology for the notational and mapping representation, analysis and evaluation of the visual data found within the urban environment, capable of allowing the observer to continuously discover relations and interactions, which would not otherwise be obvious in a first visual examination.

The elements that could be gathered to study the urban landscape in its entirety are numerous. This paper focuses on the elements that describe the visual result of the constructed space and make a reference to basic methods and techniques that could be used in collecting, analysing and evaluating such elements.

2 Greece's urban landscape

There is extensive literature on the causes of and the staging towards the current situation of the urban landscape in Greece, particularly in Athens; the prevailing, if not unique, cause is the presence of the blocks of flats [1, 2, 3, 4]. We shall cite a few extracts from recent publications, which we believe offer an overall picture of the concerns of the scientific community on how the Greek urban landscape has evolved.

Vlastos and Siolas [5] observe that "Greek towns, while having duplicated, on many occasions, the European standard, are distinguished by some specific development features: in principle, development around traditional centres has always taken place without specific urban planning, without scheduling, practically in the usual manner of legal and arbitrary private constructions. This has led to the creation of an urban space continuum, with fragmented cores lacking structural cohesion and, of course, structural facilities. Also, the historical core of Greek towns, traditionally poorer than that of European towns, has suffered the influence of the Greek modern urban development; however the "modernisation" elements that are visible in other European cities have failed to leave particular traces at the architectural level."

Vertical arrangements are known to have created a new relief and detached people from the ground and the corresponding activities related to it.

Voulgaris [4] observes melancholically that "Gardens, the uncovered spaces with the bougainvillea and the jasmines that contributed to the formation of a closed housing shelter were replaced by the compulsory uncovered space which some Building Regulation imposed to be arranged in a predetermined area of the plot and the building block in general. Its outcome was both impressive and depressive, given that the alteration of both the character and the aesthetic of our country's settlements resulted in the fragmentation of the Greek family, of social life, of neighbourhoods and of the town's political function, of democracy in its most essential form, participative democracy, given that families were isolated

Management Information Systems, C. A. Brebbia (Editor)

inside the house, in an introvert function". In the same publication [4], he observes that, "67.5 percent of Greece's total number of 1,775,000 primary residence houses are apartments in blocks of flats. However, 62 percent of their residents say to be less than happy with this solution and consider their homes to be average or unacceptable. Moreover, almost half of the households (48.7%) express their desire to change residence, if they can meet the requirements of doing so, and move towards the suburbs (74% overall and 80% in the case of Athens).

A daring and radical intervention, which could definitely solve the problem and would give the chance of changing things from scratch, is an unfulfilled dream, for obvious social and economic reasons, as well as for reasons of proprietorship.

Aravantinos [6] observes that, "The interventions that could take place in order to reform and restore an already constructed environment are generally of a limited character and aim to somehow 'tidy up' the situation, without imposing radical changes or addressing essential problems".

On this issue, Aliefs [7] observes that, "Until now, efforts aimed to address the problem of poor workmanship in constructions do not have impressive results and, for the most part, have not achieved its elimination. This means that such efforts are somewhat constrained by certain limits. Such limits are either intentional and selective, on the part of anyone trying to address the problem, or even unavoidable, in the sense that they are expressed by human weaknesses. The latter are the result of human characteristics, especially the fact that man knows little of the anatomy and physiology of his nervous, muscle and skeletal system. This finding, i.e. the lack of capability to permanently solve the problem of poor workmanship should not leave indifferent anyone wishing to deal with constructions and the urban environment in general."

Karamanou et al. [8] suggest that, "the utilisation of the construction reserves is a basic guideline in the field of residential development and perhaps the only solution for urban centres that are saturated in terms of building constructions and lack every other means of renewal."

It is obvious, from all the above observations, that there is a need to directly monitor and manage the urban landscape, which has no margin and luxury of further postponement.

3 Visual element analysis

3.1 Necessity of visual elements

The analysis and registration of the visual elements in an urban landscape can contribute to structural interventions in urban planning, such as:

1. Identifying geographical areas that are suitable for urban development and controlling the capacity and capability of certain areas threatened by additional burden due to urbanisation.
2. Analysing the existing road network in urban areas and spatially arranging and planning new networks or modifying existing ones.

Management Information Systems, C. A. Brebbia (Editor)
© 2004 WIT Press, www.witpress.com, ISBN 1-85312-728-0

3. Formulating decisions as to which elements should remain and which ones should be modified in a region or neighbourhood, depending on the historical and cultural context of the area.
4. Controlling undesirable land uses susceptible of creating visual environmental problems (shops, numerous sign-posts, orientation problems, traffic accidents due to poor visibility).
5. Spatially arranging and specially planning building constructions, in order to achieve harmonisation in the materials and colours used, as well as in the magnitudes of the elements that are usually visible, or to prevent undesirable natural phenomena, e.g. shades, intense illumination, increase of the wind speed due to the artificial relief, etc. [9].

It is obvious from the above that the registration and analysis of the visual resources of the urban environment may prove to be useful in many applications, either in order to determine the extent and the magnitude of the various problems within the urban area or to become a useful aid for further analysis and management.

3.2 The role of a GIS

A Geographical Information System shows a great contribution to this kind of analysis. As the main powerful system for spatial analysis and with its advantage of using and managing tremendous volumes of a spatial and attributes data from a variety of databases, GIS is irreplaceable for such use. Further to its abilities, GIS provides tools and modules for spatial analysis, for modelling and for decision making in order to help the urban area planners to proceed. Special modules for visual analysis are programmed and developed.

A GIS specifically for visual resources analysis can play the role of a special GIS, a Visual Environmental Geographical Information System. A system like this may play the role of the advisor and a presenter for many applications such as real estates, land planning and land developing, urban design and many more.

This specific GIS may also include data of not visual resources content in its spatial database (demographic data, housing distribution, economic data, future plans data, transportation and infrastructure data), in order to combine them with the spatial visual resources data for supporting decision makers.

4 Methodological approaches to the analysis of visual resources in urban landscapes

There are many known methodological approaches aiming to register and analyse the visual elements of the urban landscape. Perhaps the most eminent method is that of Kevin Lynch, in his work entitled “The Image of the City”, which has influenced architects and landscape experts worldwide [10].

Lynch’s work introduces the interest in the mental representation of the environment (imageability) through human experience. Using questionnaires, it examines and analyses how people perceive, utilise and recall their surrounding area. It is through these memories and impressions that the work aims to determine the sensitivities, the problems, the trends and the critical points of a given area in order to take them into account during further urban planning efforts. His research incorporates two main practices: (a) the *in situ* registration of the urban environment by skilled and experienced observers; it should be noted that Lynch developed one of the first cartographic systems for representing the urban area, and (b) the long interviews of permanent residents of the city and of people working in the area, who are asked to describe, draw and submit their personal views on the problems of the area. Comparison of the two results and synthesis of the information allows the formulation of certain findings on the sensitivity and critical points of the area.

Despite the interest arising from the analysis of the mental representation of the environment, the method has not been used systematically in Greece for designing the urban environment. A work was recently published, which applies Lynch's methodological approach successfully, in the framework of the study on the unification of archaeological sites in the centre of Athens [11].

Gordon Cullen's work entitled "Concise Townscape" has also been of influence for scholars and scientists in the ways and methods they use to analyse the urban landscape [12]. One of his main contributions was the opinion that the visual perception of space, whether constructed or open, relies on aesthetic principles. In this sense, as the author points out, the perception of a tall building lying among other buildings of equal height is different from its perception when the height of the surrounding buildings is smaller. In the first case, we simply have a tall building. In the second, the building looks like a tower, in other words it dominates the landscape. Also, it should not be ignored that people tend to compare objects on the basis of their own size. Therefore, in reality there is an "art of correlation", as there is the art of architecture. As Cullen points out, our aim would be to take all the elements that make up a town, such as buildings, trees, open areas, roads, road traffic, advertisements and everything that is related to them and combine them in such a way as to represent the "drama"; as he stresses, a "town is a dramatic event for the environment".

A further main contribution to the analysis of the urban landscape is offered by Appleyard at al. (1964) in their work entitled "The view from the road" [13], in which they refer to the dynamic perception of human environment, and to the critical role of this event in urban planning. Given that the users of a city are usually in movement, either as pedestrians or as passengers of a transport means, their perception of space, its features and complexity depends on the sequence of the different impressions they retain during their movement. These consecutive illustrations require special graphical techniques and systems of registration, which the abovementioned work does not include.

5 Visual element notational and mapping techniques

One of the fundamental beliefs of experts working in the field of landscape architecture is that during the notational and mapping procedure and analysis of the visual elements, new relations, juxtapositions and sequences emerge, which are otherwise concealed in a first visual examination.

Management Information Systems, C. A. Brebbia (Editor)

In order to be able to describe and represent the features and properties of the area examined in the best possible way, the processes used for the inventory have to be planned and prioritised in advance. Stages that are critical in every registration process are: (i) determining the observation posts that will be used; (ii) the elements that will be collected for describing the visual environment, which vary depending on the purpose and aim of the application; (iii) the presentation means that will be chosen to document the description; and (iv) the method that will be used for the evaluation of the visual quality.

5.1 Selection of the observation posts - Inventory means

A successful practice for selecting the observation posts is the shooting of photographs using a camera (analogue or preferably digital), from crossroads or areas presenting movement or gatherings of people, such as shopping centres, churches, parks, museums, etc. The observation posts and the orientation of shootings are subsequently marked on a topographic map of the area.

Another method of selecting the observation posts is the shooting of pictures from randomly selected locations of the city. Finally, a modern technique that could document the visual experience of a moving observer is the continuous video shooting of a course.

Regardless of the technique used, a basic requirement for the proper selection of the observation posts is for the researcher to have in advance a clear overview of the area surveyed. This can be achieved with the use of topographic diagrams, aerial photography and preliminary *in situ* visits.

5.2 Visual elements

In addition to the pictures taken from the selected observation posts, the observer must take note also of other features and visual elements of the environment, such as those detailed in Table 1 (Table 1):

- visibility elements (viewing angle, viewer distance, view elevation);
- the form of paths (grid, irregular, radial);
- the degree of enclosure (or the non-enclosure) of the landscape;
- the architectural patterns (horizontal and vertical distribution of buildings, shape, colour, materials);
- roadside vegetation elements (vertical distribution of trees, density, foliage);
- open areas and their uses (pedestrian streets, squares, parks, parking areas, roads, etc.);
- activity patter or population movement, as observed on the roads at the time of surveying.

A factor that is often further examined is illumination, both natural and artificial, given its important role in the picture of a city. Because natural illumination depends on a number of variables, e.g. atmospheric conditions, time of the year or of the day, etc., it is analysed as such by studying the shades and the reflection of light [9]. The vertical distribution of buildings, their orientation, as well as the existence of highly reflective materials (e.g. metal, glass, water) are fundamental elements of a survey in this context.

Management Information Systems, C. A. Brebbia (Editor)

Table 1: Visual features and elements.

VISIBILITY ELEMENTS		STREET TREES	
	VIEWING ANGLE		**HEIGHT**
	VISTA		SMALL
	WIDE ANGLE		MEDIUM or MIXED
	PANORAMA -		TALL or MIXED -
	VIEWER DISTANCE		**DISTRIBUTION**
	CLOSE		DENSE at regular intervals
	INTERMEDIATE		SPARSE at regular intervals
	FAR		SPARSE at irregular intervals
	VIEW ELEVATION	**OPEN SPACE**	
	INFERIOR		**SCALE** (extent of open area in relation to a person)
	NORMAL	**2:1-1:2**	INTINATE / ENCLOSED
	SUPERIOR	**1:2-1:4**	MODERATE
	PATHS PATTERN	**>1:4**	MONUMENTAL
	GRID		**LAND USES**
	IRREGULAR		PEDESTRIAN USE
	RADIAL		VEHICULAR
ENCLOSURE		**ARCHITECTURAL STANDARDS**	
	DEGREE OF ENCLOSURE		**BUILDING DISTRIBUTION**
	2:1		ALLIGNED
	1:3		UNEVEN
	1:4		MISSING
	>1		**ARCHITECTURAL PATTERN**
	DISTANT VIEW ENCLOSURE		SHAPE
	OPEN END		COLOUR / TEXTURE
	CLOSED END		FACE DETAILS (presentation)
		ACTIVITY PATTERN AT THE TIME OF RECORDING	
			NO PEOPLE
			NOTICEABLE AMOUNT
			RELEVANT AMOUNT

At nights, a totally different landscape emerges. Problems that are easily perceived in the daylight, such as tall buildings, criticised shapes and patterns

Management Information Systems, C. A. Brebbia (Editor)

disappear in the dark and are replaced by luminous localised or linear elements that mark off areas, which might otherwise be less than noteworthy in the daylight. Elements examined for the study of artificial illumination are the reflection of certain materials, the location, the duration and the intensity of the light source.

As mentioned already, the above visual elements can be enriched depending on the purpose of the application and are noted during their registration on tables well designed or diagrams, while the observer can also add personal remarks, as he or she deems appropriate.

5.3 Evaluation

The visual elements that have been used for the inventory, using the abovementioned techniques, can be combined and represented on thematic and combination maps depicting the information on the area, in detailed and concise form. Statistical processing of these elements can subsequently lead to results on the features and properties of the urban area examined. However, such results do not constitute an evaluation per se. As Varelidis notes in this respect, humans are not statistical items and cannot be determined on the basis of statistical figures [3].

However, there is still a need for measurement techniques. Aliefs, in his publication on poor workmanship in constructions, notes that, "the study of marginal events depends on our capability of referring to unitary notions. For example, in order to evaluate the development in an occurrence of poor workmanship, we must be able to measure it" [7].

This measurement is feasible only when using questionnaires, in which residents, workers and visitors of an area make their own evaluations and express their own views and proposals for the area in question. Referring to people's perceptions, we can identify the elements and standards of the environment, which are a source of problems, so as to take appropriate measures and use them as negative examples in future plans for the city.

6 Application

A visual element registration study carried out in a part of the Zografos area in Attica is described below (Map 1, Map 2). The object of the study was to collect visual elements of the area, so as to represent them on thematic and combination maps and to proceed to their statistical processing as necessary, with the aim to identify the visual characteristics and standards that prevail in the area. Evaluating the elements was not among the objects of our work. We plan, however, to pursue our effort in other areas, so as to allow for an evaluation and contrastive analysis between different areas. The application is commented in order to allow the technical details that are mentioned, to be used in similar efforts.

The background used for the selection of the observation posts was the Roadmap of Athens (Kapranidis and Fotis editions [14]). The area, owning to its dense construction, was divided into small quadrangular sub-areas (100 m^2

Management Information Systems, C. A. Brebbia (Editor)
© 2004 WIT Press, www.witpress.com, ISBN 1-85312-728-0

approximately), an order of magnitude that has been observed to ensure homogeneity in terms of characteristics. Prior to recording in each sub-area, the nodes were numbered and noted, together with all possible directions of observation for each node, in a clockwise manner (Map 2). According to our definition, nodes are mainly crossroad points and, where deemed appropriate, characteristic constructions such as churches, public buildings, big hotels (taking their entrance as starting point), so as to include also areas crowded with people. The directions of the roads were taken as directions for observation.

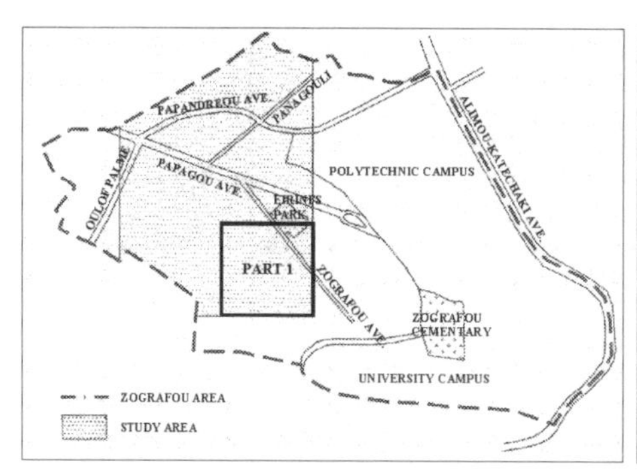

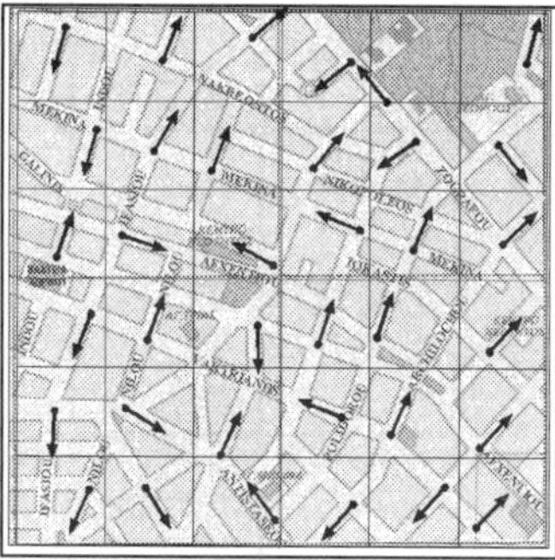

Map 1: Zografos municipality Area. Map 2: Study area (part 1).

Photo: Two typical examples of the view in the study area.

Following the numbering of all possible observation posts, we randomly sampled a post within each square of the pattern (Map 2). The layered sampling method was chosen, in lieu of the random sampling in the entire area, in order to distribute the posts throughout the area and better represent the characteristics. This was the main reason of paying special attention to the size of the squares on the pattern, i.e. to be able to represent the entire quadrangular sub-area with the selection of a single observation post.

Using the findings, we created colour thematic maps for each individual characteristic. In order to present the information concisely, we created symbolic maps in the form depicted in Table 2 (Table 2), due to the large number of the

Management Information Systems, C. A. Brebbia (Editor)

combinations that existed. Nowadays, presenting and processing large amounts of data is an easy task, with the use of GIS technology, as it has already being mentioned (Maps 3, 4, 5).

Table 2: Collected visual data from part 1 of study area.

a1 b3 c2 d1 e3 f2 g1 h1 i1 j2	a2 b2 c3 d2 e2 f1 g2 h1 i3 j2	a2 b1 c3 d2 e3 f1 g1 h1 i3 j2	a2 b1 c3 d2 e1 f1 g3 h1 i1 j2	a1 b3 c3 d1 e4 f1 g1 h1 i1 j1	a2 b2 c3 d2 e2 f1 g3 h1 i1 j2
a3 b2 c3 d2 e1 f2 g3 h2 i3 j2	a1 b3 c3 d1 e2 f2 g1 h2 i2 j2	a1 b1 c3 d1 e1 f2 g1 h2 i2 j1	a1 b1 c3 d2 e1 f2 g1 h2 i2 j2	a1 b3 c3 d1 e2 f2 g1 h2 i2 j1	a1 b1 c3 d2 e3 f1 g1 h2 i2 j1
a2 b2 c2 d2 e3 f2 g1 h2 i1 j2	a1 b3 c3 d1 e2 f2 g2 h2 i2 j2	a1 b1 c3 d2 e2 f2 g1 h1 i2 j2	a1 b3 c3 d2 e1 f2 g1 h2 i2 j2	a1 b2 c1 d1 e3 f1 g2 h2 i2 j2	a1 b2 c3 d3 e3 f1 g2 h2 i1 j1
a2 b1 c2 d2 e1 f2 g2 h2 i3 j2	a1 b2 c3 d2 e2 f2 g1 h2 i2 j1	a1 b2 c2 d2 e2 f2 g1 h2 i2 j2	a1 b2 c3 d2 e3 f2 g1 h2 i2 j1	a1 b1 c3 d2 e4 f2 g1 h2 i2 j2	a2 b2 c3 d2 e2 f2 g2 h2 i3 j1
a2 b2 c2 d2 e1 f2 g2 h2 i3 j2	a2 b2 c1 d2 e1 f2 g2 h2 i2 j1	a1 b1 c3 d2 e2 f2 g1 h2 i1 j1	a2 b3 c3 d2 e1 f2 g2 h2 i3 j1	a1 b2 c3 d2 e3 f2 g1 h2 i3 j1	a1 b3 c3 d1 e2 f2 g1 h2 i2 j1
a1 b3 c1 d1 e3 f2 g1 h2 i2 j1	a1 b3 c1 d1 e3 f2 g1 h2 i2 j1	a2 b1 c3 d2 e1 f2 g2 h2 i2 j1	a1 b3 c1 d2 e3 f2 g1 h2 i2 j1	a1 b2 c1 d2 e3 f2 g1 h2 i2 j1	a1 b2 c1 d2 e3 f2 g1 h1 i1 j1

a1: viewing angle: 0 - 90 **a2**: >> >> : 90 - 180 **a3**: >> >> : 180 - 360	**e3**: degree of enclosure: 1:4 **e4**: >> >> : 1:5 **f1**: street trees: even **f2**: >> >> : uneven
b1: viewer distance: close **b2**: >> >> : intermediate **b3**: >> >> : far	**g1**: open space: < 1:2 **g2**: >> >> : 1:2 - 1:4 **g3**: >> >> : > 1:4
c1: view elevation: inferior **c2**: >> >> : normal **c3**: >> >> : superior	**h1**: land use: pedestrian use **h2**: >> >> : vehicular
d1: paths pattern: grid **d2**: >> >> : irregular **d3**: >> >> : radial	**i1**: aligned building distribution **i2**: uneven building distribution **i3**: missing building distribution
e1: degree of enclosure : 1:2 **e2**: >> >> : 1:3	**j1**: medium noisy area **j2**: heavy noisy area

The area of Zografos surveyed shows that the standards prevailing are those usually found in Athens.

The intense relief alters the observer's relative altitude quite frequently. In many cases, the range of visibility is constrained abruptly (and dangerously in terms of traffic) owning to the high inclination (ascent). There are however, many observation posts offering overview pictures of the area, with a long range of visibility.

The widths of roads are small compared to the heights of buildings, thus creating high levels of landscape enclosure, with the exception of avenues and open areas. Multi-storey buildings prevail in an irregular vertical distribution pattern, while there are few remaining detached houses.

Management Information Systems, C. A. Brebbia (Editor)

There are very few and quite small open areas, with the exception of Eirini Park and the area of the Technical University Campus. Some tree planting efforts are also apparent in the area, especially on the very narrow pavements, within the small squares and on traffic islands of main roads. However, these elements do not prevail in the visual environment.

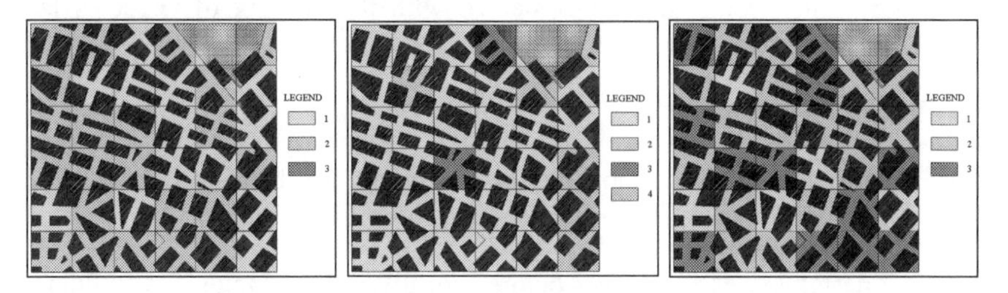

Map 3: Map of paths pattern (1=Grid, 2=Irregular, 3=Radial).
Map 4: Map of enclosure degree (1=1:2, 2=1:3, 3=1:4, 4=>1:5).
Map 5: Map of buildings distribution (1=aligned, 2=uneven, 3=missing).

7 Conclusions

The registration and analysis of visual elements can contribute to the optical enhancement of the urban environment. If every municipality or community had already registered the visual elements within its jurisdiction and had created a properly designed database for this purpose, it would have been able to monitor the urban environment, in a direct and diachronic manner, in addition to identifying visually sensitive points as well as the area's problems. This could have led to the adoption of appropriate management and prevention measures, thus possibly avoiding aesthetic cruelties in the urban environment. Today, we can only hope for a gradual restoring and improvement, which, in order to be achieved, requires detailed knowledge of the current situation and of our construction reserves. A database suitably adapted to current conditions, designed for the purpose of registering and monitoring the visual elements of the urban environment, is deemed necessary, nowadays more than ever.

The vertical development of Greece's urban areas has created a new artificial relief, which must be surveyed also in terms of its aesthetic aspects. Items that are added to building storeys, such as solar water heaters, television antennas and the various materials that are arbitrarily disposed of, are by no means decorative for this new relief and alter the city's horizon line. Walls blackened by negligence and the city's smog, the dusty and neglected window shutters, the intense and non harmonised colourfulness of shades, all make up the everyday landscape of the residents of Greece's cities. The individual proposals, which have been submitted at times for addressing the problem, have not been implemented. The potential benefits from creating gardens on the roofs of buildings, for both the aesthetic and the microclimate of cities make a sound and feasible proposal, which may be implemented sometime in the future.

Management Information Systems, C. A. Brebbia (Editor)
© 2004 WIT Press, www.witpress.com, ISBN 1-85312-728-0

The visual pollution of the urban area, a problem that could be addressed with regulatory measures, as is the case in other countries [15], cannot be omitted from the visual resource analysis process. Items subject to registration must include details such as the ones mentioned above, in order to build an integrated database of the visual environment. Designing such an integrated database, which could also be combined with the drafting of Greece's land registry, is of particular research interest for the future.

References

[1] Ananiadou, M., Tzimopoulou. "Designing the Landscape and Urban Outdoor Areas in Thessaloniki", *Technika Chronika*, A, vol. 12, no. 4, 1992 (in Greek).
[2] Anastasakis, M.S. "Athens before World War II: Planning and Building a City", *Technika Chronika*, A, vol. 10, no. 1, 1990 (in Greek).
[3] Varelidis, K. "Regulations and Housing Quality. Relation or Containment?", *Technika Chronika* (bimonthly), year 63, conference on "Human functions in modern homes", March-April, 1994 (in Greek).
[4] Voulgaris, A. "Residences, Now and Then", *Technika Chronika* (bimontly), year 63, conference on "Human functions in modern homes". March-April, 1994 (in Greek).
[5] Vlastos, Th., Siolas, A. "The Contribution of Transportation Networks in the Strategy of Articulating and Reuniting Urban Entities for Restructuring Towns: The case of Western Athens", *Technika Chronika*, A, vol. 14, no. 1, 1994 (in Greek).
[6] Aravantinos, A. *Urban Planning: Towards a sustainable development of the built environment*, Published by SYMMETRIA, Athens, 1997 (in Greek).
[7] Aliefs, A. "Constraints in Addressing the Problem of Poor Construction Workmanship", *Technika Chronika*, A, vol. 14, no. 1, 1994 (in Greek).
[8] Karamanou, Z., Koukopoulos, S. Nomikos, M. "Utilising the Construction Reserve. A Different View of Residential Development", *Technika Chronika*, A, vol. 11, no. 2, 1991 (in Greek).
[9] Smardon R., Castello T., Egging H., "Urban Visual Description and Analysis", *Foundations for Visual Project Analysis*, John Willey & Sons, N.Y.,1986.
[10] Lynch K., *The Image of the City*, MIT Press, Cambridge, 1985.
[11] Economou, D. "Mental Representations of the Athens City Centre: First Survey Findings", *Technika Chronika*, A, vol. 10, no. 1, 1990 (in Greek).
[12] Cullen G., *The concise townscape*, Architectural Press, London, 1976.
[13] Appleyard D., Lynch K. and Meyer J.R., *The view from the road*, Cambridge: MIT Press,1964.
[14] Kapranidis S., Fotis, N., *Roadmap of Athens, Piraeus and the Suburbs*, Ellinikes Touristikes & Hartografikes ekdosis, Athens 1995, p. 127 (in Greek).
[15] Smardon R., Karp J., *The Legal Landscape*, VNR N.Y.1992.

Management Information Systems, C. A. Brebbia (Editor)

Soil erosion management at a large catchment scale using the RUSLE-GIS: the case of Masinga catchment, Kenya

B. M. Mutua & A. Klik
Institute of Hydraulics and Rural Water Management, Universität für Bodenkultur (BOKU), Austria

Abstract

Kenya is one country suffering heavily from land degradation due to increasing anthropogenic pressure on its natural resources. As is common to many tropical countries, Kenya suffers from a lack of financial resources to research, monitor and model sources and outcomes of environmental degradation for large catchment domains. In order to evaluate viable management options, soil erosion modelling at the catchment scale needs to be undertaken. This paper presents a comprehensive methodology that integrates an erosion model, the Revised Universal Soil Loss Equation (RUSLE) with a Geographic Information System (GIS) for estimating soil erosion at Masinga catchment, which is a typical rural catchment in Kenya. The objective of the study was to map the spatial mean annual soil erosion for the Masinga catchment and identify the risk erosion areas. Current land use/cover and management practices and selected, feasible, future management practices were evaluated to determine their effects on average annual soil loss. The results can be used to advice the catchment stakeholders in prioritising the areas of immediate erosion mitigation. The integrated approach allows for relatively easy, fast, and cost-effective estimation of spatially distributed soil erosion and sediment delivery. It thus provides a useful and efficient tool for predicting long-term soil erosion potential and assessing erosion impacts of various cropping systems and conservation support practices.
Keywords: ArcView GIS, RUSLE, catchment, soil erosion, modelling, Masinga, Kenya.

Management Information Systems, C. A. Brebbia (Editor)

1 Introduction

Expansion of subsistence farming practices in the form of field crop agriculture and pasture within rural areas is contributing significantly to ecological alteration in many tropical countries. One of the most destructive and insidious processes, steadily increasing as a result of anthropogenic activity in these areas, is soil erosion (Landa *et al* [1]).

Kenya is one country suffering heavily from land degradation due to increasing anthropogenic pressure on its natural resources. This is more so mainly in the catchment where the rural communities encroach to open up new land for agricultural activities and settlement.

Cultivation of steep slopes within the wet highlands as a result of the population pressure on the land, and intense grazing in the semi-arid lowlands are some of the major factors enhancing soil erosion within Masinga catchment, the area of the present study. Future pressure to expand agricultural and grazing operations within Masinga will unquestionably accentuate the already high rate of soil erosion and this problem can only be effectively addressed through long-term strategic planning, based on a sustainable management approach at the catchment level.

As is common to many tropical countries, Kenya suffers from a lack of financial resources to research, monitor and model sources and outcomes of environmental degradation. Until recently, there did not exist a reliable or financially viable means to model and map soil erosion within large remote areas. However, an increase in the reliability and resolution of remote sensing techniques, modification and advancement in catchment scale soil erosion modelling techniques, and advances in Geographical Information Systems (GIS), represent significantly improved tools that can be applied to both monitoring and modelling the effects of land use on soil erosion potential.

1.1 Objective

This study is undertaken to explore the application of geographic information systems (GIS) technology as a tool in catchment management planning at Masinga. The general objective is to develop readily transferable RUSLE-GIS based procedures to be used by land resource managers to evaluate different land use and management practices in terms of soil loss potential. The primary objective is to map the spatial mean annual soil erosion over the catchment, identify and prioritise areas with high erosion risk in order to direct the implementation of viable management options.

2 Study area

The Masinga catchment area is some 6,255 km^2 in extent, lying to the east of the Aberdares mountains and south of Mount Kenya. It lies between latitudes 0° 7' South and 1° 15' South and longitudes 36° 33' East and 37° 46' East. Its location on the Kenyan map is shown in fig.1.

The geology of Masinga area can be broadly divided into volcanic rocks in the north and west, and pre-cambrian basement complex in the south-east. The landform in the catchment ranges from steep mountainous terrain with strong relief in the west, to undulating plains with subdued relief in the south-east. The elevations above means seal level (asl) in the mountainous terrain range from 2500 to 4000 m and for the undulating plains from 900 to 1200 m. The soils are generally Lithosols and Histosols (FAO classification) at the highest altitudes in the Aberdares with Humic Andosols at slightly lower elevations. Over much of the rest of the basalt foot slopes, deep fragile clays (Eutric Nitosols) predominate. On the basement complex, the soils are mostly coarser textured and shallower and are classified as Acrisols, Luvisols and Ferralsols.

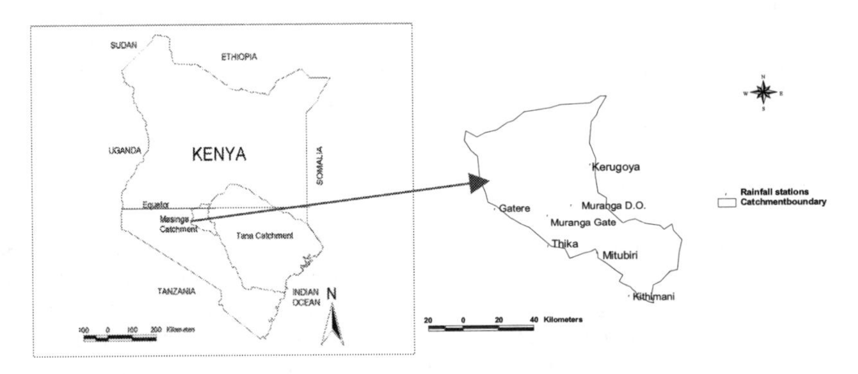

Figure 1: Location of the study area in Kenya.

The catchment falls within four agro-climatic zones, ranging from semi-arid in the east to humid near the western side. The annual rainfall is bimodal with short rains occurring from September to November and the long rains from March to May. The mean annual rainfall vary from about 600 mm on the easterly boundary to over 2000 mm on the Aberdares Mountains. The maximum temperatures vary from 25.5° C to 31.0° C generally being experienced in February or March, prior to the onset of the main rain season (long rains). Minimum mean temperatures of 21.0° C to 24.0° C occur in the month of July.

The catchment has an estimated population of 2 million people (1998 census) with most people engaged in agricultural activities. Almost all the cultivation takes place in the south-east, north-west and generally in the western areas. There is scattered cultivation in the eastern half of the area with slopes of 15% where the soils are Vertisols and where severe erosion is taking place. The remainder of the area is used for grazing with large numbers of cattle, sheep and goats being herded on the area, which is almost completely denuded of grass and with very little cover.

3 Materials and methods

3.1 GIS and soil erosion modelling

In evaluation of viable management options, soil loss modelling at the catchment scale needs to be undertaken. While soil loss data gathered from experimental plots can be extrapolated over the entire catchment, results may be misleading for catchments with heterogeneous land use patterns and physical characteristics. The problem may be addressed either by establishing numerous observation plots over the catchment to capture variation in soil loss across the area, or by employing modelling techniques. The first option is usually a costly and time-dependent proposition particularly where financial and human resources are limiting. Modelling is therefore preferred since soil loss under alternative management scenarios can be rapidly estimated at minimal cost.

In the present study, the RUSLE model was applied within a GIS environment. The analytical and manipulation tools within the GIS allowed for the quantification of the parameters from available data sets. Using the vector-based GIS programme (ArcView), a complete coverage of soil erosion within Masinga catchment was estimated under the current land use/cover and management practices.

3.2 The soil erosion model structure

The Revised Universal Soil Loss Equation (RUSLE) is among the most extensively used empirical soil erosion models. It is the present state of the art in soil erosion modelling. Basically RUSLE, which lumps the interrill and rill erosion together, is a regression equation. RUSLE is an erosion prediction model designed to predict the long-term average annual soil loss from specific field slopes in specified land use and management systems. The RUSLE model is expressed as:

$$A = LS * R * K * C * P \tag{1}$$

Where A is the estimated average annual soil loss (t ha^{-1} yr^{-1}); LS is the combination of the slope steepness and slope length factors (unitless); R is the rainfall erosivity factor (KJ mm m^{-2} h^{-1} yr^{-1}); K is the soil erodibility factor (t ha^{-1} KJ^{-1} mm^{-1} m^2 h); C is the cover and management factor which estimates the soil loss ratio (SLR) at seasonal intervals throughout the year, accounting for effects of prior land use, canopy cover or crop, surface cover, surface roughness and soil moisture; P is the support practice factor, calculated as a SLR, which accounts for tillage techniques, strip cropping and terracing.

3.3 Data source and model factor generation

Four primary data themes were required to develop the RUSLE factors. These were the digital elevation model (DEM) which is a three-dimensional raster representation of the topography, the climatic data (precipitation), the soil type coverage and the land use coverage and conservation practices. The DEM was required to derive the slope length (L) and the slope steepness (S) factors. The climatic data was required to develop the rainfall erosivity (R) factor. The soil type coverage was required to develop the soil erodibilty (K) factor and the land use coverage was used to develop the crop management (C) and conservation practice (P) factors.

Management Information Systems, C. A. Brebbia (Editor)

The data for this study was acquired from various sources. Among the sources of data was the reconnaissance survey that was carried out in the study area to establish among others the land use and management practices. The survey was carried out between July and September 2003. In addition to the survey and the field data collection, more data was collected from relevant Ministries, Departments, Research Centres and Agencies in Kenya. Other data was acquired from relevant existing databases.

3.3.1 The Digital Elevation Model (DEM)

A DEM of scale of 1: 250,000 and grid resolution of 90 metres covering the study area was purchased from the United States Geological Surveys (USGS). The DEM was projected using the Universal Transverse Mercator 37 North (UTM-37N) reference system. Most digital elevation models often contain pits or local depressions. These pits can reflect real life conditions but are often the results of poor input data or interpolation errors. Since the contributing area is based on routing the cumulative area downslope, these pits can act as sinks if not removed and therefore hinder flow routing. In this study, a depressionless DEM was created using the "Fill the Sink" command from ArcView "Hydro" extension menu that allows the removal of the sinks or pits from the DEM.

The GIS technology allows calculation of the Slope length and steepness, the LS-factor for the rill and interrill erosion through the estimation of the upslope contributing area per unit contour width by computing the flow direction and flow accumulation. Generally the upslope contributing area per unit contour width is taken as the sum of the grid cells from which water flows into the cell of interest (Mitasova *et al* [2]). Generation of the L-factor for this study was conducted using the upslope drainage area substitution method suggested by Desmet and Govers [3]. The unit contributing area for the slope length value was substituted as the slope length within the equation provided by Foster and Wischmeier [4], with each of the grid cells within the DEM considered as a slope segment having uniform slope. Substituting the values for cell outlet and cell inlet into the Foster and Wischmeier [4] equation gives the slope length L component as:

$$L_{i,j} = \frac{A_{i,j-out}^{m+1} - A_{i,j-in}^{m+1}}{\left(A_{i,j-out} - A_{i,j-in}\right)(22.13)^{m}} \tag{2}$$

Where $L_{i,j}$ is the slope length factor for the cell with coordinates (i, j), $A_{i,j-out}$ is the contributing area at the outlet of the grid cell with coordinates (i, j)(m^2) , $A_{i,j-in}$ is the contributing area at the inlet of the grid cell with coordinates (i, j)(m^2 m^{-1}), and m is the slope length exponent of the RUSLE S-factor.

In reality, a zone of deposition or concentrated flow would generally occur when a slope becomes long. The limiting slope length for interrill and rill erosion is about 120 to 150 m. Since RUSLE is only suitable for estimating erosion due to interrill and rill processes, there was a need to determine an upper bound on the slope length by limiting the flow accumulation. This was done by

Management Information Systems, C. A. Brebbia (Editor)

constraining the maximum flow accumulation to a given number of cells. To determine the number of cells, an upper bound slope length of 150 m was assumed and the DEM resolution of 90m x 90m was used.

The slope steepness S-factor was computed from the DEM using the ArcView's spatial analyst extension. The map algebra in the ArcView was applied to combine the upslope contributing area component ($L_{i,j}$) with the slope factor using equation:

$$LS = \left(FAccum \, * \, CS / 22.13 \right)^{0.4} * \left(\sin slope / 0.0896 \right)^{1.3} * 1.6 \qquad (3)$$

Where *FAccum* is the flow accumulation expressed as the number of grid cells; *CS* is the cell size (resolution of the DEM (90m)); *Sin slope* is the slope grid that had to be converted into radians; 0.4 and 1.3 are slope exponents and 1.6 a multiplying factor adapted from the values estimated by Mitasova *et al* [2].

3.3.2 Rainfall runoff factor R

The rainfall erosivity factor R was estimated by first calculating the storm erosivity indices using the product of the measured rainfall kinetic energy and its 30-minute intensity (EI_{30}) (Renard *et al* [5]). The method documented by Wischmeier and Smith [6] was applied to generate the erosivity indices for the few stations with autographic records within Masinga catchment. The storm erosivities were then accumulated for each year to give the R-factor for each of these few stations. A regression equation relating the R-factor and mean annual rainfall was developed.

Data for seven rainfall stations in Masinga with mean monthly rainfall amounts recorded over more than 20 years was collected from the Department of Meteorological Station, Dagoretti corner, Kenya. These seven rainfall stations are randomly distributed over the catchment, fig.1. A data file for each of the seven stations presenting mean annual rainfall was generated by averaging recorded amounts of rainfall over consecutive years. The R values were then estimated using the developed regression equation:

$$R = 0.56896 * P - 1.011 \qquad (4)$$

Where *R* represents the rainfall erosivity factor (KJ mm m^{-2} h^{-1} yr^{-1}) and *P* is the measured mean annual precipitation (mm).

To enable the mapping of the R-factor in a spatial domain, it required first to prepare the precipitation surface for the catchment. The data from the seven stations was geo-referenced using their latitudes and longitudes and presented in an excel table. The data was converted into a "dbaseIV" and then exported to the ArcView software. A point theme was created using the ArcView GIS software and the exported data was joined to the attribute table of this point theme. The point theme was converted to a shapefile and added as a new view using the ArcView procedure.

A surface rainfall map for the catchment was created by interpolating the rainfall point grid using the Inverse Distance Weighting (IDW) interpolation

Management Information Systems, C. A. Brebbia (Editor)
© 2004 WIT Press, www.witpress.com, ISBN 1-85312-728-0

method, a procedure supported by ArcView. Using the Analysis tool and the map calculator in the ArcView, eqn (4) was then applied to the interpolated surface rainfall grid and a resulting spatial R-factor map was generated.

3.3.3 Soil erodibility factor K

The Exploratory Soil Map of Kenya of scale 1: 1,000,000 obtained from Kenya Soil Survey was first digitised. To enable mapping of different soil types and their associated parameters in a spatial domain, the soil map of Masinga catchment was clipped from the digitised soil map of Kenya. A map of this scale was used because there were no soil maps of finer resolution covering the entire catchment, while technical and fiscal constraints could not permit a full-scale soil survey of the area. The map obtained thus contained the major soil mapping units based on FAO-UNESCO [7] classification.

Assignment of erodibility values to the FAO-UNESCO soil classification scheme for the catchment was conducted using a table of soil types and associated characteristics. In this study, the K-factor values were derived from existing literature from previous studies carried out in Kenya by Sombroek *et al* [8]. By assuming that the resolution of the K values matched that of the soil mapping units, the mean K factors for each soil type were assigned. A soil erodibility map of the catchment, expressed in class ranges was generated. The K-factor shape file was converted to a grid file.

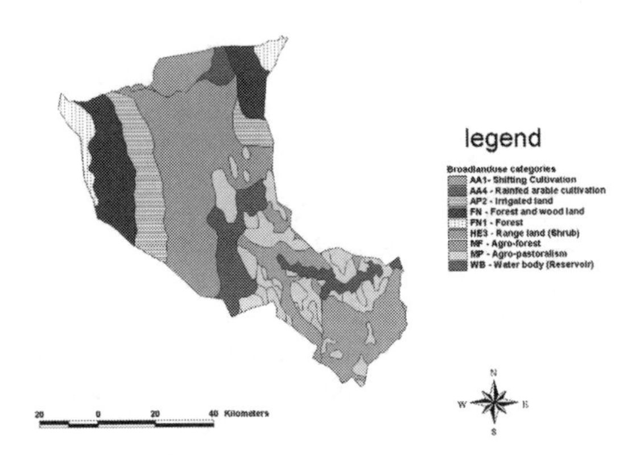

Figure 2: Broad land use /cover categories based on FAO-UNESCO classification.

3.3.4 Cover management factor C

In RUSLE, the C factors represent weighted average soil loss ratios (SLRs) that are determined from a series of sub-factors that include prior land use, canopy cover, surface cover and surface roughness (Renard *et al* [9]). However, it was not possible in this study to determine the magnitude of these sub-factors because the land use/cover data available for Masinga catchment only defined broad land use categories, fig. 2. The land use map for the catchment was

Management Information Systems, C. A. Brebbia (Editor)

generated by clipping it from the main land use map of Kenya that was prepared using remote sensing techniques and previous field surveys. The major land use/cover types were identified and classified and their associated C-factor values assigned. The C-factor for each land use/cover category was estimated based on values determined for similar crops and land cover for Kenya (Angima *et al* [10]). The C-factor map was converted from a shape file to a grid file.

3.3.5 Support practice factor P

To create a spatial map for the P-factors in this study, the digitised map of the soil units and land use/cover polygons was used. A spatial percent slope map was created and each of the polygon units was assigned a P value based on its percentage slope. The P values for contour cultivation and their corresponding percent slopes from results of Wenner [11] were assumed. The forestland was assigned a P-factor of 1. The P-factor shape file was then converted into a grid file.

4 Results and discussion

The RUSLE data themes and their associated attribute tables were integrated in the Arcview 3.2 GIS to determine the average annual soil loss for the catchment. The grid themes were overlaid together to generate a spatial mean annual soil erosion map, fig.3, using the ArcView's map algebra calculation. From the results, it can be depicted that the present land utilisation with lack of soil conservation measures, a rather low standard of husbandry on arable land, gross overstocking and lack of management on range areas is resulting in high soil erosion. The critical areas that require urgent soil and water conservation management are easily identified from the spatial erosion map.

The mean annual soil loss generated by the RUSLE model is subject to error due to inaccuracies inherent in each data layer, and the limitations of the methods used to derive the component factor values. Verification was essential to gain an appreciation of the model's predictive capabilities.

The field verification indicated that locations of very low soil loss potential were well estimated, while locations of low potential were not as accurate. However, areas of moderate, high and extreme soil loss potential were the most successfully identified. From this field validation, the RUSLE model's ability to predict soil erosion potential within the Masinga catchment is viewed as effective, when considering the purpose of this model as a conservation tool.

The GIS database and soil erosion potential map generated in this study provide valuable planning aids for managers in Masinga. With limited financial resources for land restoration or the implementation of best management practices (BMPs), it is imperative that each activity be undertaken in an area where the greatest impact can be made on mitigating soil erosion.

The database created in this research can be queried to identify those land areas currently under agriculture, with a high or extreme soil erosion potential ranking. The model can be re-run, adjusting the *C*- or *P*-value for these areas, to assess the effects of implementing BMPs leaving crop residue on the fields after harvest, or employing contouring or terracing techniques on the resulting soil

Management Information Systems, C. A. Brebbia (Editor)

erosion potential values. Financial resources can then be allocated accordingly. Similar scenarios, such as assessing the effects of clearing forested lands to meet increasing agricultural needs, can be readily accommodated with the model.

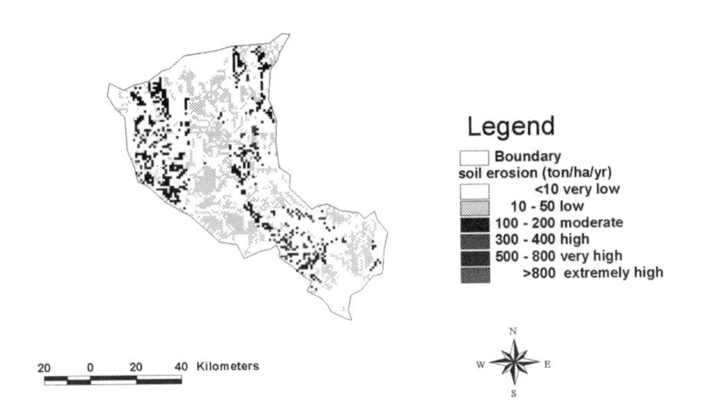

Figure 3: Spatial mean annual soil erosion within Masinga based on current land use and management practices.

5 Conclusions

This study presents a first attempt in the application of GIS technology to model soil erosion in Masinga catchment. It demonstrates the integration of an empirical model, the RUSLE within the GIS environment to estimate average annual soil losses in a spatial domain. The GIS was used to prepare required spatial data, extract input parameters for the model, execute the model computations and display results.

Refinements to the slope length and rainfall erosivity factors of the RUSLE model were introduced to improve its applicability in an area of mountainous terrain and a tropical climatic regime. By employing the upslope drainage substitution method to generate LS-values, the RUSLE is adapted to function on a semi-distributed basis, which more comprehensively considers the cumulative effects of overland flow on erosivity.

The model is based on modest data requirements, a common limitation in developing nations. Its practical utility is based upon providing a means for evaluating and comparing factor changes among alternative land areas, rather than on its prediction of absolute soil erosion for a particular grid cell. It provides a means to isolate and describe areas that are vulnerable to soil erosion lending immediate application to conservation planning. The tool can be operated locally by land managers in Masinga catchment, with available software and data.

Management Information Systems, C. A. Brebbia (Editor)

References

[1] Landa, R., Meave, J. & Carabias, J., Environmental deterioration in rural Mexico: an examination of the concept. *Ecological applications*, **7(2)**, pp. 316-329, 1997.

[2] Mitasova, H., Hofierka, J., Zlocha, M. & Iverson, L.R., Modelling topographic potential for erosion and deposition using GIS. *Geographical Information Systems*, **10(5)**, pp. 629-641, 1996.

[3] Desmet, P.J.J. & Govers, G., A GIS procedure for automatically calculating the USLE LS factor for topographically complex landscape units. *Soil and Water Conservation*, **52(5)**, pp. 427-433, 1996.

[4] Foster, G.R. & Wischmeier, W.H., Evaluating irregular slopes for soil loss prediction. *American Society of Agricultural Engineers*, **17(2)**, pp. 305-309, 1974.

[5] Renard, K.G., Foster, G.R., Weesies, G.A., McCool, D.K. & Yoder, D.C., (eds). *Predicting Soil erosion by Water*, A guide to Conservation Planning with the Revised Universal Soil Loss Equation (RUSLE). U.S. Department of Agriculture-Agriculture Handbook No. 703, 1997.

[6] Wischmeier, W.H. & Smith, D.D., (eds). *Predicting rainfall erosion losses,* USDA Agricultural Research Service Handbook 537. USDA, Washington, D.C., 1978.

[7] FAO-UNESCO, *Soil Map of the World, Revised Legend*, FAO World Soil Resources Report No. 60. Food and Agricultural Organisation of the United Nations, UNESCO, Rome, Italy, 1998.

[8] Sombroek, W.G., Braun, H.M.H. & van der Pouw, B.J.A., *The exploratory soil map and agro-climatic zone map of Kenya*, Report No. E.I, Kenya Soil Survey, Nairobi, 1982.

[9] Renard, K.G., Foster, G.R., Weesies, G.A. & Porter, J.P., RUSLE Revised Universal Soil Loss Equation. *Soil and Water Conservation*, **46(1)**, pp. 30-33, 1991.

[10] Angima, S.D., Stott, D.E., O'Neill, M.K., Ong, C.K. & Weesies, G.A., Soil erosion prediction using RUSLE for central Kenyan highland conditions. *Agricultural Ecosystems and Environment*, **97**, pp. 295-308, 2003.

[11] Wenner, C.G., Soil Conservation in Kenya, SWC Branch, MoALDM, Nairobi, Kenya.

Management Information Systems, C. A. Brebbia (Editor)

Section 6
Transportation

Statewide adaptive translative transportation geotechnical DBMS (GDBMS) framework for Virginia, USA

J. Yoon
Department of Civil and Environmental Engineering, Old Dominion University, Norfolk, Virgina, USA

Abstract

A Geotechnical Engineering Database Management System (GDBMS) Framework for transportation infrastructure planning and design was conceptualized, developed and implemented to facilitate the access to and utilization of historical and current geotechnical data specific to the Virginia Department of Transportation's (VDOT) Hampton Road Third Crossing (HR3X) and Woodrow Wilson Bridge (WWB) projects at Washington, D.C. The GDBMS framework implementation is largely based on an Internet-based, adaptive translative framework concept to allow actual heterogeneous geotechnical data to be used "as-is" without retrofitting them for the GIS spatial database, so therefore provides rapid implementation and flexible modification and expansion. The Framework automatically identifies the type and characteristic of the "as-is" geotechnical data and subsequently translates it dynamically to various output data formats to make it available to the users via standard web browser interfaces. Furthermore, the GDBMS framework facilitates additional engineering design and analysis capabilities including generation of dynamic cross-sectional borehole fence diagrams over the Internet through a standard GIS user interface for road and bridge site selection and design.
Keywords: adaptive framework and modeling, distributed GIS, geotechnical database management system (GDBMS), borehole fence diagram.

1 Introduction

In 2002, a pilot study to implement an Internet-based, spatiotemporal Geotechnical Database Management System (GDBMS) Framework was

Management Information Systems, C. A. Brebbia (Editor)

conducted for the Virginia Department of Transportation (VDOT) using a distributed Geographical Information System (GIS) methodology for geotechnical data archival and retrieval.

Previously, a feasibility study [1] was conducted to identify a number of GDBMS implementation alternatives, and the GIS-based strategy was recommended as the most viable implementation model in terms of ease of use, cost-effectiveness, flexibility and future expandability.

Main objectives of this pilot GDBMS Framework study were to conceptualize, design, develop and implement an Internet-based GDBMS Framework to facilitate VDOT geotechnical engineers to access and utilize historical and current geotechnical data specific to the Hampton Road Third Crossing (HR3X) project. Hampton Road Third Crossing project is aimed to relieve congestion at the existing Interstate 64 Hampton Roads Bridge Tunnel in Norfolk, Virginia. Proposed Third Crossing bridge site is located in the near west of the Craney Island Dredged Material Management Area (CIDMMA) where the U.S. Army Corps of Engineers (USACE) has the jurisdiction over. The CIDMMA is a 2,500-acre, confined dredged material disposal site approved by the Congress under the River and Harbor Act of 1946, and it had been receiving maintenance, private, and permit dredged material from numerous dredging projects in the Hampton Roads area since 1959.

HR3X project is a new multi-billion dollar project, and preliminary engineering activities have already started in the fall of 2002 with General Engineering Consultant (GEC) to design five discrete sections of the proposed bridge, tunnel and connecting roads. Geotechnical engineering component of the HR3X project would require multiple deep channel drilling explorations that may cost $10,000 per day. Reflecting to such high-cost project requirements, one of main objectives of the GDBMS Framework pilot study was to develop and implement a GDBMS Framework component specific to the HR3X project by using existing VDOT's geotechnical data as well as applicable Corps' CIDMMA data. Thus resulting GDBMS Framework implementation would be capable of facilitating dynamic interpolation and generation of cross-sectional boring fence diagrams in the deep channel section instead of recursively spending time and money all over again to collect new in-field deep channel drilling geotechnical data.

In 2003, site-specific HR3X GDBMS Framework pilot study implementation was substantially expanded to facilitate a Virginia statewide-scale GDBMS Framework. To demonstrate its new scalability, a new project site, Woodrow Wilson Bridge (WWB) Route 1 Interchange (R1I) site, Washington, D.C. was added to the statewide GDBMS Framework [2].

2 Organization of geotechnical DBMS (GDBMS) framework

GDBMS Framework concept lies in a "container" programming approach enmeshing two core components of GIS and geotechnical data, and an external application, gINT® software [3] over the distributed network system, i.e., Internet. Here, the concept of a "container" reflects a number of key innovations

Management Information Systems, C. A. Brebbia (Editor)
© 2004 WIT Press, www.witpress.com, ISBN 1-85312-728-0

such as a standard, non-proprietary, open-system architecture [4], an easily updateable and expandable architecture [5] that enables seamless access to and retrieval of geotechnical data from both database (= attributal/temporal) and GIS (=map/spatial).

GDBMS Framework consists of four modular components; (i) geotechnical data, (ii) adaptive translative Framework layer, (iii) analysis and design application and (iv) HyperText Transfer Protocol daemon (httpd), i.e., an ArcIMS [6] GIS-enabled web server as an Internet GDBMS Framework engine.

Data used in GDBMS Framework are categorized into two types – geotechnical data and spatial map data. Geotechnical data portion was compiled from existing and current VDOT geotechnical data and Corps' CIDMMA geotechnical data in a gINT® or equivalent format that contain standard boring, STP (Standard Penetration Test), CPT (electric Cone Penetrometer Test), DMT (Dilatometer Test), VibraCore (sediment sampling) logs.

Spatial map portion of the GDBMS Framework data was prepared from various spatial database sources including VDOT GIS layers and USGS color digital orthophotographs. These two main data components of the GDBMS Framework were then georeferenced and integrated by an adaptive translative Framework layer to provide a standard web browser-based user interface commonly accessible by Internet Explorer or Netscape. Figure 1 illustrates the concept behind the statewide GDBMS Framework.

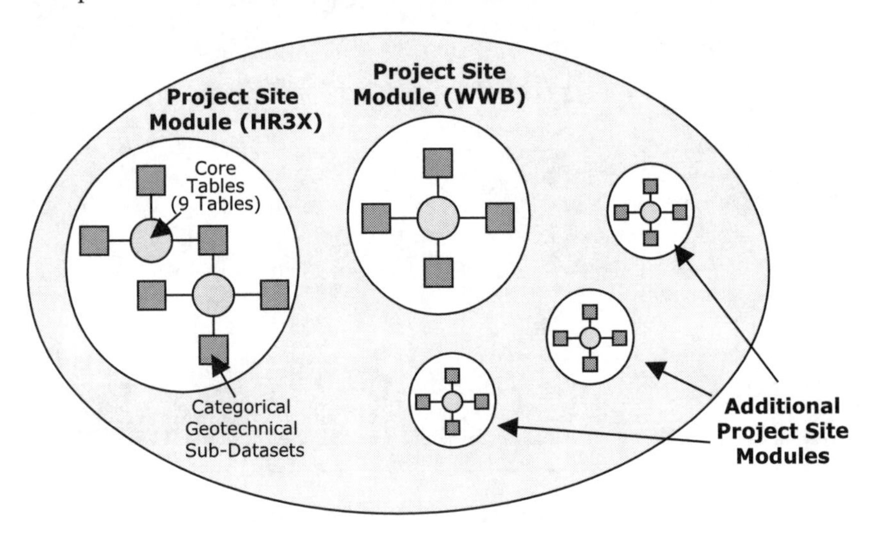

Figure 1: Concept of scalable statewide GDBMS framework.

GDBMS Framework utilizes a distributed Geographic Information System (GIS) approach to bind its module components. GIS is, in a simplistic manner, a suite of computer programs and hardwares that facilitates capture, storage, query, retrieval and analysis of spatial and attributal information. [6][7] Furthermore, in an engineering application context, a GIS can be viewed as a special case of

Management Information Systems, C. A. Brebbia (Editor)

information systems where the database consists of observation on spatially distributed features, activities or events, which are definable in space as point, lines or areas, both in visual (i.e., map) and in textual (i.e., tabular) display orientation. Thus GIS manipulates data about these points, lines, and areas to retrieve data for ad hoc queries and analyses to solve problems, come up with the answers, or try out a possible solution [8]. As a result, GIS often provides easy modification and flexibility, and native porting and linkage to other software applications in case of a specific need arises. [9]

Figure 2: GDBMS Framework user interface.

ArcIMS (Internet Map Server) [10] is an extension of a standard ESRI ArcGIS GIS application suite. Traditionally, the modus operandi of ArcGIS applications has been either stand-alone system or client-server architecture over a dedicated LAN (i.e., thin client) orientation. However since late 90's, a heavy emphasis has been given to Internet-based GIS application through ArcIMS. Figures 2 and 3 show the GDBMS Framework user interfaces.

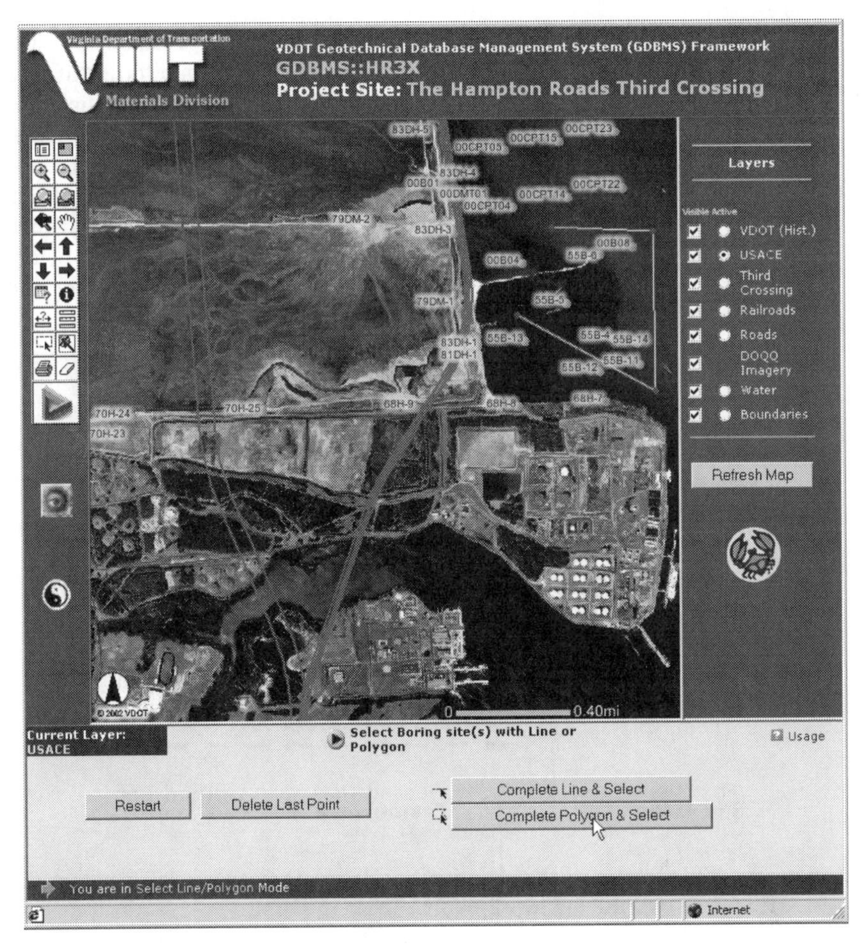

Figure 3: Dynamic boring sites display and selection.

3 Adaptive translative framework and its layers

In regard to characteristics of data, geotechnical data and spatial map data, these two data types are intrinsically heterogeneous. Linking and dynamically cross-referencing both data in a useable context was accomplished by dynamically translating these data with the adaptive translative Framework layers within the GDBMS Framework as shown in Figure 4.

Benefits of the adaptive translative Framework layer include the elimination of retrofitting existing geotechnical data and the reduction of the project time required for development and implementation. Most common problem encountered in integrating existing data to a "system" is the requirement of modifying or retrofitting existing data to a new or "ideal" format retrospectively. Such retrofitting efforts usually demand exorbitant amount of resources and time, at the same time, do not guarantee any future expandability or flexibility of

Management Information Systems, C. A. Brebbia (Editor)

the "system." To obviate such problem, GDBMS Framework was conceptualized and designed to identify particular data origin and format using a number of data signature filters. Then subsequently translate the identified data dynamically or "on-the-fly" to a standardized or common data format without physically retrofitting data itself so that engineers can use the data and information for actual analysis and design rather than spending time and efforts to interpret such heterogenous data each time. This adaptive translative Framework approach is a tremendous advantage over typical proprietary engineering database systems in terms of usability, flexibility, stability and future expansion of the implemented system.

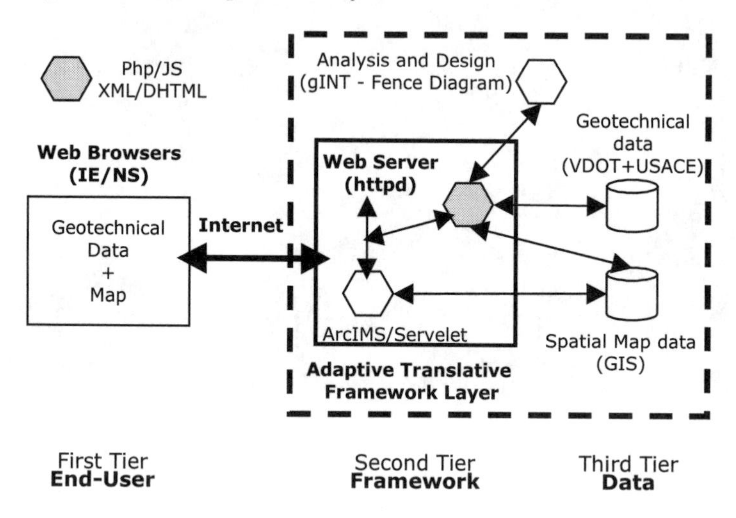

Figure 4: Schematic of the adaptive translative Framework layer within the GDBMS Framework.

Output data of dynamically translated geotechnical data by the GDBMS Framework include boring, standard penetration test (STP), electric cone penetrometer test (CPT), dilatometer test (DMT), sediment sampling (VibraCore) logs, location-specific gINT® project file translated in a standard data format and Excel CSV (Comma Separated Values) format as shown in Figure 5.

Furthermore, the GDBMS Framework layer facilitates additional design and analysis capabilities including dynamic cross-sectional borehole fence diagram generation over the Internet using gINT® software residing in the GDBMS Framework server (Figure 6). Therefore it effectively provides a valuable geotechnical analysis and design tools to the users, relieves the users from the steep learning curve for using gINT® software, at the same time eliminating expensive and complicated software requirement at each end-user's computer. All these output data are available to the users via a standard web browser interface, such as Internet Explorer or Netscape.

Management Information Systems, C. A. Brebbia (Editor)
© 2004 WIT Press, www.witpress.com, ISBN 1-85312-728-0

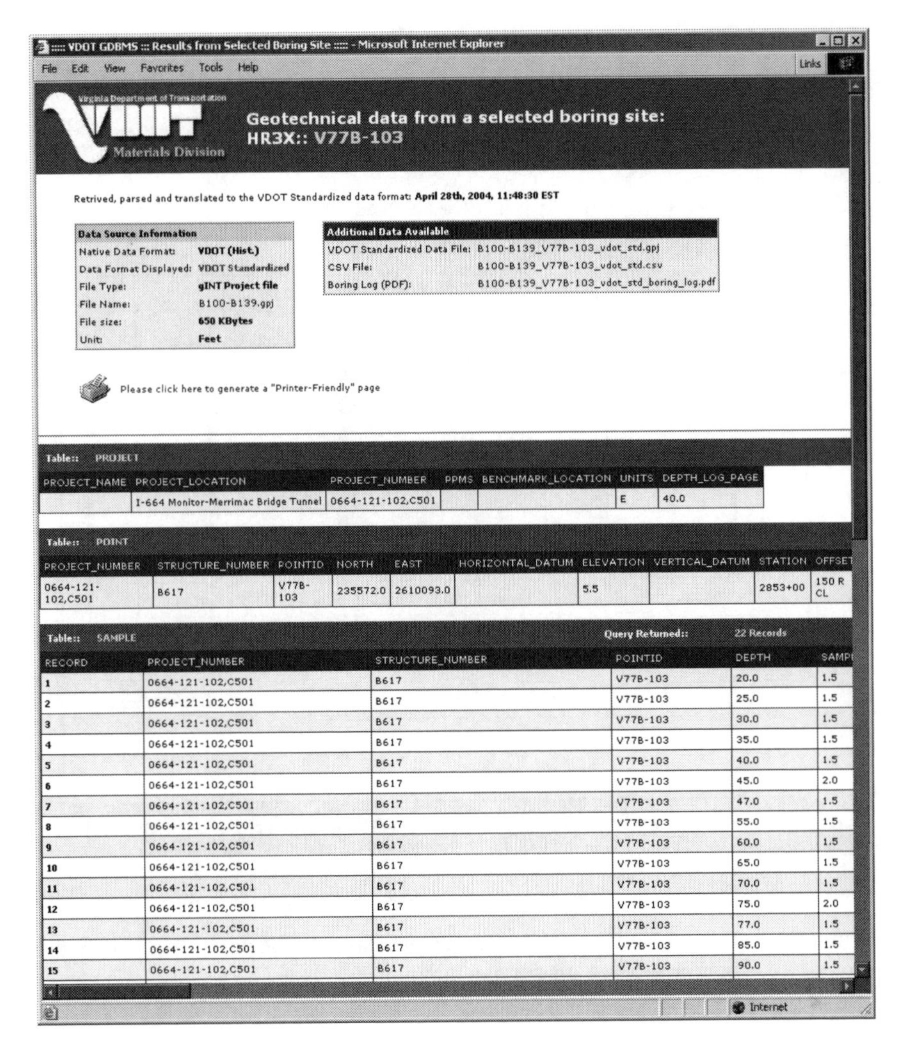

Figure 5: Dynamic adaptive translation to standardized outputs.

4 Data ROI and open system architecture concept

Majority of current engineering GIS applications require all the pertinent data, regardless of data types and formats, to be explicitly included in the their native, intrinsic GIS spatial database format, and are mainly used for retrieval and display of such data solely within the GIS. Thus, enormous amount of preparation, modification, conversion and pre-processing of source data into a specific format understood by the GIS application is essential requirement in GIS-based application. Usefulness and effectiveness of an engineering GIS application is directly proportional to the flexibility, size and quality of its database.

Management Information Systems, C. A. Brebbia (Editor)

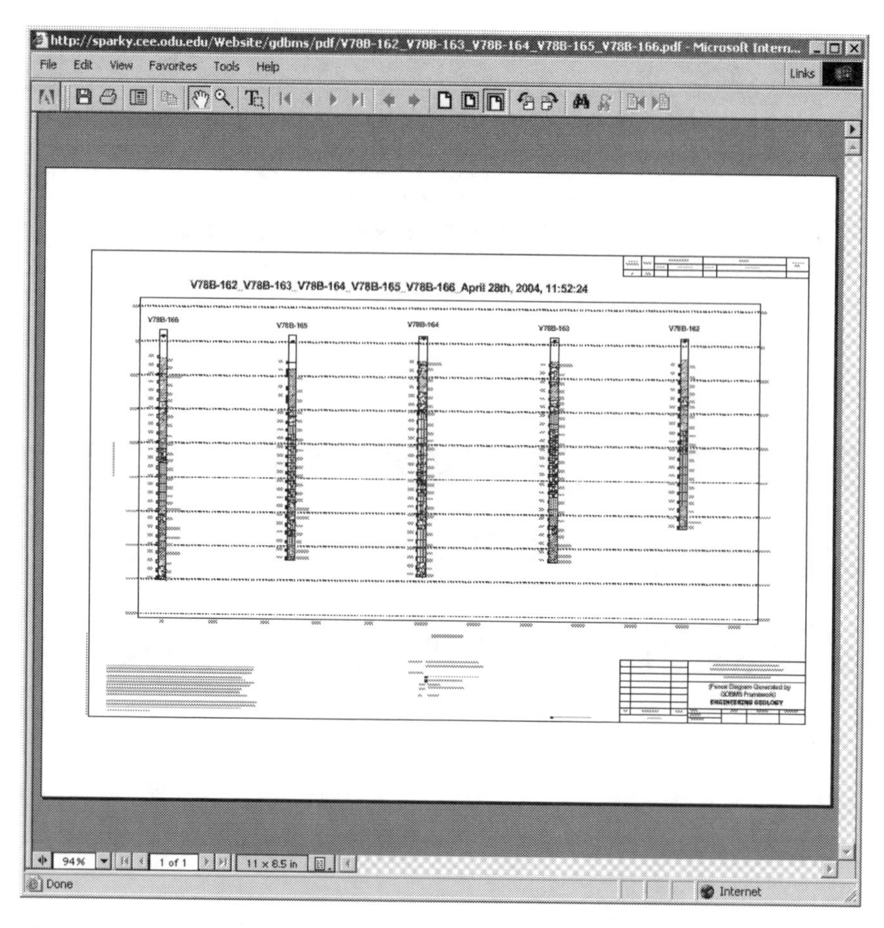

Figure 6: Design and analysis Framework layer - dynamic cross-sectional borehole fence diagram generation.

Consequently, time and resource required for such data conversion into GIS data format has been a major huddle for any engineering GIS application in term of ROI (Return On Investment) turnaround, particularly in time-sensitive project management situation – it easily takes years of data preparation and once it is completed, usually the data becomes already obsolete.

Instead of binding all pertinent geotechnical data into the GIS, the GDBMS Framework flexibly links geotechnical data to skeleton GIS internal database, so that managing additional geotechnical data after system implementation become a matter of copying and dropping data files "as-is" to GDBMS Framework directory – flexibility and expandability -- therefore provides a rapid implementation of useable data through GIS methodology.

The adaptive translative Framework layer was based on the open system architecture principle with standard XML (Extensible Markup Language) [11],

Management Information Systems, C. A. Brebbia (Editor)

ArcXML (Arc Extensible Markup Language) [12], Php (Hypertext Preprocessor) [13], DHTML (Dynamic Hyper Text Markup Language) [14], Javascript script programming languages to ensure flexibility, expandability, portability and scalability of the Framework. All these programming languages are public domain, and all are freely available. Choice of script programming languages for the GDBMS Framework was to create a non-proprietary implementation so that in case of any future need, the GDBMS Framework can be easily modifiable and portable without repeating a new development and implementation cycle.

5 Summary and conclusion

An Internet-based, spatiotemporal Geotechnical Database Management System (GDBMS) Framework was conceptualized, designed, developed and implemented using a distributed Geographical Information System (GIS) methodology to facilitate accessing, retrieval and utilizing historical and current geotechnical data specific to the Virginia Department of Transportation's (VDOT) Hampton Road Third Crossing (HR3X) project. This adaptive translative GDBMS Framework was designed and implemented from the user's perspective instead of the developer's so that it would satisfy followings characteristics; usability and ease of use, compatibility (with existing implementation), reliability and robustness, and flexibility and expandability.

Utilization of heterogenous geotechnical data was facilitated through dynamic, adaptive translative GDBMS Framework layer, not through static, proprietary, hard-coded applications. Such static application approaches often impose precondition for existing geotechnical data to be retrofitted or re-processed, and certainly incur unnecessary time, efforts and high resources commitment. On the contrary, the adaptive translative Framework layer allows actual geotechnical data to be used "as-is" without retrofitting, therefore it rapidly provides valuable data for intended analysis and design.

Although this pilot study emphasizes the GDBMS Framework methodology to the Hampton Road Third Crossing (HR3X) and Woodrow Wilson Bridge (WWB) projects at Washington, D.C. projects, the very Framework methodology can be applicable to any future geotechnical and other spatial database managements and engineering design and analysis in general.

References

[1] Ishibashi, I and Yoon, J., Feasibility Study of VDOT Geotechnical Database Management System (GDBMS), VTRC Contract Research Report, Virginia Transportation Research Council, Charlottesville, VA, February, 2002.

[2] Yoon, J. Implementation of Statewide VDOT Geotechnical Database Management System (GDBMS) Framework: The Woodrow Wilson Bridge (WWB) Route 1 Interchange (R1I) Site, VA, VTRC Contract Research Report, Virginia Transportation Research Council, Charlottesville, VA, October 2003.

Management Information Systems, C. A. Brebbia (Editor)

[3] Geotechnical Computer Applications (GCA), 2002, gINT 5.0, Geotechnical and Geoenvironmental Software System, Windsor, CA.
[4] Mallach, E.G., 2000, *Decision Support and Data Warehouse Systems*, McGraw-Hill Publishing Company, New York, NY.
[5] Umar, A., 1997, *Object-Oriented Client/Server Internet Environments*, Prentice Hall Professional Technical Reference Series, New York, NY.
[6] Bernhardsen, T., 2002, *Geographic Information Systems: An Introduction, 3rd Ed.*, Norwegian Mapping Authority (Distributed by John Wiley & Sons).
[7] Clarke, K.C., B.O. Parks, and M.P. Crane, 2002, *Geographic Information Systems and Environmental Modeling*, Prentice Hall Publishing, New York, NY.
[8] Duecker, K.J., 1979, Land resource information systems: a review of fifteen years' experience, Geo-Processing, 1(2):105-128.
[9] Connolly, T.M. and C.E. Begg, 2001, Database Systems: A Practical Approach to Design, Implementation, and Management, 3rd Ed., Addison-Wesley Publishing, New York, NY.
[10] East, R., R. Goyal, A. Haddad, A. Konovalov, A. Rosso, M. Tait, and J. Theodore, 2001, *The Architecture of ArcIMS, a Distributed Internet Map Server Advances in Spatial and Temporal Databases*, Lecture notes in computer science. no.2121: 387-403, Springer-Verlag.
[11] World Wide Web Consortium (W3C), 2001, *Extensible Markup Language (XML), XML 1.1 W3C Recommendation*, November 2001, Institut National de Recherche en Informatique et en Automatique, Keio University.
[12] ESRI Press, 2001, *ArcXML Programmer's Reference Guide, ArcIMS 3*, ESRI Press, Redlands, CA.
[13] Gutmans, A and Z. Suraski, 1997, *Php: Hypertext Preprocessor*, Apache Software Foundation (ASF).
[14] World Wide Web Consortium (W3C), 2000, *Dynamic Hyper Text Markup Language (DHTML) W3C DOM level 2 (DOM2)*, Massachusetts Institute of Technology, Institut National de Recherche en Informatique et en Automatique, Keio University.

Management Information Systems, C. A. Brebbia (Editor)

Tracing traffic dynamics with remote sensing

S. P. Hoogendoorn & H. J. van Zuylen
Transport & Planning Department, Delft University of Technology

Abstract

Science has not yet been completely successful in explaining the various puzzling but interesting phenomena of congested traffic flows. The reasons that these phenomena are poorly understood are mainly due to the lack of adequate data needed to develop this theory. Common measurement methods collect only 'snapshots' of the situation (i.e. at a limited number of cross-sections, or a single, instrumented vehicle). Such data do not provide sufficient information to study the dynamics of individual drivers in their continuously changing traffic and roadway environment.

This paper presents a new innovative traffic data collection system to detect and track vehicles from aerial image sequences. Besides the longitudinal and lateral positions as a function of time, the system can also determine the vehicle lengths and widths. Before vehicle detection and tracking can be achieved, the software handles correction for lens distortion, radiometric correction, and orthorectification of the image. Vehicle detection occurs approximately each tenth of a second, after which vehicles are tracked both forward and backward in time (10 frames in each direction). The resulting data redundancy is then used to improve accuracy and reliability of the traffic data. Post-processing furthermore entails using Iterative Extended Kalman filtering to improve data quality.

The software was tested on data collected from a helicopter, using a digital camera gathering high-resolution monochrome images of a Dutch motorway near the city of Utrecht. From the test, it is concluded that the techniques for analyzing the digital images can be applied automatically without many problems, while nearly 100% of all vehicles could be detected and tracked.

1 Introduction

The objective of the present study is to develop a data collection method to collect vehicle trajectories (longitudinal and lateral position of the centre of the

Management Information Systems, C. A. Brebbia (Editor)

vehicle represented by a rectangle as a function of time) and individual vehicle characteristics (vehicle length and width) in particular during congested traffic flow conditions. Given the fundamental requirements of underlying research into driver behaviour, specific demands to the monitoring system hold for the temporal and spatial resolution. For one, the final system must have a resolution of at least 40 cm. Since the roadway length that can be observed by a single camera is determined by resolution of the camera, the used high resolution B&W digital camera (resolution of 1300 pixels × 1030 pixels), could observe 1300 × 0.4 = 520 m roadway length.

Given average headways between vehicles, and their average speeds, it was decided that the time between two observations should not exceed 0.1 s. It can be shown that in case the specifications above are met, the locations of the vehicles can be determined with an accuracy of 1/4 pixel (= 0.1 m). The resolution of the speeds that are determined from the vehicle positions is thus 1 m/s.

2 Airborne traffic data collection

Data was collected from a helicopter, using a digital camera system. The camera that was used in the end, provided greyscale images at a resolution of 1300 by 1030 pixels with a maximum frequency of 8.6 Hz. The area that each pixel represents is determined by the specifications of the camera (light sensitive chip and lens) and the height at which the images are collected. To decrease the probability that clouds obstruct the observations, it was decided to not fly higher than 500 m. It was decided to use a camera with a 2/3" chip, and a lens of 16 mm. A Personal Computer equipped with a frame grabber was attached to the camera enabling real-time storage of the digital images. The camera was attached to the helicopter and was fixed. No gyroscopic mounting was used to attach the camera, assuming that the resulting vibrations and movements of the helicopter would not influence the quality of the collected data too much.

The data was collected from a height of 500 m, implying that approximately 500 m of roadway could be observed at a single time instant. During the measurement, the helicopter tended to slowly drift from its location.

3 Process overview

The collected raw data consist of large sequences of digital greyscale high-resolution images of the roadway and the traffic it carries. The approach to process these data and determine the individual traffic data consist of the following steps, described in the ensuing of the paper:

1. Image processing (correction for lens distortion, radiometric correction, and orthorectification / geo-correction).
2. Determining background image by removing dynamic objects from reference image.
3. Vehicle detection and tracking.
4. Data post-processing (geo-referencing, handling data redundancy, and filtering).

Management Information Systems, C. A. Brebbia (Editor)

4 Image processing

The first step entails applying image processing techniques, with the aim to 'standardize' the images. This implies that the intensities in the all images need to be comparable, all images refer to the same plane of observation, and that all images have the same orientation. To this end, the following techniques are used respectively:

1. *Radiometric correction* ensures that the effects of changing lighting conditions (e.g. due to clouds, etc.) are reduced.
2. *Lens distortion correction* removes the so-called pincushion effect from the images. Even though the pincushion effect is very moderate, removing it appear crucial to ensure correct orthorectifcation.
3. *Orthorectifcation* or *geocorrection* of the images effectively corrects for the movement of the helicopter (changes in height, rotation, pitch, and yawn).

See [1] for a detailed description of radiometric and lens distortion correction. Let us consider orthorectification in some detail. In an aerial photograph of a rectangular object, the image of this object will only be rectangular is the camera is located exactly above the middle of the rectangle (neglecting the lens distortion discussed above). Otherwise, the perspective of the image will be distorted, depending on the location and the angle of the camera. On top of this, the size of the rectangle will depend on the height at which the images are collected, and will the image be rotated around the vertical axis.

During orthorectification, perspective distortion, scale and rotation of the images are adjusted such that the objects on the image are projected at the same location as the same objects in the *reference image R*. This reference image *R* is determined *from several images that are collected at different time instants*, and thus reflects a relatively part of the roadway (compared to the individual images). These images are stitched together using dedicated image processing software to form the reference image.

Orthorectification needs *control points*, which are points in the image that are visible in both the reference image and the processed image. In theory, only 4 control points are needed. However, due to the fact that the determination of the location of the control points cannot be achieved with 100% reliability, 10 to 30 points were used instead of four, which also gives an indication of the accuracy of the process.

To handle the fact that the control points in the reference image and the processed image may be far apart, a special process was developed that uses the information of the control point location in image *I*-1 to determine the location of these points in image *I*. This process starts from reference image *R*. If *R* is not the first image of the sequence, the process is also performed backwards (for images *R*-1, *R*-2, etc.). Two sets of control points have been used.

The first set, the so-called *characteristic control points* are used for coarse matching. These are around these points are unique for the entire image and can thus be used to match to images where the amount of perspective distortion is large (i.e. the objects on the image and the objects on the reference image are far apart). Typical objects in these sets are lanterns, gantries, etc. The second set – the *roadsurface control points* – contains points of the roadway surface, i.e. on the reference plane such as the lane markings. These points are not unique and are used for finematching only.

The two phases of the process are:

1. *Coarse matching* finds the control points in the characteristic set of the reference image R in image I. This is achieved using an iterative approach that maximizes the cross-correlation coefficient of the pixels around the control points in the characteristic set. The search window contains 50 × 50 pixels. The accuracy of coarse matching is approximately 1 pixel. The results of coarse matching are used to determine the transformation of image I-1 to image I.
2. *Fine matching* uses the same approach as coarse matching. In this case however, the road surface control points are used instead of the characteristic control points, while the search window is only 7 × 7 pixels large.

5 Removing dynamic objects from reference image

Vehicle detection is based on removing a *background image B* (i.e. the empty road) from the individual vehicles, thus leaving the vehicles (dynamic objects). The approach to removing dynamic objects is simple, and based on the assumption that the road will be empty most of the time (referred to as the *histogram approach*). Considering the median intensity of a pixel in a sequence of images thus reflects the intensity of the road surface, i.e. the background, at that particular location. Application of this procedure to all pixels in the reference images effectively removes the dynamic objects from the scene. Fig. 1 shows an example of applying the approach to the data collected.

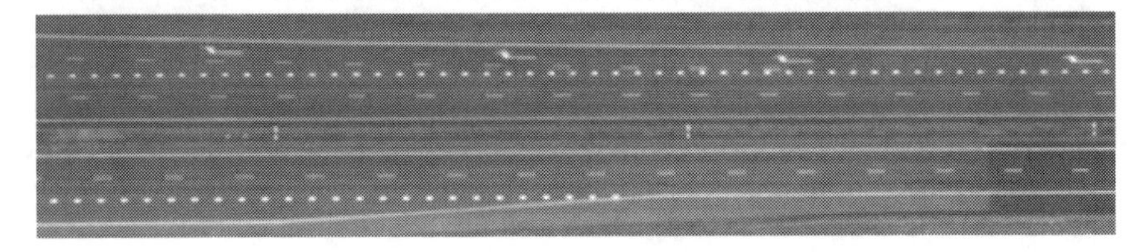

Figure 1: Empty road determined for A27 motorway using histogram approach.

6 Vehicle detection and tracking

Having determined the background image B, the next step of the approach is the detection and tracking of the vehicles. During vehicle detection, the location of the vehicle, and its dimensions is determined. Vehicle tracking entails following the detected vehicle both backward and forward through a series of *geocorrected* images.

Management Information Systems, C. A. Brebbia (Editor)
© 2004 WIT Press, www.witpress.com, ISBN 1-85312-728-0

6.1 Vehicle detection

For any image, vehicle detection is based on the difference between the current image I and the background image B. A first approximation is to use a threshold value to decide whether a pixel represents a vehicle or not. If so, neighboring pixels can either be identified as a vehicle or not. In practice, a number of complicating factors will occur:

1. Both light and dark vehicles will cast shadows, which are generally darker than the roadway surface.
2. Light vehicles have dark spots (windshields, etc.).
3. On occasion, a small vehicle completely drives in the shadow of a big vehicle (e.g. a truck or a bus). As a result, the shadow of the small vehicle disappears. Furthermore, the intensity of the vehicle itself may be close to the intensity of the background image.

The biggest problems are caused by vehicles that have the same intensity as the roadway surface or vehicles that have the same intensity as their shadow. Different approaches have been implemented to resolve these issues (morphological grayscale operations, binary morphological operations, split and merge image segmentation, etc.). We refer to [2] for a detailed description. When the vehicles are detected, the positions as well as the length and width of the vehicles could be established with relative ease.

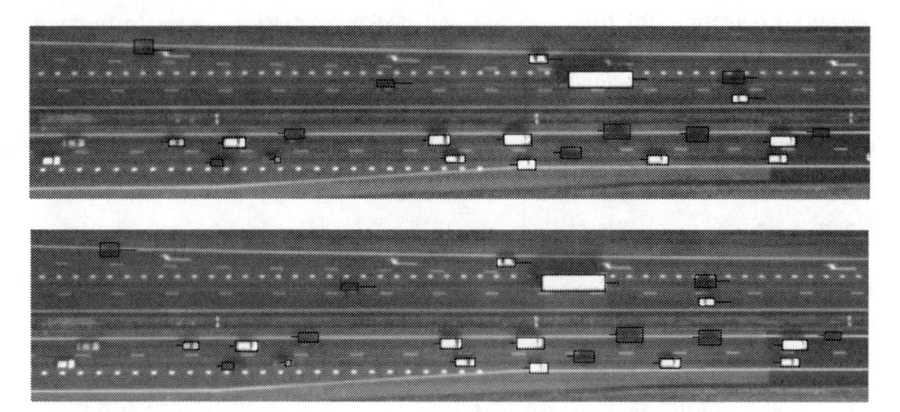

Figure 2: Vehicle detection and tracking results.

6.2 Vehicle tracking

The aim of vehicle tracking is to follow the vehicles detected in an image, i.e. to determine their position in the other images. In most cases, tracking is done using an approach similar to the control-point approach used in the orthorectification step (coarsematching and finematching). Its application yields a unique label for all vehicles detected during the vehicle detection step, enabling determination of the vehicle trajectories. To improve the accuracy, the original subimage of the detected vehicle (i.e. determined during detection) as well as its

Management Information Systems, C. A. Brebbia (Editor)

subimage in the previous image *I*-1 were jointly used to determine the position of the vehicle in the current image *I*.

As a final note, it is mentioned that *vehicle detection* is performed for each frame *I*, that is at each 10th of a second. Vehicle tracking entails both forward and backward tracking of all detected vehicles in the next / previous 10 images. As a result, in each image, a vehicle may result in multiple data points: one from vehicle detection (given it is detected), and multiple data points from tracking. Redundant data is used to improve the accuracy of the traffic data.

7 Data post-processing

This post-processing step entails four aspects:

1. For each detected vehicle, combine the redundant data to a unique vehicle trajectory.
2. Translate image coordinates into real-life coordinates.
3. Check and correct data manually (manually detect missed vehicles, remove false vehicle detections).
4. Improving data accuracy by Iterative Extended Kalman filtering.

7.1 Combining redundant data

In theory, for each vehicle 21 data points are determined in the previous steps: one from detection, 20 from forward and backward tracking. Due to missing detection, in practical situations this number may be substantially less. Using dedicated cluster analysis techniques, these points are clustered, identifying eventual outliers. For the remaining points, the median value is considered and considered to be the best estimate for the vehicles position (lateral and longitudinal), as well as its dimension.

7.2 Translation into real-life coordinates

In the final step, the image vehicle coordinates are translated into real-world coordinates. Scaling and translation determine both the longitudinal position of the rear bumper of the vehicle relative to an arbitrary location on the roadway, and the lateral location of the vehicle relative to the right lane demarcation. To this end, maps of the roadway are used.

7.3 Manual data checking and correction

A dedicated tool has been developed, which lets the used check the resulting detection and tracking results. False detections and incorrect tracking can be removed. The tool also offers the user the opportunity to manually detect vehicles. The system in turn takes care of the required vehicle tracking.

Management Information Systems, C. A. Brebbia (Editor)
© 2004 WIT Press, www.witpress.com, ISBN 1-85312-728-0

7.4 Iterative Extended Kalman filtering

In the current version of the software, data is collected with an accuracy of 1 pixel (i.e. roughly 40 cm in real-life coordinates), which implies that the speeds can be determined with accuracy in the order of 4 m/s. For most applications, this will be too coarse.

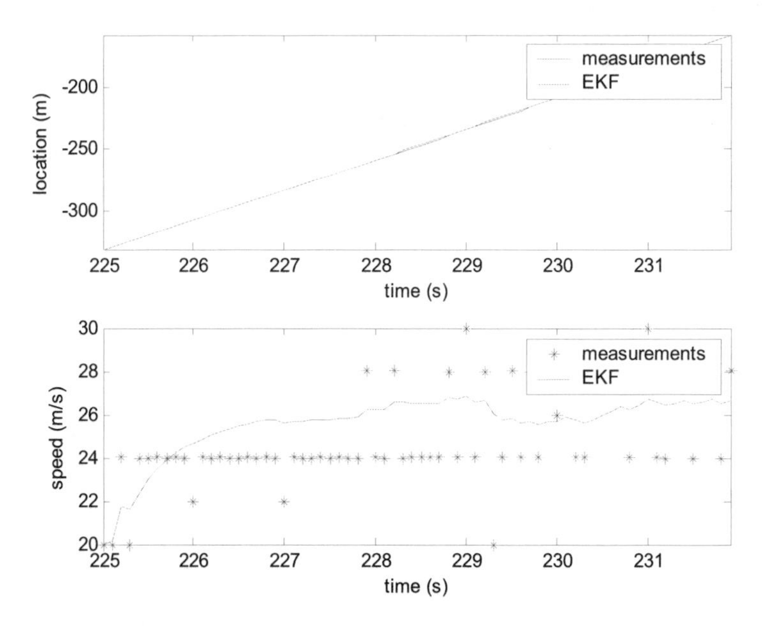

Figure 3: Application of the Extended Kalman filter to individual vehicle measurements (up: vehicle positions; down: longitudinal speeds).

To resolve this problem, an Iterated Extended Kalman Filter was developed that combines rudimentary knowledge of vehicle dynamics and driver behaviour with data from the data collection system. The filter consists of two elements: a so-called state-space equation (the model) describing the vehicle dynamics, and a measurement equation describing how the model output relates to the data. In the situation presented here, a simple car-following rule is used to predict the acceleration of a vehicle as a function of the distance headway and speed of the preceding vehicle, i.e.

$$\frac{d}{dt}r_i = v_i, \text{ and } \frac{d}{dt}v_i = \kappa(v_{i-1} - v_i) + \varepsilon$$

where ε is some Gaussian noise term describing the errors in the model. Since only the locations are measured direction, the measurement equation is given by

$$y_i = r_i + \eta$$

Management Information Systems, C. A. Brebbia (Editor)

where η is also some Guassian noise term. The Iterative Extended Kalman Filter is then used to combine both equations and to determine an estimate for the location, speed and acceleration. A similar approach is used for the lateral dimension.

Figure 3 illustrates application of the approach for one vehicle, which has been detected and tracked. The figure shows how the filter effectively smoothes the speed measurements.

7.5 Example results

Figure 4 below shows results of application of the approach outlined in this paper. The data in the figure where collected at weaving area on the A2 motorway near the Dutch city of Utrecht.

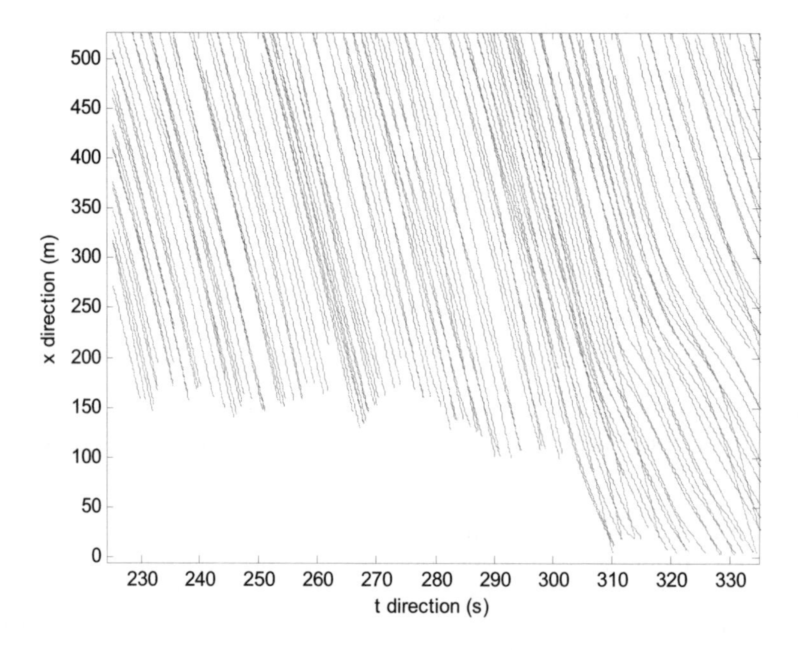

Figure 4: Example vehicle trajectories for situation on A2 motorway near Utrecht, The Netherlands.

8 Conclusions

This paper describes a new data collection system that was developed to determined individual vehicle trajectories from high-resolution grayscale images. These images were collected using a digital camera mounted underneath a helicopter, and stored on a personal computer. Having applied a number of photogrammetric operations on the images, nearly all vehicles could be detected

Management Information Systems, C. A. Brebbia (Editor)
© 2004 WIT Press, www.witpress.com, ISBN 1-85312-728-0

and tracked, yielding both vehicle positions, and vehicle dimensions. The spatial resolution of the raw data was 40 cm; the temporal resolution is 0.12 s. Using the measurement set-up and detection and tracking algorithms described in this paper, it was possible to track vehicles on an area of 500 m length. The accuracy and reliability of the data was increased substantially by multiple detection and tracking of the vehicles, and by using an Iterated Extended Kalman filter.

Acknowledgements

The research described here is part of the research programme "Tracing Congestion Dynamics – with Innovative Traffic Data to a better Theory", sponsored by the Dutch Foundation of Scientific Research MaGW-NWO. It is co-sponsored by the Traffic Research Center AVV of the Dutch Ministry of Transportation, Public Works and Water Management.

References

[1] Hoogendoorn, S.P. et al (2003). Microscopic Traffic Data Collection by Remote Sensing. Transportation Research Records 1855, pp. 121-128.

[2] Brackstone, M. and McDonald (1999), M. Car-following: a historical review. Transportation Research Part F, 2, pp. 181–196.

Management Information Systems, C. A. Brebbia (Editor)

Web based system for transportation and GIS database management

S. S. Pulugurtha & S. Nambisan
Transportation Research Center, University of Nevada, Las Vegas, USA

Abstract

The graphical display format provides a very powerful mechanism to effectively communicate vast amounts of information on roadway projects and related databases. An application tool in an Internet Server environment, combined with a Geographic Information System (GIS), can be used to disseminate such spatial information to dispersed users effectively, efficiently, and in a timely manner. Such a system is a powerful tool to communicate and exchange information amongst various entities, which include the general public, elected officials, managers, and staff of public and private sector organizations. This paper discusses the functionality, development, and capabilities of the system.
Keywords: Geographic Information Systems, GIS, Internet, web, transportation, map, display, communication.

1 Introduction

The general public, elected officials, managers and staff of the public and the private sector organizations are interested in the transportation system, its relationship to landuse, its functionality and other databases because of its tremendous impact on public safety, quality of life, business and the economy of the region. The spatial nature of the road network makes the use of maps a natural tool to display information. The graphical display format provides a powerful mechanism to effectively communicate vast amounts of information on roadway projects. It also helps serve as an index to additional information in databases related to the graphical display. Such information is frequently exchanged and communicated among various agencies and organizations. Over the last decade Geographic Information Systems (GIS) programs have gained wide acceptance for managing such activities [1].

Management Information Systems, C. A. Brebbia (Editor)

Previously, information in graphical format has been exchanged as hard copies or electronic copies (mailing or personal delivery), which is a time consuming process especially when multiple users are involved. The GIS technology has helped improve the efficiency of such activities. The Internet provides an efficient and quick way to access, share, analyze, communicate and exchange information across communications networks. A system that combines these two technologies offers tremendous potential for the transportation community. The dissemination of spatial information pertaining to the transportation network and other databases in a graphical format using the Internet requires the development of an application tool to work in an Internet Server environment combined with GIS or other systems which can handle spatial data. The level of detail is to be established based on the needs and requirements of the end user. In addition, the end user need not have GIS data and software installed on the local computer.

This paper discusses the functionality, development and capabilities of an Internet-based transportation projects and GIS database management tool. The focus is precisely to demonstrate the development of such as a tool, discuss its advantages, and illustrate the use of this tool.

2 Functionality and development

The common formats in which transportation data exist currently include spreadsheets, databases, ARC/INFO coverages, ArcView shape files, and AutoCAD drawings. To the extent possible, the system utilizes existing data in their current format so as to reduce the time and resources needed to recreate such information. The system is developed to display spatial data files graphically on the Internet utilizing data stored in such standard GIS coverages and spreadsheets. The data most critical to the development of the system include the representation of the street network and its attributes, parcel level data, and data from various other organizations/divisions whose projects may have a bearing on the quality of life and activities in the region as related to the transportation network. The framework for the system development consists of three elements:

1) Create GIS coverages and generate database
2) Develop server-side application, and
3) Develop client-side user interface.

2.1 Create GIS coverages and generate database

The system is based on a street centerline (SCL) coverage maintained by the Clark County Geographic Information System Management office (GISMO), a county agency. The SCL represents the road network with key pieces of information which include the geography of the network, link identification number, road names, address ranges for links, and lengths of links. Information pertaining to projects and other related databases are periodically (daily, weekly - as appropriate) obtained in the form of spreadsheets from the appropriate divisions and agencies. The project coverages created using ARC/INFO are related using unique project identification number (ID) for each project. These IDs use links on the SCL coverage as the basis to identify specific projects. In turn, the project IDs are related to the spreadsheet or database created using software such as Microsoft Access, Microsoft Excel or dBase. Shapefiles are generated using an Arc Macro Language (AML) program, which joins the project coverages and databases. The next step is to convert the data files into formats that could be viewed using web browsers.

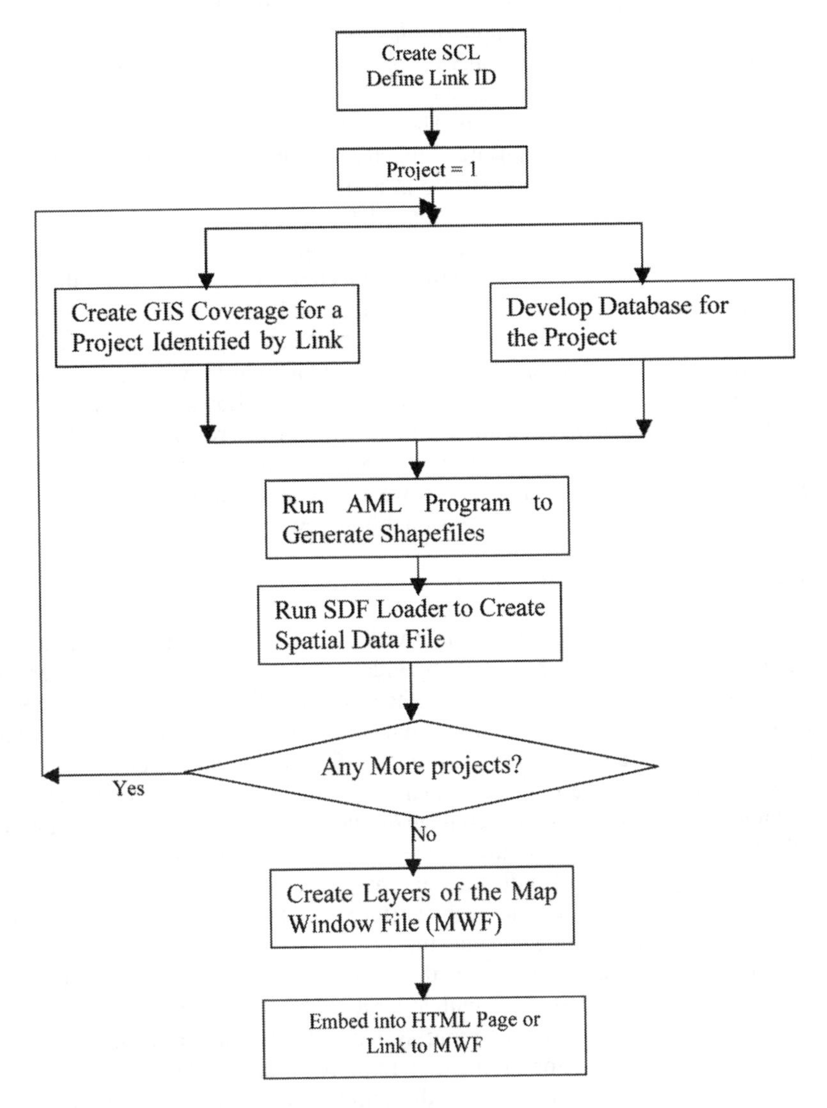

Figure 1: Methodology - flowchart.

Management Information Systems, C. A. Brebbia (Editor)
© 2004 WIT Press, www.witpress.com, ISBN 1-85312-728-0

2.2 Develop server-side application

The shapefiles are processed on the Internet server by the Spatial Data File (SDF) loader. The SDF loader is used to convert the shapefile into a SDF and Spatial Index File (SIF). An off-the-shelf web authoring software (MapGuide Author) product is used on the server to read and author the files generating layers with symbology, colors, and features to differentiate the projects. The maps are created by combining the data resources, SDFs, and databases. Various search functions are developed to help the end user perform several operations. Such operations range from simple display tools such as pan and zoom, to other tools that facilitate analyses, queries or automation of office activities and practices. The maps are stored as map window files (MWF), which contain the complete specifications of the map. These maps are stored at locations accessed by the Internet server. The maps may directly be embedded or linked from a web page. This methodology is summarized in Figure 1.

The system is built using an 866 MHz Pentium III processor running Microsoft Windows 2000 server as the operating system with 512 MB RAM, 144 Gbyte hard disk, and a T-1 Internet connection. It has to be noted that hard disk requirements depend on the data for storage and retrieval. The hardware and the Internet connection will certainly affect the system performance. Software needed include the following: 1) Autodesk's MapGuide Server, 2) AutoDesk's MapGuide Author [2], 3) Spatial Data File (SDF) Loader and Viewer, 4) Coldfusion [3], 5) HTML [4], 6) JavaScirpt [5], 7) Microsoft Office 2000 [6] suite containing Microsoft (MS) Access, MS Excel, and MS Word programs, 8) Netscape Navigator (version 4.0 or newer) or Communicator (version 4.0 or newer), and 9) Internet Explorer (version 4.0 or newer) as the web browsers.

2.3 Develop client-side user interface

A client accesses the system by using a web browser (e.g., Netscape Navigator, Communicator or Internet Explorer) by either clicking on a link to the website for the system or typing in the URL for the same. The first time user is prompted through a series of screens (text based) that walks the user through the hardware and software required to use the system, and disclaimer and policies on the use of the information. A map of Clark County is displayed when the end user with the required plug-ins opens the web page or clicks on the link upon accepting the disclaimer and policies on the use of the system. The plug-ins, created to work with the web browsers to handle data and maps are available at no cost and can be downloaded from the product website (there is a link from the system for the first time user) and operated inside the user's local personal computer. The minimum required specifications for the client-side (end user) local computers are any 486 or Pentium based PC with at least 16MB RAM, either Windows 3.1, 95, 98, NT, 2000 or XP as operating system, 10MB hard disk and Netscape Navigator or Communicator or Internet Explorer (version 4.0 or newer). The user initiates the request to provide the required data on a map through the Internet using the services of the web browser on the local computer. This request is sent across the Internet to the web server. The request is then sent

Management Information Systems, C. A. Brebbia (Editor)

through the web server to MapGuide Server software installed on the server. The request is processed on the server and results sent to the computer on the client-side. Figure 2 summarizes these processes. The user interface is designed to provide various analytical and display capabilities. These are summarized next.

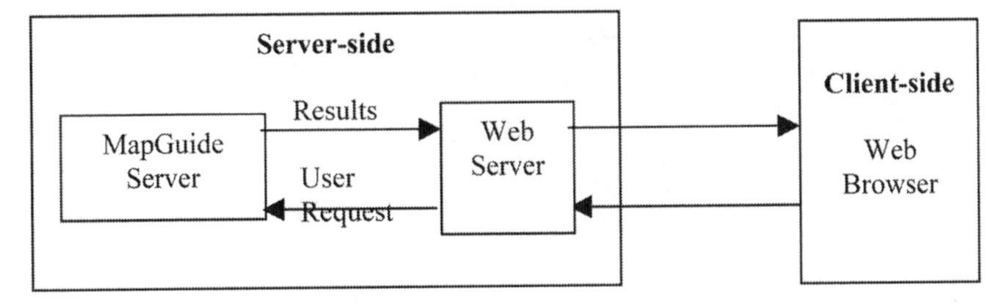

Figure 2: System functionality.

3 System capabilities

The system is used to display information, and allows the user to perform queries, analyze, automate office activities and export data. The capabilities of the system can be divided into two categories: 1) built-in capabilities, and 2) added search functions.

3.1 Built-in capabilities

The system has many built-in functions. Some of the capabilities based on these functions include: 1) easy layer turn on/off, 2) easy zoom-in/zoom-out, and pan 3) zoom go-to, 4) information at the tooltip, and, 5) creating reports and labels. These functions are shown clearly on the left side of the screen. These capabilities have been demonstrated and illustrated in the literature [7, 8]. For example, Figure 3 shows the capability of the system to provide project related information at the tooltip.

3.2 Added search functions

To quickly narrow down to the area or project of importance several search functions have been created utilizing JavaScript [5] and Active Server Pages (ASP). They include search functions to find an address, intersection (set of cross streets), find an owner of a parcel, find a specific parcel, locate a firemap, search for Township-Range-Section (TRS) or zip code. Links to these buttons are displayed on the left side of the screen. When the end user clicks on the search button for a project type (say, find projects in a specific township-range-section), the system prompts for the user to enter the township (Figure 4), range and section details. The JavaScript then activates the server. Based on the selection, the system zooms in to the location according to the pre-set scale. Figure 5 shows the Clark County Public Works (CCPW) projects, traffic signal priority needs projects, traffic capacity improvement projects, roadway capacity

Management Information Systems, C. A. Brebbia (Editor)
© 2004 WIT Press, www.witpress.com, ISBN 1-85312-728-0

improvement projects, and 5 Year no-cut projects in a selected area (township-range-section T21S-R62E-S08).

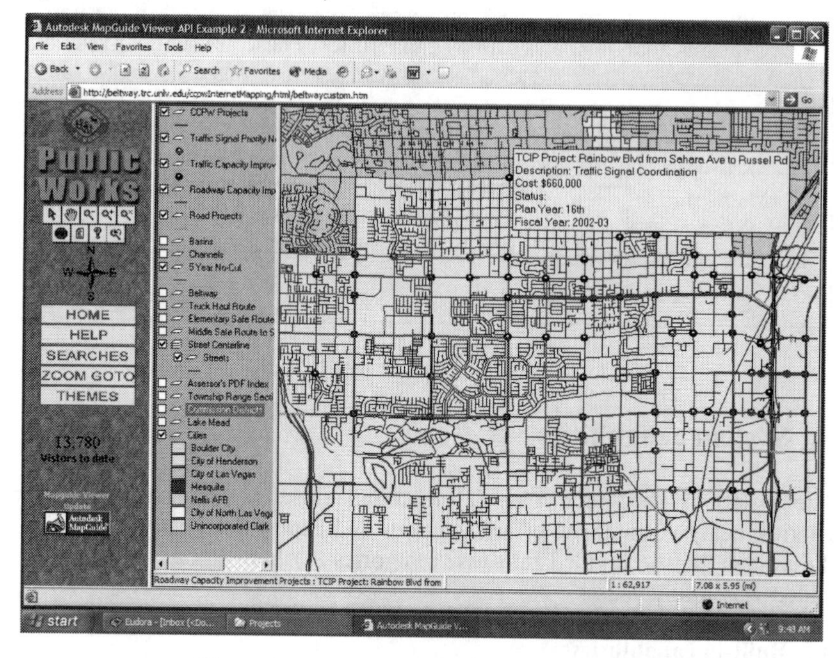

Figure 3: Project information at tooltip.

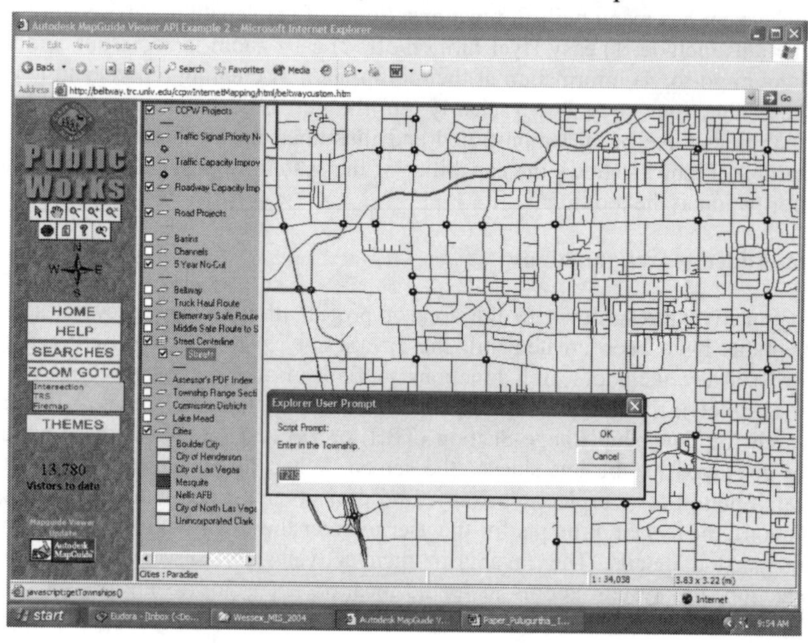

Figure 4: Zoom go-to to search for projects in a township-range-section.

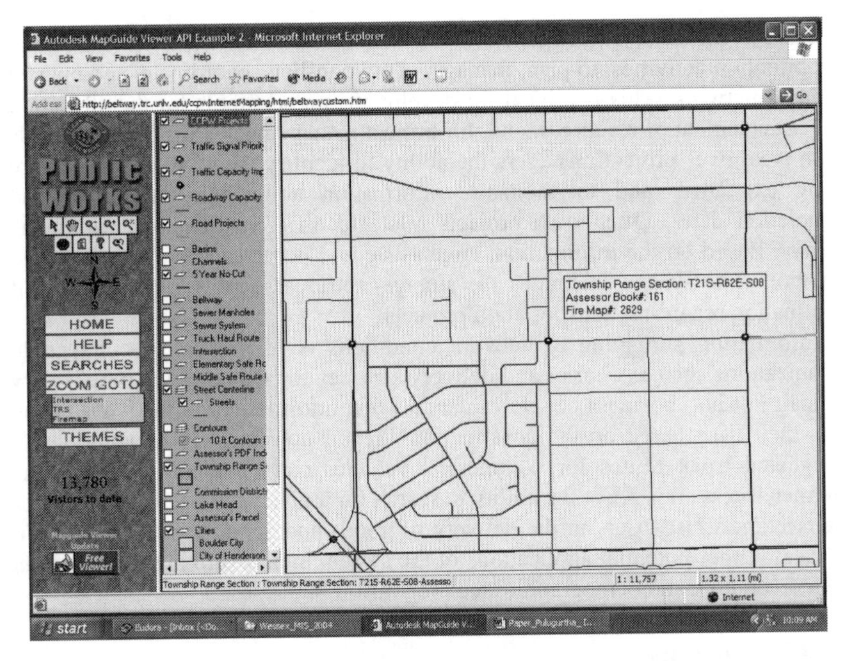

Figure 5: Projects in a township-range-section.

4 Benefits to users

The system assists the end users in many ways. For example, a user can zoom into a particular region to identify projects underway in the area. This could be in the area around an intersection, on a given street, in a township range or area around a parcel. The system assists elected officials and public works managers allocate resources, conduct long-term planning, and identify potential inequalities in expenditures and projects across neighborhoods. The system provides visual displays of ongoing activities in the area so that citizens can grasp the magnitude of what is being done and can see the progress being made. Access to the information and database enable diverse group of citizens to interact with the local government.

Roadway congestion and the associated delays affect travelers from the general public. Additionally, special events that draw several thousands of people in attendance or weather related events (flash flooding) could also cause traffic delays. Delays and events such as flash flooding also have safety implications for the road users. Providing road user's information, in a timely manner, on such projects and events could help the public better plan their trips and potentially reduce such delays. It also enhances the safety of the public. In turn, reduced delays affect economic diversification, development efficiency, productivity, and vehicle emissions. For real-time management, the available information has to be dynamic in nature.

Management Information Systems, C. A. Brebbia (Editor)

The system assists local agencies, who manage construction, maintenance and rehabilitation activities, to plan, manage, avoid conflicts, coordinate, and operate the road network. This is critical for developing support tools for the allocation and management of resources, and, for project management. Identifying projects in an area gives project managers the ability to identify where the resources are being expended, and to exchange information about the start dates and completion dates. Often such projects relate to different divisions within the agency. Based on the information, emphasis could be placed on areas that need attention. The system facilitates the storage, retrieval, analysis and display of information regarding transportation projects

Information pertaining to network conditions is of interest to commercial organizations such as courier and delivery services for which travel time is very critical. Further, business people can access this information on the website and plan their trips based on the construction / repair activities. Information about designated truck routes for commercial vehicles can be identified using the Internet access site. Also, the ability to search for a given address, street name or intersection assists users in the delivery of goods and services, and para-transit services. Other potential applications of the system include linking the data with socio-economic and demographics data for various analyses.

5 Conclusions

This paper has summarized the development, functionality, and capabilities of an integrated GIS and Internet system for transportation and landuse management. The tool was developed to support activities of Clark County Department of Public Works in Las Vegas, United States. It has been in place since 1998 and is being widely used by the Public Works staff and several other local agencies for information dissemination and exchange. The system was developed using data stored in tabular or spreadsheet format as well as those stored as GIS coverages. The system is used to display information, perform queries, analyze, automate office activities and export data. The capabilities of the system include built-in capabilities such as easy layer turn on/off, zoom-in, zoom-out, pan, zoom go-to, generate reports, and created query tools to search for a project or projects in a specific location. The Internet applications are easily accessible to the general public, elected officials, as well as to private sector organizations and public agencies.

Acknowledgments

The authors thank the Clark County Department of Public Works and the Applied Research Initiatives Program at University of Nevada, Las Vegas (UNLV) for financial support. The authors are grateful to N. T. Rajah, Thomas D. Miller, several CCDPW staff and students of UNLV Transportation Research Center for their help, guidance and support, which have been critical for the success of project.

Management Information Systems, C. A. Brebbia (Editor)

References

[1] Environmental Systems Research Institute (ESRI), Inc. Understanding GIS - The ARC/INFO Method, *John Wiley Sons, Inc.*, New York, 1997.

[2] Autodesk. Autodesk MapGuide: User's Guide. *Autodesk, Inc.*, Cupertino, CA, 2000.

[3] Allaire. Coldfusion: User Guide. *Allaire Corp.*, Cambridge, MA, 1997.

[4] Castro, E. HTML for the World Wide Web. *Peachpit Press*, 2nd Edition, Berkeley, CA, 1997.

[5] Negrino, T. Javascirpt for the World Wide Web. *Peachpit press*, 2nd Edition, Berkeley, CA, 1998.

[6] Busby, M. Learn Microsoft Office 2000. Wordware Publication, Plano, TX, 2000.

[7] Miller, Thomas D., Srinivas S. Pulugurtha, Shashi S. Nambisan, Jeffery J. Jensen and Chang-Hwan Park. An Internet Based Transportation Projects Management System. Compendium of Technical Papers (CD-ROM), *69th Annual ITE and District 6 ITE Conference*, Las Vegas, NV, August 1-4, 1999.

[8] Nambisan, Shashi S., Srinivas S. Pulugurtha, Jeffery J. Jensen, Robert Puterski, Thomas D. Miller, and Chang-Hwan Park. Prototype Internet Applications for Transportation System Management. *Transportation Research Record Series1660,* Journal of Transportation Research Board, National Research Council, Washington, D. C., p (122-131), 1999.

Management Information Systems, C. A. Brebbia (Editor)
© 2004 WIT Press, www.witpress.com, ISBN 1-85312-728-0

Decision support for operational ambulance control

T. Andersson
Department of Science and Technology, Linköping University, Sweden

Abstract

Operational ambulance control naturally includes the dispatching of ambulances to calls, but equally important, and an even more complicated task, is the need to maintain a sufficient level of preparedness in the area of responsibility. Preparedness is a measure of the ability to serve potential patients with ambulances in a swift and efficient manner. Here, a number of enhancements to the existing information systems are presented, including way of calculating and visualising the preparedness in a geographical information system (GIS). Furthermore, a simulation module and real-time decision support tools for automatic dispatching and dynamic ambulance relocation, based on mathematical modelling and heuristic solution algorithms, are developed.
Keywords: ambulance logistics, optimisation, preparedness, decision support tools, geographical information systems.

1 Introduction

SOS Alarm [11] is the company in Sweden responsible for receiving all calls to the national emergency number, 112, and also for controlling all ambulance movements. The operations are run from a SOS central of which there is one in each county (administrative district) in Sweden. One of the services offered by SOS Alarm is called *ambulance logistics*, and the customers are the county councils. An ambulance can be requested by the public or by the health care. A SOS operator receives the request and prioritises it according to three degrees:

- Prio 1: Urgent, life threatening symptoms.
- Prio 2: Urgent, not life threatening symptoms.
- Prio 3: Non-urgent calls.

Management Information Systems, C. A. Brebbia (Editor)
 ISBN 1-85312-728-0

A Prio 3 call is often a transport to or from the hospital and the patient's home, or between hospitals. These requests are often made by the health care but can also be made by the public at a special ambulance order number.

Orders received more than a day before they are to be executed can be used to create a transportation schedule. Today, the planning mostly consists of checking that it is possible to perform the pre-ordered transports and still have enough ambulances left to cover for incoming urgent calls, i.e. without compromising the *preparedness* in the county.

Calls that are received the same day as they are to be executed may be urgent or non-urgent. A SOS operator decides which is the case when the call is prioritised. An ambulance controller (that may or may not be the same person as the operator) then has to decide which ambulance to assign to the incoming call. This decision depends on a number of different parameters; e.g. the location of the ambulances, their equipment and also the preparedness in the area. Preparedness refers to the ability to get one or more ambulances to a potential call site, i.e. where the patient is located, within a reasonable time. At a certain location, the preparedness depends on how many ambulances that can reach the location within a certain time frame, as well as their travel time. Of course, the preparedness also depends on the expected need for ambulances in the area; e.g. in a densely populated area it may be necessary to have more than one ambulance close by in order to get an adequate preparedness.

To a Prio 1 call, the closest ambulance (or ambulances if more than one is needed) is always assigned. This assignment may however cause a drop in the preparedness, probably for the zones close to the ambulance that was assigned to the new call. This might make it necessary for the controller to relocate another ambulance to the affected areas, in order to cover for the ambulance that is busy. For Prio 2 and 3 calls, the controller may choose to assign another ambulance than the closest one in order to preserve the preparedness, if this can be done without compromising the safety of the patient.

Keeping an adequate preparedness is one of the most complex tasks for the ambulance controller. It requires knowledge of where call sites are likely to appear and of how fast the ambulances can travel through different parts of the county, as well as knowledge of where the ambulances currently are located and if they are available. Today, most ambulances in Sweden have GPS receivers and transmit their position and status to the SOS central. However, to know where ambulances might be needed in the future, and how fast they can get there requires experience.

2 Information systems in the SOS central

The main computer system in a SOS Central is called CoordCom, originally developed by Ericsson [6] in 1987. CoordCom supports communication by radio, telephone and data, and is also used to book, edit and store information about ambulances and calls. Information about a new call might, for example, be called in through the public switched telephone network, in which case CoordCom automatically checks, and enters the address from where the call

Management Information Systems, C. A. Brebbia (Editor)
© 2004 WIT Press, www.witpress.com, ISBN 1-85312-728-0

originated. The SOS operator prioritises the call and enters additional information into CoordCom, and CoordCom indicates for the ambulance controllers that there is a new call to be served. The ambulance controller assigns an ambulance to the call, and the system sends a data message, using the Mobitex network [10], to the ambulance personnel. When the ambulance is on its way to the call site, the ambulance personnel acknowledges this by sending a data message back to the central. The Mobitex network is also used to transmit the position of the ambulances, which is received using the global positioning system (GPS), at regular intervals. A new system that will replace CoordCom is being developed by Ericsson, and will be taken into service in the beginning of 2005.

The ambulance controllers also have a geographical information system (GIS) to support them in their work. The old GIS was recently replaced by the ResQMap system, developed by Carmenta [5]. ResQMap is connected to CoordCom and can graphically display the information that is stored in CoordCom, e.g. the location of calls and ambulances. Today it is not possible to issue any kind of order in the GIS, due to the inability of CoordCom to process them, but with the arrival of the new communication and call handling system, the ambulance controller will be able to perform more of the work tasks directly in ResQMap.

Although the current systems, and the forthcoming system, provide substantial support to the ambulance controllers, there are a number of desirable features that they do not currently offer, including:

- **Calculation and visualisation of preparedness.** To state the preparedness in ambulance logistics is a way of describing the current operational situation. If the preparedness is "good", the ambulance controller feels confident that it is possible to efficiently serve a normal amount of calls in the near future. This situation might however change rapidly; suppose that there is a serious traffic accident, forcing the controller to dispatch a large number of ambulances to the call site. These ambulances will probably be the ones that are closest to the call site, making the surrounding area devoid of available ambulances. Suddenly the preparedness is "bad", at least around the call site. Naturally, "good" and "bad" are quite vague expressions, and does not give a clear picture of the operational situation. Therefore an application capable of calculating and visualising the preparedness would be valuable.
- **Simulation.** Simulation of ambulance control would as a minimum include the generation of calls and simulated ambulance movements. This would make it possible to use the simulator as an educational tool, by letting an ambulance controller make all the decisions, such as the assignment of calls to ambulances and the dynamic relocation of ambulances. In order to use simulation to evaluate more strategic questions, e.g. fleet sizing and ambulance station location, it is

Management Information Systems, C. A. Brebbia (Editor)

necessary to simulate the ambulance controller as well, which means finding algorithms for automatic assignment and automatic relocation.

- **Automatic dispatch**. In order to simulate an ambulance controller, a decision support tool capable of selecting which ambulance that should serve a new incoming call is needed. The tool should of course pick the most suitable ambulance, from a patient safety and a preparedness perspective. This also makes the tool useful as decision support in operational ambulance control, where it can give the controller suggestions on possible assignments.
- **Automatic relocation**. In order to keep a high level of preparedness in the area of responsibility, an ambulance controller may choose to relocate ambulances to cover zones with a low preparedness. A decision support tool that can judge when a relocation is necessary and which ambulances that should be relocated is also needed to simulate the work performed by an ambulance controller. Just as for the automatic assignment tool, automatic relocation can be used in operational ambulance control.

3 Enhancements to current systems

3.1 Calculation and visualisation of preparedness

In order to find a quantifiable measure for preparedness, we first divide the area of consideration into a number of different zones. To each zone j, a weight c_j is associated, which mirrors the demand for ambulances in the zone. The weight can for example be proportional to the number of calls served in the zone during a specific time period, or to the number of people currently resident in the zone. The preparedness in zone j can then be calculated as:

$$p_j = \frac{1}{c_j} \sum_{l=1}^{L_j} \frac{\gamma^l}{t_j^l} \tag{1}$$

where L_j is the number of ambulances that contribute to the preparedness in zone j, t_j^l is the travel time for ambulance l to zone j, γ^l is the contribution factor for ambulance l and the following properties hold:

$$t_j^1 \leq t_j^2 \leq \ldots \leq t_j^{L_j} \tag{2}$$

$$\gamma^1 > \gamma^2 > \ldots > \gamma^{L_j} \tag{3}$$

Thus, the preparedness is calculated by letting the L_j closest ambulances to zone j contribute to the preparedness with an impact that is decreasing as the travel time to the zone increases. A more thorough discussion on the preparedness measure (1), and alternative ways of calculating the preparedness, can be found in Andersson *et al* [2].

Given a certain fleet of ambulances with a known status and location, the preparedness in each of the zones can be calculated. The most obvious area of use for this is in the operational control, where the preparedness dynamically can

Management Information Systems, C. A. Brebbia (Editor)

be updated. The ambulance controller can check when the preparedness is low in a zone and do something to correct this, e.g. relocate an ambulance to cover the affected zone.

To benefit fully from the preparedness measure, it should be presented in a way that makes it easy for the ambulance controller to evaluate the situation. This can, for example, be implemented by letting different colours represent different levels of preparedness in the GIS.

3.2 Simulation

Today, the GIS is capable of, among other things, to visualise and manage information about calls and ambulances. This makes it an ideal platform for an ambulance control simulator, especially since the ambulance controllers are already used to the system interface.

In order to simulate calls, a call generator that imitates the information that is usually received from CoordCom is needed. Calls can be generated stochastically according to historically verified distributions, which is proper if the simulator is to be used for evaluating strategic decisions. A constructed sequence of calls may be a preferred input if the simulator is used as an educational tool, since this makes it possible to build scenarios to educate and test the skills of the ambulance controllers. Of course, the constructed sequence may also be a set of historical calls.

For the simulation to work, it is also necessary to supply the GIS with ambulance information, and for the ambulances to automatically find and travel a route from an origin to a destination when ordered.

3.3 Automatic dispatch

Sometimes it is easy to decide which ambulance to assign to a new call, e.g. for a Prio 1 call that requires only one ambulance, the ambulance with the shortest expected travel time to the call site is always dispatched. However, the controller must still ensure that the ambulance carries the necessary equipment and that the ambulance personnel are qualified to handle the call. If the call is not as urgent, an ambulance operator may choose to dispatch an ambulance with a longer travel time, if this assignment means that the drop in preparedness will be less significant. The controller may also reassign an ambulance already on its way to a call site, if the new call is more urgent.

All of the decisions above have to be simulated by an automatic dispatch algorithm, for a simulator to be credible. Since the implementation of the preparedness measure (1) keeps track of which ambulances that are close to a zone, it is easy to find the closest ambulance to a certain zone. The check, if a certain ambulance may handle a certain call, can be made beforehand and stored together with the ambulance information. To check which ambulance to dispatch to a Prio 2 or 3 call, an algorithm has been developed that checks all available ambulances within a certain travel time from the zone, and picks the one whose unavailability causes the least drop in the preparedness as calculated by (1).

Management Information Systems, C. A. Brebbia (Editor)
© 2004 WIT Press, www.witpress.com, ISBN 1-85312-728-0

By letting ambulances on their way to a Prio 2 or 3 call still contribute to the preparedness, it is also possible to assign these to calls that are more urgent, e.g. an ambulance on its way to a Prio 3 call, can be assigned to a new Prio 2 or Prio 1 call. To ensure that the waiting times for the less urgent calls do not grow beyond what is practically feasible, pseudo priorities are used when making the assignments. The pseudo priority for a call changes if the call has not been served within a certain time, e.g. a Prio 3 call that has not been reached by an ambulance in T_3 minutes changes pseudo priority from 3 to 2. This means that an ambulance that is on its way to serve this Prio 3 (pseudo Prio 2) call cannot be reassigned to a Prio 2 call, but still to a new Prio 1 call. A similar approach is used in Weintraub *et al* [12] where requests for service vehicles used for repairing breakdowns in the electricity system move up in priority the longer they remain in the service queue. Gendreau *et al* [7] use a slightly different approach and only picks ambulances already on their way to a call if it is the only available ambulance that can reach the more urgent call site quickly, or if there is another ambulance that can cover the less urgent call within a certain time. It may be noted that the real priority of a call, and thus not only the pseudo priority, may change if a patient has to wait for medical care.

3.4 Automatic relocation

An ambulance controller may relocate an ambulance if there is reason to believe that there exists a location where the ambulance is more likely to be able to serve a new call in a shorter time. This relocation will then increase the general preparedness. In order to simulate this decision, it is assumed that relocating one or more ambulances is beneficial if the level of preparedness, as given by (1), drops below a certain threshold value, P_{min}. This gives rise to the following optimisation problem:

Minimise
{the maximum relocation time among the relocated ambulances}
subject to
{all zones must have a preparedness level of at least P_{min} after the relocation is complete}
{at most R_{max} ambulances may be relocated}

A mathematical representation, and a solution algorithm for the problem can be found in Andersson *et al* [4]. Today, the ambulance controllers solve the problem above manually, although the objective and the constraints are not quantified. In order to simulate the relocation decisions, quantifications are necessary. The decision to solve the problem is triggered by any zone having a level of preparedness below P_{min}, and the objective is to correct this situation as quickly as possible. The constraint that not more than R_{max} ambulances may be relocated is necessary to ensure that it will be possible to use the solutions to the problem in practice; no ambulance controller would approve of a solution that involved too many relocations. If more than two or three ambulances are to be relocated, it is hard for the controller to evaluate the solution, and might feel that it is not worth the trouble to execute it.

Management Information Systems, C. A. Brebbia (Editor)
© 2004 WIT Press, www.witpress.com, ISBN 1-85312-728-0

4 Evaluating the system enhancements

Once the system enhancements have been developed, they have to be calibrated, verified and evaluated. The calibration phase includes collecting necessary data and finding values for parameters. For the calculation of the preparedness, this means constructing the zones, finding proper weights and obtaining expected travel times between the zones. Each one of these tasks is a research challenge in itself, but fortunately this also means that there often has been some prior work in the area, some of which is described in Goldberg [8]. The construction of the zones and the travel time modelling often comes hand in hand if previous developed models are to be used for obtaining the travel times. One way of finding expected travel times is to use an equilibrium based traffic model; see for example Alsalloum *et al* [1], where traffic models are used to obtain travel times for ambulances. This type of model is often used for evaluating alternative modifications in the traffic infrastructure. The solution from the traffic model is a state of equilibrium where no traveller can find a shorter route from their origin to their destination. The model primarily determines the traffic flow on links in a traffic network. The travel time on a link in the model is expressed as a function of the flow on this link; typically, the function prescribes that when the flow increases the travel time increases. The travel time, as well as the traffic flow, for each link in the network in the equilibrium situation can then be extracted from the computed equilibrium solution. The traffic equilibrium can be run for different traffic situations and travel demands, making it possible to obtain travel times for, for example, different time periods during a day.

The demand for ambulances in a zone, or the probability that an ambulance will be needed in a zone, can be found in different ways. Common methods for forecasting demand in other areas include causal methods, time series models and judgemental methods. In causal methods the forecasted demand is expressed as a function of independent variables such as e.g. population and employment in the zone. Kamenetzky *et al* [9] describe the development of a causal model for estimating the demand for pre-hospital care in Southwestern Pennsylvania. Time series models are based on historical data, which is analysed to find out how the demand varies with time. Common components in these models are trends, cyclical variations and seasonal variations. In a demand function for urgent ambulance health care, the seasonal variations will be prominent, since the demand in a zone might vary substantially depending on the time of the day, which day of the week it is, or which month it is. Judgemental methods are mainly built on knowledge and qualified guesses of the present and of future events; e.g. the knowledge that a big sports event will take place in a zone will of course affect the expected demand for ambulances. In order to obtain a fair estimation of the expected operational demand for ambulances in a zone, a combination of the three types of methods may be needed.

When all the necessary data has been collected, proper values for the parameters have to be found. In the case of the preparedness measure (1), there are two different kinds of parameters; the weights c_j and the contribution factors

Management Information Systems, C. A. Brebbia (Editor)
© 2004 WIT Press, www.witpress.com, ISBN 1-85312-728-0

γ^l. A way of calibrating these, as it was done for the county of Stockholm in Sweden, is described in Andersson *et al* [3].

Verifying that the preparedness measure (1) actually mirrors what the ambulance operators refer to as preparedness, can only be done by the operators themselves. This can be facilitated by visualising the calculations in the GIS and let the operators determine whether the calculations are correct. In this manner, it will also be made clear if the operators have any use of the measure, i.e. they evaluate the system enhancement at the same time as they verify that it works as intended. However, if the evaluations of the automatic dispatch and the automatic relocation applications show that they have a positive impact on the call service times, this also verifies that the preparedness measure is useful.

The call service times are the most important quality measure of the ambulance logistics service, and they are calculated as the time span from when the call is received by the SOS operator until the ambulance personnel reach the patient. In Andersson *et al* [4] it is shown that the use of dynamic relocations, which are based on the preparedness measure, decreases the call service times. This verifies that the preparedness measure, and the automatic relocation application, works as intended.

It should also be verified that the decisions made, or proposed, by the two automatic suggestion applications are attractive from a user perspective, i.e. that the ambulance operators agree that the decisions are logical. If they do not look like decisions that could have been made by one of the operators, the users will probably never trust the applications enough to use them.

The verification that a simulation model actually simulates the system it is built for, is usually done by feeding the model with historical data and checking that the simulated actions and results correspond to the historical outcome. However, that would be difficult in this case, since there are numerous situations where the decisions made by the simulator may not be the same as the ones made historically, without the simulated decisions being wrong. For example, for a new Prio 2 call there are two ambulances equally suited to serve it; the simulator picks one while the ambulance operator in reality chose the other one. At the time the decision is made, neither the operator nor the simulator can know the full consequences of the assignment, which may differ substantially depending on where the next set of calls sites are located. They can only pick the ambulance that they think is best suited for serving the call, the operator bearing in mind the need for an adequate preparedness, while the simulator uses the preparedness calculations to reach its decision.

The verification that the simulator works therefore has to be done by verifying that the components work as intended, i.e. that the call generator can produce valid scenarios, that the ambulances travel as they should and that the automatic dispatch and relocation applications make logical decisions.

5 Final remarks

In this paper, a number of enhancements to the GIS in the SOS centrals in Sweden has been described and discussed. These include a way of calculating

the preparedness in different zones in the area of responsibility, as well as a simulation module and two decision support tools for operational ambulance control.

Currently SOS Alarm collaborates with Carmenta and Linköping University in a project aiming to install the enhancements as external and internal modules to the GIS in the SOS central in Stockholm. Computational testing and theoretical reasoning has shown us the benefits of the new applications, but it is not until the intended users have evaluated them, that their true worth will be clear.

It may be noted though, that the developed applications are not primarily intended for the ambulance controllers who are top of the class with years of experience. They usually do not need this type of decision support, even if their work performance should not suffer from having it. The enhancements are primarily intended for the less experienced controller who has not yet been able to acquire and memorise the vast amount of information required to make a quick and faultless decision every time there is a call for an ambulance. As described earlier, the new applications are also intended for evaluating strategic decisions, which are generally made by the county councils. The intended users in this case are still SOS Alarm employees, since SOS Alarm provides the city councils with the background information necessary for making the decisions.
Even if it the effects from the new applications proves to be small, i.e. the reduction of the call service times are only reduced marginally, a mean reduction of a few seconds for a set of calls can be the difference between life and death when studying as single call. This undoubtedly makes it worthwhile to continuously improve the quality of the ambulance logistics service.

References

[1] Alsaloumm, O., I. & Rand, G., K., A goal-programming model applied to the EMS system at Riyadh City, Saudi Arabia. Working Paper LUMSWP2003/035, Lancaster University Management School, Lancaster, UK, 2003.

[2] Andersson, T., Petersson, S. & Värbrand, P., OPAL – Optimized Ambulance Logistics, *Proc. of TRISTAN V: The Fifth Triennial Symposium on Transportation Analysis*, 2004.

[3] Andersson, T., Petersson, S. & Värbrand, P., Calculating the preparedness for an efficient ambulance health care, Working Paper, Dep. of Science and Technology, Linköping University, 2004.

[4] Andersson, T., Petersson, S. & Värbrand, P., Dynamic ambulance relocation for a higher preparedness, *Proc. of the 35th Annual Meeting of the Decision Sciences Institute*, 2004.

[5] Carmenta, www.carmenta.se.

[6] Ericsson, www.ericsson.com.

[7] Gendreau M., Laporte G. & Semet F., A dynamic model and parallel tabu search heuristic for real-time ambulance relocation. *Parallel Computing*, **27**, pp. 1641-1653, 2001.

Management Information Systems, C. A. Brebbia (Editor)

[8] Goldberg, J., Operations research models for the deployment of emergency service vehicles. *EMS Management Journal*, **1**, pp. 20-39, 2004.

[9] Kamenetzky, R., Suman, L. & Wolfe, H., Estimating the need and demand for prehospital care. *Operations Research*, **30(6)**, pp. 1148-1167, 1982.

[10] Mobitex Association, www.mobitex.org.

[11] SOS Alarm, www.sosalarm.se.

[12] Weintraub, A., Aboud, J., Fernandez, C., Laporte, G. & Ramirez, E., An emergency vehicle dispatching system for an electric utility in Chile. *Journal of the Operational Research Society*, **50**, pp. 690-696, 1999.

Management Information Systems, C. A. Brebbia (Editor)

Author Index